Probability
and
Statistics

Probability and Statistics

A. M. Mathai

Associate Professor of Mathematics and Statistics
McGill University, Montreal, Canada

P. N. Rathie

Professor of Statistics, Instituto de Matematicá e Estatística,
Universidade Estadual de Campinas, Brazil

First published in India 1977 by
THE MACMILLAN COMPANY OF INDIA LIMITED
Delhi Bombay Calcutta Madras

First published in the United Kingdom 1977 by
THE MACMILLAN PRESS LTD
London and Basingstoke
Associated companies in New York Dublin
Melbourne Johannesburg and Madras

ISBN 978-1-349-02769-9   ISBN 978-1-349-02767-5 (eBook)
DOI 10.1007/978-1-349-02767-5

Preface

This book is an introduction to basic statistical concepts and their applications. The only prerequisite is some knowledge of differential and integral calculus, which can be picked up in fifteen one-hour lectures. All other mathematical requirements are given in the book itself, in the form of technical notes.

The book is intended to give a clear picture of basic statistical concepts. Since the theory of sets is usually taught in secondary schools, only the concepts needed for a meaningful discussion of basic probability theory and statistical populations are discussed; the utmost care is taken to make the treatment rigorous at the mathematical level assumed.

Every concept is properly defined and illustrated with examples and counter examples. The important points are specifically made in ' comments ' following the examples. A sufficient number of problem for a complete understanding of the topics discussed is given at the end of each section, and an additional set of problems appears at the end of each chapter, some of which supplement the theory in that chapter. Answers to these problems are provided at the end of the book. Also a number of statistical decision problems are included following the discussion of mathematical expectation.

Acceptance sampling and quality control problems are discussed after introducing probability models, with a view to giving some practical applications of the models. A brief introduction to the topic of entropy is also given. Other applications have been omitted to keep the book to a manageable size. All the important results, correspondences and relationships are given in tabular form for quick reference. Most of the illustrations are taken from daily life, keeping in mind the age group of students for whom the book is intended.

It is suggested that the book be used for a twenty-five-week course of three hours per week, given as a first course in statistics for arts, science, commerce and engineering students. This could usefully be supplemented by an additional one-hour problems session each week for those students who have difficulty with the subject.

A. M. MATHAI
P. N. RATHIE

Contents

Part III STATISTICAL INFERENCE

Chapter 8 Statistical Estimation 241

Chapter 9 Tests of Statistical Hypotheses 264

List of Symbols

ϵ—element of
$\notin$—not an element of
$a_1, \ldots, a_n$—set of numbers $a_1, \ldots, a_n$
ϕ—null set
$\leqslant$—less than or equal to
$\subset$—contains
$\not\subset$—not contained in
Σ—sum
π—product
$\overline{X}$—mean
G.M.—geometric mean
H.M.—harmonic mean
S.D.—standard deviation
R.M.S.D.—Root mean square deviation
M.D.—mean deviation
m'_r—r-th moment about the origin
m_r—r-th central moment
M'_r—r-th absolute moment
$\cup$—union
$\cap$—intersection
$\overline{A}$—complement of A

$(n)_r$—permutations
$\binom{n}{r}$—combinations
$\approx$—approximately equal to
$P(A)$—probability of the event A
$P(B \mid A)$—probability of B given A
$H(p_1, \ldots, p_n)$—entropy
$|.|$—absolute value

s.v.—stochastic variable

$E(.)$—mathematical expectation

$\mu_{(r)}$—factorial moments

var$(.)$—variance

$M(t)$—moment generating function

$\phi(t)$—characteristic function

γ_2—kurtosis

cov$(.)$—covariance

ρ—correlation coefficient

L—likelihood function

λ—likelihood ratio

Part I

Statistical Populations
and
Sampling

CHAPTER ONE

Statistical Populations and Sampling

1.0 SETS

Definition. *A collection or an aggregate of well-defined objects is called a set.* Those objects which belong to the set are called *elements* of the set. Sets are usually denoted by capital letters A, B, C, etc., and their elements by small letters a, b, c, etc. Set, aggregate, group, collection, etc., are synonyms in ordinary language but in mathematical language they do not have the same meaning. ' Well-defined ' here means that any object may be classified as either belonging to a set A or not belonging to the set A.

Notations. 1. $a \in A$ (a is an element of the set A, where $\in$ is a Greek letter 'epsilon').

2. $b \notin A$ (b is not an element of the set A).

3. $A = \{0, 1, -1\}$ (A is a set with the elements 0, 1 and -1). Here $0 \in A$, $1 \in A$, $-1 \in A$ but, for example, $2 \notin A$).

4. $B = \{3 \leqslant x \leqslant 5\}$ (the set of real numbers between 3 and 5 including 3 and 5).

5. $C = \{(x, y) \mid x+2y=3\}$ (the set of all values for x and y such that $x+2y=3$).

6. $D = \{(x, y) \mid x, y \in E\}$ (the set of all pairs of elements (x, y) where x and y are both elements of the set E).

Example 1.1 Let A be a set of all real numbers between 10 and 15. Then there is an infinite number of elements in this set A, since there is an infinite number of real numbers between 10 and 15.

Example 1.2 Let B be the set of assumptions for a particular mathematical statement. Then there is a finite number of elements in the set B.   The elements are some abstract quantities, namely, assumptions for a mathematical statement.

Example 1.3 Let C be the set of all real numbers x such that $x^2 = -4$.   We know that there is no real number whose square is -4 and hence the set C has no elements.   Such a set is called a null set or an empty set or a vacuous set and is usually denoted by ϕ (phi).

Example 1.4 Let D be the set of books, pens, desks and students in a class-room.   This set has different types of objects as elements. If Mr. Fox is a student in the class then Mr. Fox is an element of D, so also is a desk in the class-room.

Example 1.5 Consider a spinner with its circular dial marked with numbers 0 to 100, with zero and 100 coinciding.   Let E be the set of possible points on which the indicator can stop when rotated. Since the indicator can stop anywhere from 0 to 100 there is an infinite number of points in the set E.   Further, every point in the interval 0 to 100 is in E and hence E can be called a *continuous set* in the sense that it represents a continuous interval.   Sets which have only a set of individually distinct elements are called *discrete sets*.   For example, a set of students in a class is a discrete set whereas the set of lifetimes of electric bulbs which can be produced by a particular method is a continuous set unless the method is so perfect that bulbs with a fixed lifetime can always be produced.

Comment. It may be noted that the element of a set can be animate or inanimate objects, real or abstract quantities and a set can be real or hypothetical, discrete or continuous, with a finite or an infinite number of elements.

1.1 A Statistical Population

Definition.   *A set whose every element is characterized by* k —*characteristics is called a* k —*variate population*, where k is a positive integer.   If $k=1$ then the population is univariate; if $k=2$ then it is called a bivariate population.   Later we will consider statistical populations defined by stochastic (random) variables.   It may be noticed that this notion of 'population' is not always the same as the notion of population in ordinary conversation.   It is the reference set based on which statistical hypotheses are tested and statistical decisions are made. Usually a statistical population is a set of numbers or a set of measurable quantities.

Example 1.1.1 Let S be the set of height measurements of all the students in a university at a particular time. Then S can be called a univariate finite population. Here every element is characterized by one characteristic, namely, 'height measurement'.

Example 1.1.2 Let S_1 be a set of height and weight measurements of all the citizens in a city at a particular time. Then S_1 can be called a bivariate finite population. Here each element in S_1 consists of two numbers, namely, the height and weight measurements of a particular citizen in the city under consideration. In other words, each element in S_1 is characterized by two characteristics. S_1 can also be considered to be a bivariate finite population consisting of two univariate populations, namely, one univariate population of height measurements and the other one of weight measurements.

Example 1.1.3 Let S_2 be the set of life spans of all television picture tubes of a particular category. Then S_2 can represent a univariate continuous population. The life span of any particular tube can be anywhere from zero to infinity. In other words, theoretically, an element in S_2, namely, the life span of a particular tube, can be anywhere on the continuous interval, zero to infinity.

Comment. From the above examples it is easy to see that according to the definition, a statistical population can be continuous or discrete (with individually distinct elements such as height measurements of a fixed number of people), real or hypothetical (such as 'true effects' of drugs), finite or infinite in the sense of containing a finite or an infinite number of elements. An arbitrary set may or may not be a statistical population and if it is a statistical population it may consist of one or more statistical populations.

Exercises

1.1   Count the number of elements in the following sets, if possible.

(a) $A = \{2, -1, 0, \phi\}$.
(b) $B = \{Y, 1, 4, \text{Miss Chick}\}$.
(c) $C = \{x \mid 2x + 3 = 5\}$.

     (d) $D=\{(x, y) \mid x+y=1\}$, x and y integers.
     (e) $E=\{x \mid 0 \leqslant x \leqslant 10\}$, x a rational number.
     (f) $F=\{(x, y) \mid x, y \in A\}$.
     (g) $G=\{x \mid x^2-2=0$, where x is rational$\}$.

1.2   Construct 2 examples each of a set which is (1) discrete, (2) continuous, (3) finite, (4) infinite, (5) real, (6) hypothetical.

1.3   Can the following sets be statistical populations? If so, explain what type of populations they are.

     (a)   The set of incomes of all the people in a city at a particular time.

     (b)   The profits made by a shop on 100 items of a particular commodity.

     (c)   The set of possible profit that can be made by a shop on a perishable item costing £1.00.

     (d)   The set of yields of wheat on 5 experimental plots and that of corn on 10 experimental plots.

     (e)   The set of I.Q's and length of noses of all the people in a certain age group, in a country at a particular time.

     (f)   The set of cows in a farm at a particular time.

     (g)   The set of books and chairs in a particular office at a particular time.

1.4   Is the following a multivariate population?

     (a)   The height, weight and length of right leg of all the people in a city at a particular time.

     (b)   The beauty (assuming that it is measurable) and I.Q's of all the participants of a beauty contest.

     (c)   The sales of a commodity on a particular day in 15 different shops and the numbers of traffic accidents in a country in 15 different years.

1.2 A Subset

Definition. A set B is said to be a subset of a set A if all the elements of B are also elements of A. This is written as $B \subset A$ (*B is contained in A or A contains B*).

Example 1.2.1 Let A be the set of all students in a class and B be the set of all female students in the class. Then evidently $B \subset A$. If all the students are female students then naturally B is the same as A. In other words $A \subset A$. If there are no female students in the class then evidently B is a null set ϕ, that is, $\phi \subset A$. The following general statements can be made.

For any set A, $A \subset A$ and $\phi \subset A$.

Definition. *Two sets A and B are said to be equal if all the elements of B are also elements of A and vice versa.* That is, if $A \subset B$ and $B \subset A$ then $A = B$.

Example 1.2.2 Let $A = \{x \mid 1 \leqslant x \leqslant 5\}$ and $B = \{x \mid 2 \leqslant x < 4\}$. Here, since the interval 2 to 4 is contained in the interval 1 to 5, $B \subset A$. The elements in A which are not included in B are the points on the intervals $1 \leqslant x < 2$ and $4 \leqslant x \leqslant 5$.

In general $A - B$ denotes the set of elements which are in A but not in B, when $B \subset A$.

Example 1.2.3 Let $A = \{1, 2, 3\}$ and $B = \{2, 3, 5\}$. Then $A \not\subset B$ and $B \not\subset A$. $\not\subset$ means 'not contained in'. Here A is not a subset of B and B is not a subset of A even though there are some common elements.

Comment. It is easy to notice that if B is a subset of a set A then it is possible that both A and B have infinite number of elements or B has a finite number of elements and A has an infinite number of elements. In general, any set is a subset of itself and the null set is a subset of any set.

1.21 *A Sample*

Definition. *A subset of a statistical population can be called a sample.* A sample, in general, means only a subset of any given set and when the reference set is a statistical population then the subset is a sample from a statistical population. The word 'sample' is used in statistical literature in both these senses, that is, the subset of any set as well as the subset of a particular statistical population. But usually there won't be any confusion in its usage.

Example 1.21.1 Let S be the strength (measured in a certain unit) of all steel beams that can be produced by a certain process, and let A be the strength of 10 beams produced on a certain day. Then evidently $A \subset S$, that is the 10 measurements form a sample from the population of measurements representing the strength of all such beams. Here S is a continuous set whereas A is a discrete set consisting of 10 elements.   Further, since a subset of a set is a set itself, a sample itself can be considered to be a population if it is not considered with reference to a larger set of which the sample is a subset.   In this example both the population and the sample are univariate.

Example 1.21.2 Let S be the set of annual incomes and weights of all the citizens in a particular country at a particular time. Let A be the set of annual incomes of all the citizens in a city of that country at that time and let B be the set of annual incomes and weights of 100 citizens in that country at that time.   Here S is a bivariate finite population and $B \subset S$ and hence B is a sample of size 100. But $A \not\subset S$ and hence A is not a sample of S. A can be considered to be a sample of the univariate population of annual incomes of all the citizens of that country at that time.

Comment. It should be pointed out that one cannot take a p-variate sample from a k–variate population if $k \neq p$. Further, both the population and the sample can contain an infinite number of elements and both can be continuous.   But in the discussions of statistical decision making we will be mainly concerned with statistical populations which may be continuous or discrete whereas the sample is a discrete set of a finite number of elements.

1.22 *A Representative Sample*

A subset from a given set may be obtained in different ways.   For example one can go and catch 10 birds from the totality of birds in a certain island.   What he gets, may be the weak ones which could not escape capture.   If he finds that 8 of his 10 birds are affected by air pollution, he won't be able to make a general statement that 80 per cent of the birds in that island are affected by air pollution, unless his sample of 10 birds represents in some sense the bird population of that island.   Similarly, if 10 per cent of the bullets produced by a machine at a particular hour, are defective, one cannot generalize that the machine produces 10 per cent defectives, unless the production in that hour is typical of that machine's performance.   So, in order to make generalizations, or in other words, in order to make valid inference regarding a population, based on a sample from it, the sample must be representative of the population in some sense.   The statistical theory of sampling is mainly concerned with the problem of selecting representative samples from a given population. It is important to notice that a sample from a population need not always be a representative sample. Often misleading decisions or statements are made by experimental scientists due to the fact that the sample experimental units are not representative of the population on which general statements are made.

Exercises

1.5 Write down all the subsets of the following sets
   (a) $A = \{1, 2, -1\}$.
   (b) $B = \{(i, j) \mid i, j \in \{0, 1\}\}$.
   (c) $C = \{a_1, a_2, a_3\}$.

1.6 Is the following statement correct, where A denotes a population and B denotes a sample from A?
   (a) A—the increase in weight in 100 experimental animals, due to a particular diet.
      B—the increase in weight in 10 of them.
   (b) A—the possible increase in weight in an experimental animal, due to a particular diet.
      B—the increase in weight in 5 animals due to that diet.
   (c) A—the I.Q's of 10 students.
      B—the marks obtained by 4 of the students in a particular test.
   (d) A—ability to withstand heat, and weight of 10 people.
      B—the weights of the 10 people.

1.7 Is the following a representative sample?
   (a) The sales in money value during the first week out of the sales for a month, in a shop.
   (b) The items produced in the first, third and eighth hours out of the items produced by a machine for a whole day.
   (c) The cards obtained by a person in a card game when the cards are 'well shuffled'.
   (d) The numbers obtained by spinning a 'well balanced' spinner three times.
   (e) The brick, which fell on a worker's foot while constructing a wall, out of the bricks used for the construction.

1.3 SAMPLING TECHNIQUES

In section 1.2 we considered the problem of sampling to some extent. Here we will deal with some of the methods available for getting a representative sample from a given population. Sampling is needed only in some cases where economic and other considerations compel one to study a representative sample instead of the whole population. But sampling is not needed in every problem. For example, in order to find out the chances of getting a red marble from an urn containing some red and some white marbles nobody need resort to picking up a sample of marbles and do complicated calculations when it is possible and easy to take out every marble and count them. But in order to study the average amount spent on clothing by the people in a country or to find out whether majority of the people in a country like to have a blue shirt and red cap as the official attire for their prime minister, if somebody decides to conduct a survey of each and every individual in the country then firstly, it may not be practicable and secondly, the information collected by such a complete survey may not be worth the money spent. In such a situation it is more appropriate to study a representative sample. There are other situations where a complete survey is impossible. For example, in order to find out the average life span of electric bulbs if somebody burns out all the bulbs then there won't be any bulb left for sale. When the units are destroyed by the investigation then a complete study of the population is out of question. Further, if one can get the required information from the sample, within allowable limits of error then one needs to study only the sample, rather than studying the population. Hence the techniques of obtaining a representative sample are very important in statistical investigations.

1.31 *Random Sampling Numbers*

One simple method of getting a representative sample is to use a game of chance or some mechanical devices. For example, if we want to select two students at random from a class of 50 students one way is to write their names on some cards which are similar in all respects and shuffle the cards well and take one card at random by some mechanical device. Replace it, shuffle again and select another one and thus we get two names. Here *at random* means that all the cards are given equal chances of being selected and it does not mean that the selection is done in some haphazard manner. In simple problems such simple procedures are possible but if the population size is quite large or infinite or if the population is continuous such a technique may not be practicable even if it works in some cases. Sampling is usually done with the help of a table of random sampling numbers.

A table of random numbers is a table of numbers which are generated by a mechanical or an electronic device in such a way that each number is obtained independently of the preceding number and all the numbers from 0 to 9 have an equal chance of appearing in a particular row and column of a sample page of random numbers. A sample page of random numbers is given at the end of this section. Suppose that there are 21 cards out of which 5 are to be selected at random. These 21 cards can be numbered from 00 to 20. One method of selecting a random sample of size 5 is to take a page of random numbers and start at any column and row. Suppose we start at the first row and first column, with the number 8. Then read off the numbers in the same row and continuing to the next row and so on. Divide the numbers into successive couplets. That is 87/65/77/77/65/43/43/20/08/.... Omit all numbers which are larger than 20. That is, omit 87/65/77/43/, select the card numbered 20, select 8 and so on. Continue till a sample of size 5 (a set of 5 numbers less than 21) is obtained. Other methods of using a table of random numbers are given in the tables of random numbers. A list of such tables is given at the end of this chapter. If a random sample of 2 persons is to be selected from a set of 45 persons then the same technique can be used by giving each person a number, starting from 00 to 44 and then selecting 2 numbers at random by using a table of random numbers. The technique described above fails if the population is infinite or if the population is continuous. Some techniques of sampling from theoretical populations will be discussed later.

8765	7777	6543	4320	0864	5185	6049
8998	3558	2557	6115	8672	4788	3461
3168	1403	4571	5975	0547	6522	7069
0566	7889	8456	6346	4802	1149	5951
6426	2924	9350	2275	1626	3901	5527
6651	6510	3162	9672	2834	2507	5341

9058	3339	2397	5737	8135	3872	2007
2108	1705	3813	5519	9333	4852	4185
5165	9360	4525	3886	8412	2299	0711
7313	9792	7106	6899	4005	0904	4910
0825	6773	7598	4371	1969	6340	8309
2625	8315	0941	9256	0197	9454	9652
2356	9096	1453	0549	2002	2552	4555
7323	5803	3127	8930	2057	0988	3046
1204	5567	6771	2338	9109	1448	0557
9924	9305	9229	8535	7764	6300	4065
7794	8316	6110	4426	0536	4962	5499
1210	2414	3625	6039	9664	5704	5368
9235	1568	0804	2372	3176	5549	8726
8657	1908	0566	2475	3041	5516	8557
8245	5060	3306	8367	1673	0040	1713
5618	4727	0345	5072	5418	0490	5909
1702	2928	4631	7559	2191	9751	1942
8407	9246	7654	6900	4554	1455	6009
0006	4898	4905	9804	4710	4515	9225
0940	9440	0381	9821	0202	0023	0226
3252	1443	4696	6140	0836	6976	7812
6880	9007	5888	4896	0785	5681	6466
8830	8003	6833	4837	1670	6507	8178
1080	5811	6891	2703	9594	2298	1893
2260	4889	7150	2039	9190	1230	0420
4974	5117	0091	5208	5300	0509	5809
6815	5020	1836	6856	8692	5549	4241
1974	2729	4704	7433	2137	9571	1709
9477	8617	8094	6711	4806	1518	6325
9511	6454	5966	2421	8388	0810	9198
3169	2751	5921	8672	4594	3267	7861
3445	9334	2779	2113	4892	7006	1899
2666	4370	7037	1407	8444	9851	8295
1174	0085	1260	1345	2606	3952	6558
5098	1087	6186	7274	3461	0736	4197
5915	4104	0020	4125	4146	8272	2418
4291	0248	4539	4787	9326	4113	3440
9095	2461	1557	4018	5575	9594	5170
0580	7856	8436	6293	4730	1023	5753
1210	8303	9513	7817	7331	5149	2480
1491	3956	5448	9405	4854	4259	9114
6115	5454	1570	7024	8594	5619	4214
2860	4163	7023	1187	8210	9397	7608
8010	4064	2074	6139	8214	4353	2567

1.32 *Simple Random Sampling*

Here we will discuss the problem of selecting a simple random sample from a population of a finite number of individually distinct elements. The notion of a simple random sample from a theoretical population will be discussed later.

Definition. Consider a population of size N (that is, N elements which are distinct in some sense) and let a sample of size n be taken. *If a sample of size n is obtained in such a way that all possible such samples of size n (with distinct elements) had an equal chance of being selected then the sample obtained is called a simple random sample from the population under consideration.*

If one card is drawn from a well shuffled ' deck of 52 cards then it is a simple random sample of size one. If two cards are drawn at a time from a ' *well shuffled* ' deck of 52 cards then it is a simple random sample of size 2. If another set of 2 cards are drawn without replacing the first set of 2 then we don't have two simple random samples of size 2 each because the first sample is taken from 52 cards whereas the second sample is taken from 50 cards. Thus the chances of being selected are different for these different sets of cards. If two cards are selected at random one by one *without replacement* still the 2 cards constitute a simple random sample of size 2 since every such set of 2 cards drawn in such a fashion have the same chances of being selected. Let A represent the set of two cards drawn from a well shuffled set of 52 cards one by one *with replacement* and let B represent the set of 2 cards drawn from a deck of well shuffled cards, one by one without replacement. Here both A and B individually qualify to be simple random samples but A and B are not two simple random samples of size 2 each because the chance of getting A is different from the chance of getting B. That is A is a simple random sample of size 2; B is a simple random sample of size 2; whereas A and B are not two simple random samples of size 2 each when they are considered together. These two are two different types of simple random samples. It must be noticed that a simple random sampling is different from what can be called *purposive sampling*. If a man is entrusted with the duty of selecting one girl to represent a set of 20 girls then the one that he is going to select is the one who he thought would be a true representative of the set. An element of personal bias plays a vital role in this selection. If a butcher is allowed to select two rabbits from a set of rabbits the unlucky rabbits need not be a simple random sample of size 2. These are cases of purposive sampling. In short, a simple random sample insures against personal bias while selecting a sample.

1.33 *Stratified Sampling*

A simple random sampling procedure may not be appropriate in many situations. If the population from which a sample is taken is heterogeneous in some sense then it is advisable to divide the population into homogeneous sets and treat each stratum thus obtained as a separate population or as a sub-population. For example, for a public opinion survey, students, prisoners, labourers and so on cannot be treated as a homogeneous set as a whole.

Definition. If the given population is subdivided into homogeneous (in some sense) sets (strata) and if sampling is done in each set then the procedure is called a stratified sampling.

Sometimes administrative convenience may compel one to resort to a stratified sampling. There are other situations where a simple or stratified sampling technique is not advisable. A number of other methods such as quota sampling, systematic sampling, cluster sampling and multi-stage sampling are available. For a detailed discussion of the various techniques see books on sampling, some of which are listed at the end of this chapter.

Exercises

1.8 Is a sampling procedure needed in the following situations?

(a) To determine the average age of first year Arts students in a university.

(b) To estimate the average reduction in cavities with the use of a new tooth-paste.

(c) To study the effect of a new fertilizer on the yield of a particular variety of wheat.

(d) To study the average strength of rods produced by a particular process.

(e) To find out the number of 'intelligent' students in a set of 100 students.

(f)   To find out whether majority of the people favour a legislation to nationalize gambling in a particular country.

1.9   By using a table of random numbers select 5 simple random samples of size 10 each from the following data:

−2, 7, 5, −1, 0, 4, 25, 7, 4, 1, 2, 9, 12, 15, 20, 18, 17, 19, 15, 12, 11, 10, 5, −1, 10, 11, 9, 15, 12, 14, 16, 8, 9, 7, 5, 12, 13, 18, 17, 16.

1.10   Is the following a simple random sample?

(a)   From an urn containing 20 marbles (all homogeneous) a person picks up two marbles one by one without replacement.

(b)   From the same urn in (a) a person picks up 3 marbles with replacements.

(c)   The items produced in the first and seventh hours in a production process.

1.11   How do you select a random sample of size 5 in the following case:

(a)   Thread is being produced by a machine and 5 pieces are to be taken from it for quality control inspection.

(b)   Picture tubes are produced one by one by a machine.

(c)   Sugar is being refined and the refined sugar comes out of a machine in a continuous stream.

(d)   Out of 35 executives 5 are to be selected for a committee.

1.4 REPRESENTATION OF A NUMERICAL SAMPLE

Representation of data (observations on some characteristics under investigation) is of some importance in statistical analysis and there are several methods available for the presentation of data. Instead of writing down a set of numbers, if they are arranged into some tabular forms then it may be more appealing to the eyes. If a mass data is classified into different classes then it may take less space for the presentation. Instead of a table of numbers, if the numbers are presented in the form of some charts or graphs a layman may be able to understand the data better.

1.41 Classification

Classification of given data into different convenient classes is one way of representing the data. This can be explained with the help

of the following example.   Table 1.1 gives the marks obtained by
50 students in an examination.

TABLE 1.1

60	48	37	55	72	80	72	65	68	79
65	45	39	56	71	79	55	65	69	80
62	25	20	68	73	76	58	67	54	48
60	30	22	65	75	75	60	69	58	82
50	32	52	70	78	74	62	71	80	89

If the data is presented as in Table 1.1 it may not help much if
one wants to get an idea of the distribution of marks or the numbers
of students whose marks lie in certain intervals.   The data in
Table 1.1 may be classified by taking a suitable number of classes.
The lowest mark is 20 and the highest is 89.   So if the total range
is divided into 7 equal intervals then we can represent the data as
in Table 1.2.

TABLE 1.2

Class (Marks)	Tally	Frequency (Number of students)
20—29	///	3
30—39	////	4
40—49	///	3
50—59	⊬ ///	8
60—69	⊬ ⊬ ////	14
70—79	⊬ ⊬ ///	13
80—89	⊬	5
		50

Table 1.2 is obtained by putting tally mark in the appropriate
classes corresponding to the various numbers in Table 1.1. Here the
intervals 20–29, 30–39 and so on are called the *classes*, 20, 30, 40,
etc., are called the *lower class limits* and 29, 39, etc., are called the
upper class limits.   Here there are gaps between two successive
intervals.   If we would like to have the intervals to be continuous
then the intervals can be taken as 19.5 to 29.5, 29.5 to 39.5 and so

on. Here the class width is 10 units which is often called the *true class width*, 19·5, 29·5, 39·5, etc., are sometimes called the *true lower limits* and 29.5, 39·5, etc., are sometimes called the *true upper limits*. Continuous class intervals are justifiable here because theoretically the marks can be any point from 0 to 100 if the marks are given in percentage. But in all cases the classes cannot be assumed to be continuous. For example, consider the data of traffic accidents. Table 1.3 gives the number of accidents involving cars with 1, 2, 3, 4, 5 and 6 passengers.

TABLE 1.3

Number of passengers in a car	*Frequency (Number of accidents)*
1	15
2	32
3	28
4	22
5	15
6	4
	116

Now coming back to our data in Table 1.1, suppose that a student had obtained the mark 59.6. The classification in Table 1.2 is ambiguous because 59.6 is not included in any class. Hence if the data obtained is such that the numbers are rounded off at the first decimal place, in the sense that for example a number in the interval 49.5 to 50 will be counted as 50 and a number in the interval 40 to 40.4 is counted as 40, thus the ambiguity of classification can be avoided by taking the classes as 19.5 to 29.4, 29.5 to 39.4 and so on. Similarly if the numbers in the original data are rounded off at the second decimal places then again by taking the class limits with two decimal places, ambiguity can be avoided.

While representing data the rule for deciding upon the number of classes depends upon the purpose for which the data is going to be represented. For example, if one is interested in the classification of marks in table 1.4 according to 4 grades such as grade 1 (80 and over), grade 2 (65 to 79), grade 3 (50 to 64) and grade 4 (49 or less) then the classification is as given in Table 1.4.

TABLE 1.4

Class (Marks)	Frequency (Number of students)
less than 50	10
50—64	13
65—79	22
80 and over	5
	50

Here the class widths are different for the different classes. If one would like to see more subdivision of the classes, then the data in Table 1.1 can be represented as in Table 1.5.

TABLE 1.5

Class	Frequency	Class	Frequency
20—24	2	55—59	5
25—29	1	60—64	5
30—34	2	65—69	9
35—39	2	70—74	7
40—44	0	75—79	6
45—49	3	80—84	4
50—54	3	85—89	1
			50

While classifying data the choice of the number of classes, the width of each class, the continuity of the intervals (this is important when deciding upon the true class limits) etc., depend upon the nature of the data, and the purpose for which the data is going to be classified. When data are classified it is likely that some information about the data is lost. For example, from Table 1.2 we know only that there are 3 students who got marks in the interval 20 to 29 but unless Table 1.1 (original data) is available we cannot get the exact marks obtained by the three students. If only Table 1.2 is available a fair means of assigning marks to the 3 students in the interval 20 to 29 is to give each the mid-point of the interval, namely, 24.5. The mid-point of 'true interval' is often called the *class mark*.

1.42 *Bar Diagrams*

A diagrammatic representation of data is often a visual aid for better understanding of data. There are several diagrammatic representations available of which bar diagram is a simple representation. *In a bar diagrammatic representation the data is represented by erecting bars, whose areas are proportional to the frequencies in the various classes, over the mid-points of the class intervals.* Sometimes bars whose heights are proportional to the frequencies are erected over the intervals and which often leads to misrepresentation of data unless the class intervals are all of the same length. The data in Table 1.3 is represented in Figure 1.1. The data in Table 1.2 is represented in Figure 1.2. The data in Table 1.2 with the frequencies in the classes from 50 to 59 combined, is represented in

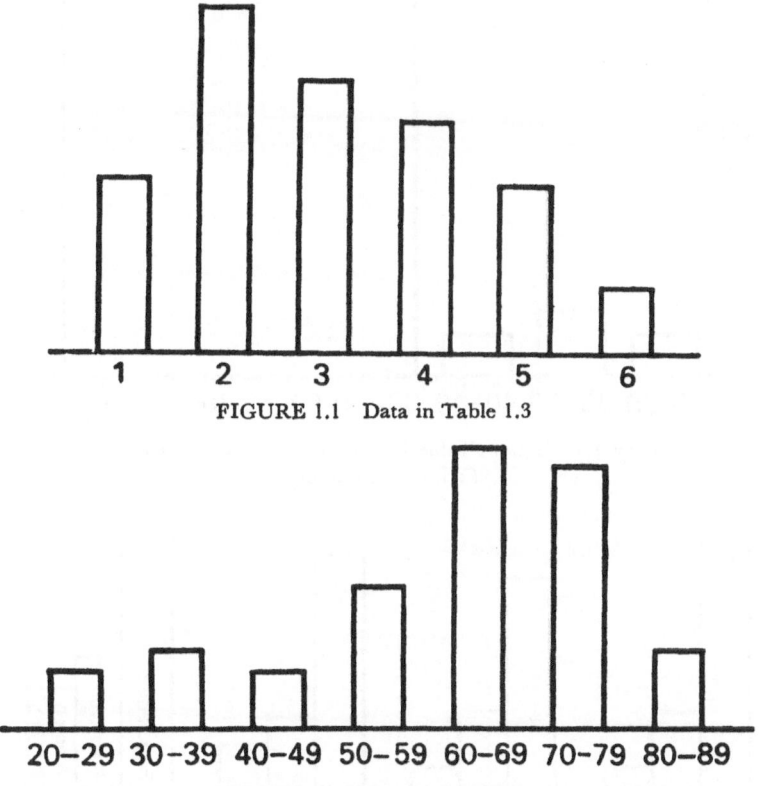

FIGURE 1.1   Data in Table 1.3

FIGURE 1.2   Data in Table 1.2

Figures 1.3 and 1.4 out of which in Figure 1.4 the height of the bar is proportional to the frequency and it is seen that it leads to misrepresentation.

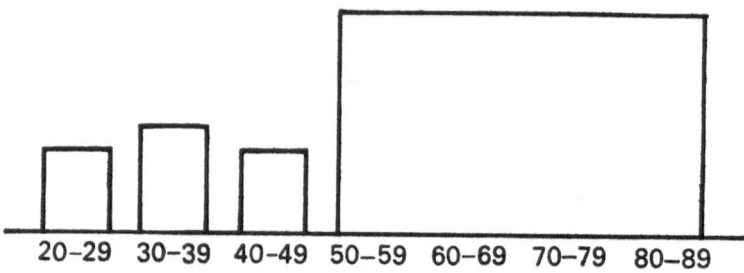

FIGURE 1.3   Data in Table 1.2 with the last 4 classes combined
(correct representation)

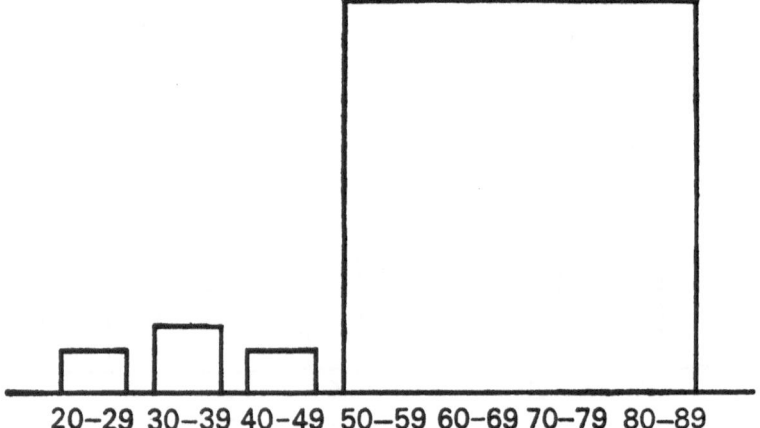

FIGURE 1.4   Data in Table 1.2 with the last 4 classes combined
(Misrepresentation)

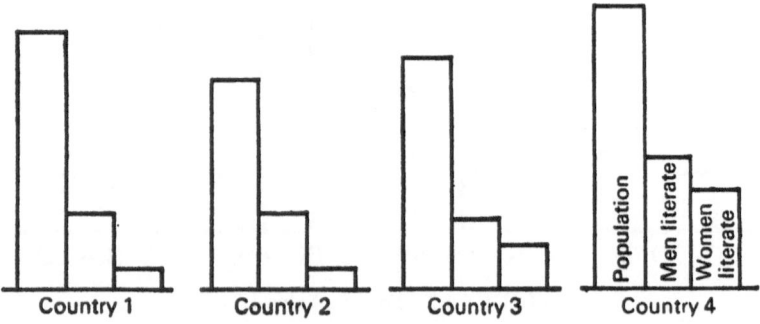

FIGURE 1.5   Literacy rates and populations of 4 different countries

Figure 1.5 gives a bar diagrammatic representation of literacy rates among men and women and the total populations in 4 different countries.

The bar diagrammatic representations can be made more attractive by different designs and colour charts for the representations.

1.43 *Histograms*

In a histogram, rectangles whose areas are proportional to the frequencies, are erected over the true intervals. A histogram is not usually used for representing data as in Table 1.3 where the intervals cannot be justifiably assumed to be continuous. Figure 1.6 gives a representation of the data in Table 1.2 in histograms.

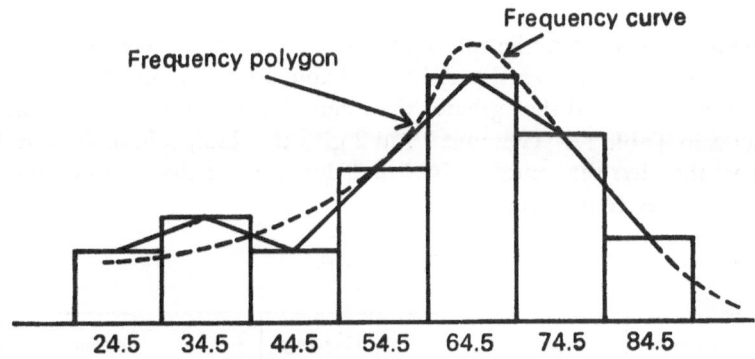

FIGURE 1.6 Histogram of data in Table 1.2
Frequency polygon and frequency
curve for the same data are also
given here

1.44 *Frequency Polygons and Curves*

Data from a continuous population are also represented by *frequency polygons* and by *frequency curves*. Instead of a histogram if points are plotted against the mid-points of the true class intervals (*class marks*) and if the points are connected by straight lines then a frequency polygon is obtained. Figure 1.6 gives a frequency polygon for the data in Table 1.2. If a histogram is smoothed by the 'best fitting' smooth curve then a frequency curve is obtained. While drawing a frequency curve the following rules are helpful.

(a) If the population, of which the given data is a sample, cannot be assumed to be continuous when a representation of the data by a frequency curve is not advisable.

(b) If the class intervals are of different lengths appropriate adjustments are to be made in order to avoid false representation.

(c) The curve need not pass through all the points through which a frequency polygon passes. Use the rule that the area of the histogram excluded is approximately equal to the area included while drawing the curve.

Figure 1.6 gives a frequency curve for the data in Table 1.2.

1.45 *Ogives*

Sometimes we may be interested in cumulative frequencies rather than the frequencies themselves. Table 1.6 gives the ' less than cumulative ' and the ' greater than cumulative ' frequencies for the data in Table 1.2. Columns 1 and 2 give the data, columns 4 and 6 give the ' less than cumulative ' and the ' greater than cumulative ' frequencies respectively.

TABLE 1.6

Class	Frequency		Frequency		Frequency
		less than		*greater than*	
		19.5	0	19.5	50
20—29	3	29.5	3	29.5	47
30—39	4	39.5	7	39.5	43
40—49	3	49.5	10	49.5	40
50—59	8	59.5	18	59.5	32
60—69	14	69.5	32	69.5	18
70—79	13	79.5	45	79.5	5
80—89	5	89.5	50	89.5	0

If the frequencies in column 4 of Table 1.6 are plotted at the upper class limits given in column 3 of Table 1.6 and when the points are connected by straight lines we get a ' less than ogive '. Similarly the frequencies in column 6 of Table 1.6 give a ' greater than ogive ' when plotted. It should be noticed that the upper and lower class limits are taken instead of the class marks and the cumulative frequencies are plotted. Figure 1.7 gives the 'less than cumulative' and the 'greater than cumulative' ogives for the data in Table 1.2.

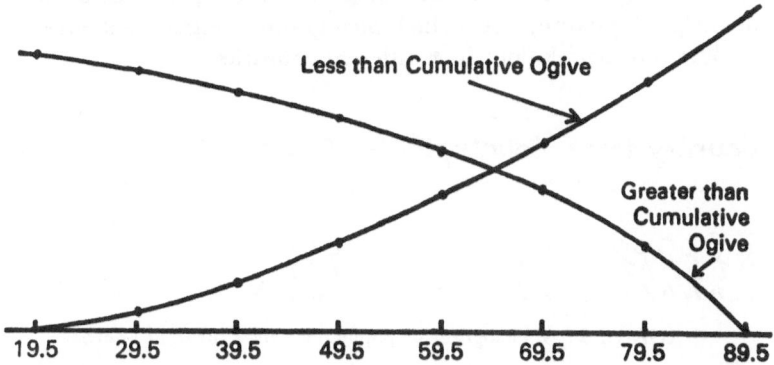

FIGURE 1.7 Less than and greater than ogives for the data in Table 1.6

1.46 *Pie Diagrams*

In this representation the frequencies in the various classes are represented by the various sectors of a circle. For example the data in Table 1.7 are represented in a pie diagram in Figure 1.8.

TABLE 1.7

Population in a country religion wise	
Protestants	50%
Catholics	40%
Others	10%

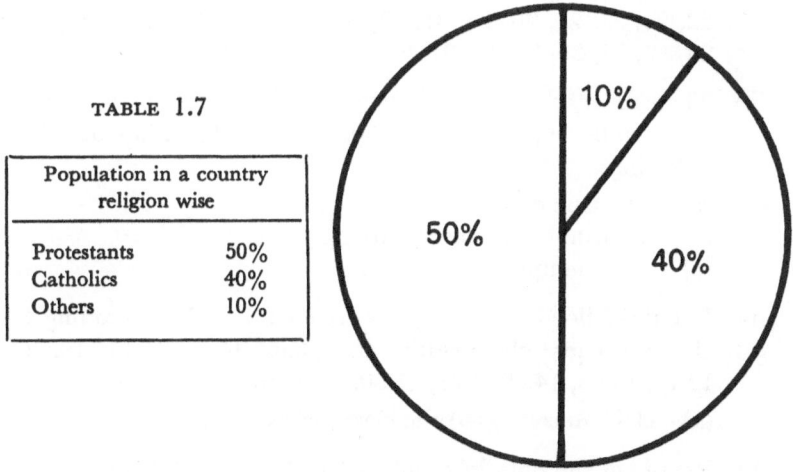

FIGURE 1.8   Pie diagram for the data in Table 1.7

1.47 *Pictograms*

In order to have an idea about the comparative strength of armies or population and so on of different countries often a pictorial representation is made. Such a diagrammatic representation with the help of pictures are called pictograms. Figure 1.9 gives a pictogram of populations in 4 different countries.

Country 1   Country 2   Country 3   Country 4

FIGURE 1.9   A Pictogram of populations in 4 different contries

Exercises

1.12 Select a simple random sample of size 40 from the following data. Classify the sample into a frequency table by selecting an appropriate number of classes.

10, 11, 12, 16, 13, 14, 16, 22, 27, 32, 35, 39, 40, 28, 29, 30, 22, 24, 26, 32, 33, 35, 26, 37, 32, 31, 29, 28, 29, 40, 40, 38, 22, 24, 23, 27, 26, 29, 31, 32, 34, 33, 31, 30, 31, 29, 27, 29, 28, 32, 34, 39, 38, 18, 21, 23, 19, 17, 18, 20.

1.13 The class marks in a frequency table (of whole numbers) are given to be 10, 15, 20, 25, 30, 35 and 40. Find out the following:
(a)   The true classes.
(b)   the true lower class limits.
(c)   the true upper class limits.

1.14 Put the following data into a frequency table containing 3 classes of equal class widths: 5.01, 6.02, 14.20, 16.51, 19.50, 12.11, 13.45, 14.02, 8.21, 11.50. Obtain the following
(a)   class intervals, (b) the class marks.

1.15 Represent the data in problem 1.14 by (a) bar diagram, (b) histogram, (c) frequency polygon, (d) frequency curve.

1.16 The following are the temperature readings in 20 consecutive hours: 60.0, 60.2, 62.0, 65.3, 67.4, 60.4, 68.2, 70.0, 70.0, 61.2, 62.4, 61.7, 63.4, 64.0, 66.2, 68.0, 69.4, 66.7, 62.0, 61.4. Classify the data by selecting intervals of width 2. Represent the data by a histogram and also plot the less than and greater than ogives.

1.17 The following table gives a classification of telephone calls received by a switch board. What is an appropriate diagrammatic representation for the data?

Number of calls per minute	Frequency (Number of such minutes)
0	10
1	25
2	22
3	15
4	8
5	4
6	2

1.18 Represent the following data, (which is a classification of 50 people) by appropriate diagrams.

Intelligence	Frequency
Above average	12
Average	20
Below average	18

1.19 The following is a classification of the employees in a factory. Represent the data by a pie diagram.

Employees	Frequency
Executive	1
Clerks	14
Technicians	45
Laborers	40

1.20 Is the following representation correct, (not misleading), which is supposed to be a pictogram of the statement that the profit in a shop on the second day is double that of the first day.

$\$$ $\mathbf{\$}$
first day   second day

1.5 TECHNICAL NOTES: Σ AND π NOTATIONS

Σ (Sigma) is a usual notation for a sum and π (Pi) is a usual notation for a product. For example, if x denotes the production of wheat (in bushels) in an experimental plot and if 5 such plots yielded 10, 15, 14, 20, 15 bushels respectively then the total yield from these 5 plots can be denoted as,

1. $\Sigma x = 10 + 15 + 14 + 20 + 15 = 74.$

   Here Σx denotes the sum of all values x takes in this particular problem under consideration. Consider the variables x, y and z where x takes the values, 2, 0, -1, y takes the values, 1, 1, -2 and z takes the values 0, 1, 2 correspondingly. Then according to the above notation we have,

2. $\Sigma xy = (2)(1) + (0)(1) + (-1)(-2) = 4.$
3. $\Sigma(x+2) = (2+2) + (0+2) + (-1+2) = 7.$
4. $\Sigma x(y+z) = (2)(1+0) + (0)(1+1) + (-1)(-2+2) = 2.$
5. $\Sigma x^2 = (2)^2 + (0)^2 + (-1)^2 = 5.$
6. $\Sigma x^2 y^2 = (2)^2(1)^2 + (0)^2(1)^2 + (-1)^2(-2)^2 = 8.$
7. $(\Sigma x)^2 = (2+0-1)^2 = 1^2 = 1.$

Another usual notation is to indicate, under the Σ sign, the values x can take so that from the Σ notation itself one will know the values x can take. For example, if x takes the values a, b, and c then $\Sigma x = a + b + c$ and this can be written as,

8. $\underset{a,\,b,\,c}{\Sigma} x = a+b+c = \underset{a,\,b,\,c}{\Sigma} a = a+b+c = \underset{a,\,b,\,c}{\Sigma} b = b+c+a =$

   $\underset{a,\,b,\,c}{\Sigma} c = c+a+b.$

   In this notation the elements are summed up taking them in a cyclic order.

9. $\left(\sum\limits_{a,\ b,\ c} a\right)^3 = (a+b+c)^3 \neq a^3 + b^3 + c^3$

$\sum\limits_{a,\ b,\ c} 5 = 5+5+5 = 15.$

Since 5 does not contain the letters a or b or c corresponding to each term of the sum we get only 5 and hence the sum is 15.

When there are a number of elements in a set the elements may be denoted with the help of some suffixes. For example, a class of 10 girls and 15 boys may be numbered as a_1, a_2,..., a_{10}, b_1, b_2, ..., b_{15} where a's denote girls and b's denote boys. This is a convenient notation. The costs of production of 20 items may be denoted by c_1, c_2, ..., c_{20} so that the cost of production of the 12th item is c_{12} and the total cost of production of all the 20 items together is,

$$c_1 + c_2 + \ldots + c_{20}$$

In this case also a $\sum$ notation can be conveniently used. That is,

10. $\sum\limits_{j=1}^{20} c_j = c_1 + c_2 + \ldots + c_{20}$

11. $\sum\limits_{i=1}^{2} a_i b_i = a_1 b_1 + a_2 b_2$

12. $\sum\limits_{j=1}^{4} a_j^3 = a_1^3 + a_2^3 + a_3^3 + a_4^3$

13. $\sum\limits_{j=1}^{3} a_j^i$ for $i=1, 2$ means,

$= a_1^i + a_2^i + a_3^i$ for $i=1, 2$.

That is, $a_1^1 + a_2^1 + a_3^1$ and $a_1^2 + a_2^2 + a_3^2$.

14. $\sum\limits_{i=1}^{2} \sum\limits_{j=1}^{3} a_i b_j = \sum\limits_{i=1}^{2} \left\{ \sum\limits_{j=1}^{3} a_i b_j \right\} = \sum\limits_{i=1}^{2} \{a_i b_1 + a_i b_2 + a_i b_3\}$

$= \sum\limits_{i=1}^{2} a_i (b_1 + b_2 + b_3) = a_1(b_1 + b_2 + b_3) + a_2(b_1 + b_2 + b_3)$

$= (a_1 + a_2)(b_1 + b_2 + b_3)$

$= \left(\sum\limits_{i=1}^{2} a_i \right)\left(\sum\limits_{j=1}^{3} b_j \right)$

15. $\displaystyle\sum_{i=1}^{2}\sum_{j=1}^{3} a_{ij} = \sum_{i=1}^{2}\left\{\sum_{j=1}^{3} a_{ij}\right\} = \sum_{i=1}^{2}\{a_{i1}+a_{i2}+a_{i3}\}$

$\displaystyle = \sum_{i=1}^{2} a_{i1} + \sum_{i=1}^{2} a_{i2} + \sum_{i=1}^{2} a_{i3} = (a_{11}+a_{21}) + (a_{12}+a_{22})$

$+ (a_{13}+a_{23})$

$\displaystyle = a_{11}+a_{12}+a_{13}+a_{21}+a_{22}+a_{23} = \sum_{j=1}^{3}\sum_{i=1}^{2} a_{ij}$

16. $\displaystyle\sum_{i,j=1}^{5} a_{ij} = \sum_{i=1}^{5}\sum_{j=1}^{5} a_{ij}$

17. $\displaystyle\sum_{i=1}^{2}\sum_{j=1}^{4} a_i(b_j+c_i) = \sum_{i=1}^{2}\sum_{j=1}^{4} (a_ib_j+a_ic_i)$

$\displaystyle = \sum_{i=1}^{2}\sum_{j=1}^{4} a_ib_j + \sum_{i=1}^{2}\sum_{j=1}^{4} a_ic_i$

$\displaystyle = \left(\sum_{i=1}^{2} a_i\right)\left(\sum_{j=1}^{4} b_j\right) + \left(\sum_{i=1}^{2} a_ic_i\right)\left(\sum_{j=1}^{4} 1\right)$

$\displaystyle = (a_1+a_2)(b_1+b_2+b_3+b_4) + (a_1c_1+a_2c_2)4.$

18. $\displaystyle\pi_{a,\,b,\,c} a = abc = \pi_{a,\,b,\,c} b = \pi_{a,\,b,\,c} c.$

19. $\displaystyle\pi_{a,\,b,\,c} ab = (ab)(bc)(ca) = a^2b^2c^2 = (abc)^2 = \left[\pi_{a,\,b,\,c} a\right]^2$

$\displaystyle = \left(\pi_{a,\,b,\,c} a\right)\left(\pi_{a,\,b,\,c} b\right)$

20. $\displaystyle\pi_{a,\,b,\,c} a(b+c) = a(b+c)b(c+a)c(a+b) = abc(a+b)(b+c)(c+a)$

$\displaystyle = \left(\pi_{a,\,b,\,c} a\right)\left(\pi_{a,\,b,\,c} \frac{(a+b)}{c}\right)$

21. $\displaystyle\pi_{a,\,b,\,c} a^3 = a^3b^3c^3 = (abc)^3 = \left(\pi_{a,\,b,\,c} a\right)^3$

22. $\displaystyle\pi_{1=i}^{5} a_i = a_1a_2a_3a_4a_5$

23. $\displaystyle\pi_{i=1}^{3}\pi_{j=1}^{2} a_ib_j = \pi_{i=1}^{3} a_i \pi_{j=1}^{2} b_j = (a_1a_2a_3)(b_1b_2)$

24. $\displaystyle\pi_{i=1}^{n} \sum_{j=1}^{2} a_i b_{ij} = \pi_{i=1}^{n} a_i \sum_{j=1}^{2} b_{ij} = \pi_{i=1}^{n} a_i(b_{i1}+b_{i2})$

$\quad = a_1(b_{11}+b_{12}) \cdot a_2(b_{21}+b_{22}) \ldots . a_n(b_{n1}+b_{n2})$

$\quad = a_1 \ldots . a_n \cdot (b_{11}+b_{12}) \ldots . (b_{n1}+b_{n2}).$

Exercises

1.21 Write down the following in full:

    (a) $\displaystyle\sum_{i=1}^{n} x_i f_i$; (b) $\displaystyle\sum_{i=1}^{n} (x_i - y_i) f_i$; (c) $\displaystyle\sum_{i=1}^{n} x_i y_i f_i$.

1.22 If $\bar{x} = \displaystyle\sum_{i=1}^{n} x_i/n$ show that $\displaystyle\sum_{i=1}^{n} (x_i - \bar{x}) = 0$.

1.23 If $\bar{x} = \displaystyle\sum_{i=1}^{n} x_i f_i / \sum_{i=1}^{n} f_i$ show that $\displaystyle\sum_{i=1}^{n} f_i (x_i - \bar{x}) = 0$.

1.24 If $a_1 = 1$, $a_2 = 2$, $a_3 = -1$, $a_4 = 0$ evaluate the following:

    (a) $\displaystyle\sum_{i=1}^{4} a_i^2$; (b) $\displaystyle\sum_{i=1}^{2} a_i - \sum_{i=3}^{4} a_i$; (c) $\displaystyle\sum_{i=1}^{4} a_i^3$

1.25 Write down the following by using Σ and π notations

    (a) $x_{11}+x_{12}+x_{13}+x_{21}+x_{22}+x_{23}$.

    (b) $x_{11} y_{11} + x_{22} y_{22} + x_{33} y_{33}$.

    (c) $x_1^2 f_1 + x_2^2 f_2 + \ldots + x_n^2 f_n$

    (d) $(x_1 - \bar{x})^2 + (x_2 - \bar{x})^2 + \ldots + (x_n - \bar{x})^2$.

1.26 Show that $\displaystyle\left(\sum_{i,\,j=1}^{n} x_{ij}\right)^2 \neq \sum_{i,\,j=1}^{n} x_{ij}^2$, in general.

1.27 Evaluate $\displaystyle\pi_{i=1}^{n} x_i^2$ where $x_1 = \tfrac{1}{2}$, $x_2 = \tfrac{2}{3}$, ..., $x_n = \dfrac{n}{n+1}$

1.28 If $f(x_j) = a^x_{\,j}$ find $\displaystyle\pi_{j=1}^{n} f(x_j)$.

1.29 If $f(x_j) = ce^{-\kappa_j/\beta}$ evaluate $\displaystyle\pi_{j=1}^{n} f(x_j)$.

1.30 If $f(x_j) = ce^{-(x_j - \alpha)^2/\beta}$ evaluate $\displaystyle\pi_{j=1}^{n} f(x_j)$.

1.6 Measures of Central Tendency and Dispersion

Description of a statistical data is by no means complete by classifying the data or presenting the data with the help of charts and diagrams. We can study the data in more detail depending upon the purpose for which the data is collected. That is, the different aspects or characteristics of the data are studied for different purposes. A technician admires a new alarm clock for its new mechanical features, a businessman may admire it for its market value, an artist admires it for the beauty in its design but all these people may be interested to see whether it keeps correct time and if not how much it gains or loses on the average per day, and whether its behaviour in losing or gaining time is erratic or not. In a statistical data, as in the case of an alarm clock, there are some aspects in which most of its users are usually interested in. They are the central tendency of the data and the dispersion in the data.

1.61 *Measures of Central Tendency*

These measures are also sometimes referred to as measures of 'location', measures of 'central values', measures of 'positions' and so on. In defining these measures, the attempt is to represent the whole data by a typical value. In general it is not possible to represent the whole data by a single value. In section 1.41 we have seen that even the classification of the data is done at the expense of some information about the whole data. So naturally these 'typical values' will represent only certain aspects of the data.

1.62 *The Mean*

Definition. A simple arithmetic mean (the sum divided by the total number) is defined as the mean value of a set of numbers. That is, the mean value of $x_1, x_2, ..., x_n$ is denoted by $\bar{x}$ and is,

$$\bar{x} = \frac{x_1 + x_2 + ... + x_n}{n}$$

This is not always the same as the word 'average' used in ordinary conversational language such as 'average person', 'average intelligence ' and so on. The mean value of the set of numbers —1, 2.2, 1, 5, 7.8 is

$$\frac{-1+2.2+1+5+7.8}{5} = 3.$$

In general we may use the notation,

$$\bar{x} = \frac{\Sigma x}{n} \qquad (1.1)$$

where Σx denotes the sum of all the n values x takes. The mean value of the set of numbers, 1, 1, 2, 1, 2, 2, 2, 3.5, 4.7, 5.5, 6.3, 8 is

$$\frac{1+1+1+2+2+2+2+3.5+4.7+5.5+6.3+8}{12}$$

$$= \frac{1\times3+2\times4+3.5+4.7+5.5+6.3+8}{12} = 3.25$$

Here, number 1 is repeated 3 times and number 2 is repeated 4 times and hence a simplification in the computation is possible. Evidently if x_1 occurs f_1 times, x_2 occurs f_2 times, ..., x_n occurs f_n times then the mean value of all these x_1, x_2, ... x_n is,

$$\bar{x} = \frac{x_1 f_1 + x_2 f_2 + \dots + x_n f_n}{f_1 + f_2 + \dots + f_n} = \frac{\Sigma x f}{\Sigma f} \qquad (1.2)$$

These values, x_1, x_2, ..., x_n can be considered to have weights f_1, f_2, ..., f_n and hence

$$\bar{x} = \frac{\Sigma x f}{\Sigma f}$$

is often called the *weighted mean value* of x_1, x_2,..., x_n with the weights f_1, f_2, ..., f_n respectively. In order to find the average marks obtained by a student in 4 different subjects it may not be advisable to add up the marks of the 4 subjects and divide by 4 if one subject was taught twice a week for 5 weeks, another one at the rate of 3 times a week for 12 weeks and the other two twice a week for 25 weeks. A typical average mark is obtained by assigning appropriate weights to the marks and then taking a weighted average. The average $\bar{x}$ can also be considered to be the *centre of gravity* of a system of weights f_1, ..., f_n acting at the points x_1, x_2, ..., x_n on a horizontal line.

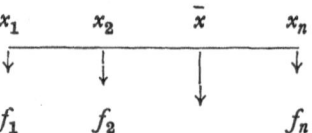

FIGURE 1.10 Weights $f_1, ..., f_n$, acting at the points $x_1, ..., x_n$

Then a simple average $\bar{x} = (x_1 + x_2 + ... + x_n)/n$ is the centre of gravity of a system of unit weights at the points $x_1, ..., x_n$.

Suppose that the weights of 4 army recruits are obtained as 75, 76, 91, 84 and the average weight is reported as $(75+76+91+84)/4 = 81.5$. After reporting, it was noticed that the weighing scale was faulty and the initial point was 5 instead of zero. That means the true weights are 5 units less than the weights reported and thus the true average is

$$\frac{(75-5)+(76-5)+(91-5)+(84-5)}{4}$$

$$= \frac{75+76+91+84}{4} - \frac{20}{4}$$

$$= 81.5 - 5 = 76.5.$$

In general if

$$x = a + bu$$

then, $\bar{x} = a + b\bar{u}$ (1.3)

where u and x take corresponding values and a and b are some constants. In the example, the true weight x was the weight u given by the balance 'minus 5'. That is

$$x = -5 + u$$

and the average was 81.5 and hence by using the formula,

$$\bar{x} = a + b\bar{u}$$

where $a = -5$ and $b = 1$ we get the true average $\bar{x}$ as

$$\bar{x} = -5 + \bar{u} = -5 + 81.5 = 76.5.$$

Formula (1.3) is a convenient and useful formula by which we can compute the average of a set of very large numbers by subtracting or adding a suitable number to all the numbers and thus reducing to simple numbers, if possible.

Example 1.62.1 Evaluate the arithmetic mean of the following numbers: 10235, 10234, 10236, 10237.

Solution. Let $x=10230+u$ then $\bar{x}=10230+\bar{u}$. Here u is $x-10230$. That is,

$$u_1=10235-10230=5; \ u_2=10234-10230=4$$
$$u_3=10236-10230=6; \ u_4=10237-10230=7$$

Hence,

$$\bar{u}=(5+4+6+7)/4=5.5$$

Therefore,

$$\bar{x}=10230+5.5=10235.5$$

Example 1.62.2 Evaluate the arithmetic mean from the following frequency table.

TABLE 1.8

Class	10—12	13—15	16—18
frequency	2	3	1

Solution. Since we do not know the exact values of the 2 observations in the first class 10–12 and so on (this information is lost due to classification) we can only assign the class marks as typical values for all the observations in the particular classes. Hence the 2 observations in the first class are given the class mark 11, the 3 observations in the second class are given the class marks 14 and the one in the third class is given the value 17. Thus the required average is,

$$\bar{x}=\frac{11\times2+14\times3+17\times1}{2+3+1}=13.5.$$

When we have a large frequency table, the computations can be made easier by using the formula (1.3). This is illustrated in the following table, for the data in Table 1.8.

TABLE 1.9

Class	*Frequency f*	*Class mark*	*u*	*uf*
10—12	2	11	−1	−2
13—15	3	14	0	0
16—18	1	17	1	1
	$6=\Sigma f$			$-1=\Sigma uf$

Here u is obtained by subtracting 14 from the class marks and then dividing by the class interval 3. That is,

$$u = \frac{x-14}{3} \text{ or } x = 14 + 3u$$

Hence, $\bar{x} = 14 + 3\bar{u}$ (by using formula (1.3))

But, $\bar{u} = (\Sigma uf)/\Sigma f = -\frac{1}{6}$

Therefore, $\bar{x} = 14 + 3(-\frac{1}{6}) = 14 - \frac{1}{2} = 13.5$.

Comment. It is easy to see that when the formula $\bar{x} = a + b\bar{u}$ is used in a frequency table, and if the classes are of equal width then a is the class mark which is subtracted from the other class marks and b is the class width.

Exercises

1.31 The depth of a mountain stream at every metre away from one shore at a particular point is given as, 0.5, 0.7, 1, 1.5, 3, 6, 10, 15, 12, 6, 3, 2, 1, 1, 0.5, 0.2. Calculate the average depth. Is this average of any use for a person who wants to cross the stream?

1.32 The total sale of a particular item in a shop, on 10 consecutive days, is reported by a clerk as, $35.00, $29.60, $38.00. $30.00, $40.00, $41.00, $42.00, $45.00, $3.60, $3.80. Calculate the average. Later it was found that there was a number $10.00 in the machine and the reports of the 4th to 8th days were $10.00 more than the true values and in the last 2 days he put a decimal in the wrong place thus for example $3.60 was really $36.0. Calculate the true mean value.

1.33 Three people A, B, C were given the job of finding the average of 5000 numbers. Each one did his own simplification. A's method. Divide the set into sets of 1000 each, calculate the average in each set and then calculate the average of these averages. B's method. Divide the set into 2000 and 3000 numbers, take average in each set and then take the average of the averages. C's method. 500 numbers were unities. He averaged all other numbers and then added one. Are these methods correct?

1.34 Is it possible to find out the average weight of the people from the following table:

Weight (kg)	Number of people
Less than 50	25
50—59	35
60—69	30
over 69	20

1.35 Calculate the mean values from the following frequency tables:

(a)

Class	Frequency
5—9	3
10—14	7
15—19	10
20—24	8
25—29	2

(b)

Class	Frequency
5—9	2
10—14	4
15—24	10
25—44	24
45—64	20

1.36 If x, u, v all can take values and if $x = au + bv$ show that $\bar{x} = a\bar{u} + b\bar{v}$, where a and b are some constants.

1.37 If possible, calculate the following from 1.35 (a) and (b).
   (a) Number of observations between 9.6 and 21.
   (b) Number of observations between 9.5 and 19.5.
   (c) Number of observations between 10 and 24.
   (d) The point below which 50% of the observations fall.
   (e) The interval on which the middle 50% of the observations fall.

1.38 *Geometric and Harmonic means*: In mathematical calculations we often use other types of averages known as geometric mean and harmonic mean. But in our discussion ' mean value ' of a set of numbers means the arithmetic mean. *The geometric mean of a set of non-negative numbers $x_1, x_2, \ldots, x_n$ is a nth root of the product.* That is, the geometric mean (G.M.) is,

$$\text{G.M.} = \left\{ x_1 \ldots x_n \right\}^{\frac{1}{n}}$$

The harmonic mean of a set of non-negative numbers is the reciprocal of the arithmetic mean of the reciprocals of the numbers. That is, if x_1, x_2, ..., x_n are the numbers, the harmonic mean (H.M.) is,

$$\text{H.M.} = \frac{n}{\sum\limits_{i=1}^{n}\left(\dfrac{1}{x_i}\right)}$$

Calculate the G.M. and H.M. of the following data.

(a) 2, 1, 3, 5.

(b) $x_1, x_2, ..., x_n$ with frequencies $f_1, f_2, ..., f_n$ respectively.

1.39 *Simple aggregate index*: In economic studies of price changes in commodities, usually index numbers are constructed so that one can have an idea about the prices of commodities in any year compared to the prices in a base year (a typical year based on which the index number is calculated). A simple aggregate index number is given by the average $\left(\dfrac{\Sigma P_n}{\Sigma P_0}\right)100$, represented in percentage, where P_0 and P_n denote the price of a particular commodity in the base year and in the current year (year under consideration) respectively. The following table gives the prices of detergents of 5 different brands in 1960 and 1968. Calculate the simple aggregate index of the prices taking 1960 as the base year.

	Brand 1	Brand 2	Brand 3	Brand 4	Brand 5
1960	0.90	0.92	0.94	0.90	0.96
1968	1.12	1.15	1.16	1.16	1.21

1.40 *A weighted aggregate index*: Since various amounts of the different commodities are produced, consumed or sold an improved index number is a weighted average taking the quantities produced or consumed or sold in the base year or in any other typical year as the weights. *Hence a weighted aggregate index is the weighted average,* $\left(\dfrac{\Sigma q_n P_n}{\Sigma q_n P_0}\right) 100$ *represented in percentage,* where q_n denotes the quantity of a particular item in the current year.

For the data in Problem 1.39 calculate the weighted aggregate index taking the quantities produced in 1968 as the weights. The output of the various brands in 1968 is given in the following table.

	Brand 1	Brand 2	Brand 3	Brand 4	Brand 5
1968	20	25	30	40	55

1.63 *The Fractiles*

The simple arithmetic mean is not often a good measure of central tendency. For example, for a person who wants to cross a river the average depth may not be of much use. For a construction company the average temperature at a particular place is not enough if they want to construct something which can withstand the heat at any time. In summer it may reach over 100°F and in winter it may go below −20°F but the average may be 70°F and if something is erected to withstand 70°F then it would be foolish. Even if in some cases an arithmetic mean is desirable we may not be able to compute it. For example, the data in problem 1.34 is classified in such a way that the end classes are not well defined, in this sense, the class marks cannot be computed without further assumptions. Also a simple arithmetic mean is affected if one or more observations are wrong in the data. To save some of these situations, there are other measures of central tendency available. They are the median and the mode.

Definition. A certain fractile point, for example, the P% point, is defined as that point below which falls P% of the total number of observations. The 10%, 20%,..., 90% points are known as the *decile points*. The 25%, 50%, 75% points are known as the first *quartile* Q_1, the *second quartile* Q_2 and the *third quartile* Q_3, points respectively. The second quartile point Q_2 is also known as the median point.

Definition. The median point is that point such that the number of observations below it equals the number of observations above it.

For convenience of calculations, if there are an odd number of numbers then the middle number is taken as the median and if there are an even number of numbers then the arithmetic mean of the two central ones is taken as the median.

Example 1.63.1 Calculate the median from the following data (a) 1, 5, 7, 15, 23.   (b) 9, 3, 2, 12.

Solution. (a) Since there are 5 observations the middle one is the third one if they are arranged according to their magnitudes. Hence the median is 7.   (b) There are 4 observations and therefore this median is the average of the middle ones.   When the numbers are arranged according to their magnitudes, that is, 2, 3, 9, 12, the central ones are 3 and 9 and hence the median is $(3+9)/2=6$. The techniques of calculating the fractiles are not that simple if some of the observations are repeated or if the data is given in a frequency table.   From a frequency table the different fractiles can be calculated by using more assumptions.

Example 1.63.2 From the following frequency table calculate (1) The median, (2) the 12th percentile point, (3) the first quartile Q_1 and (4) the second decile point.

TABLE 1.10

Class	Frequency	True lower limit	Less than cumulative frequency
5—7	1	4.5	0
8—10	3	7.5	1
11—13	8	10.5	4
14—16	10	13.5	12
17—19	7	16.5	22
20—22	1	19.5	29
	30	22.5	30

Solution. Columns 1 and 2 of Table 1.10 give the data and columns 3 and 4 are computed for calculating the required fractiles. The fractiles will be calculated by using the following assumption. Since nothing is given about the true values of the observations, the observations in any class are assumed to be uniformly or evenly distributed in that class interval.

(1) The median.   There are 30 observations and therefore this median value corresponds to the 15th observation, because in an ungrouped data we would have looked for the middle

item but since the data is given in a frequency table we are assuming that the observations are spread on a continuous line and we want to take that point on the line such that half of the frequencies are below the point and half of them are above the point. But the 15th observation is in the interval 13.5—16.5 because we have 12 observations which lie below 13.5 and 22 observations lying below 16.5. Hence we will divide the interval 13.5 to 16.5 equally among the 10 observations, that is, $3/10=0.3$ each for every observation. That is the 15th observation now corresponds to the number

$$13{\cdot}5 + (15-12) \times \frac{3}{10} = 14.4.$$

Hence a general formula for the median, calculated from a frequency table, can be given as follows.

$$\text{Median} = L_m + \left(\frac{N}{2} - f_{cm}\right)\frac{c}{f_m}$$

where L_m denotes the true lower limit of the *median class* (class in which the median lies), N is the total frequency, f_{cm} is the less than cumulative frequency up to L_m, c is the class interval of the median class.

(2) The 12th percentile point. Following the same argument as in (1) it is seen that the 12th percentile point is that point corresponding to 30 $(\frac{12}{100})=3.6$th observation. Even though there cannot be an observation which is a fraction we will calculate the percentile by using 3.6 because we are selecting a point on a continuous interval. The 3.6th observation lies in the interval 7.5—10.5. Up to 7.5 we have one observation and we will distribute the interval 7.5—10.5 for the frequency 3 and take the fraction for 2.6 units. Hence the required point is,

$$=7{\cdot}5 + (3.6-1)\tfrac{3}{3} = 10.1$$

(3) The one-fourth of the total frequency is 7.5 and hence Q_1 corresponds to 7.5, which falls in the interval 10.5—13.5. That is,

$$Q_1=10.5+(7.5-4)\tfrac{3}{8}=11.825.$$

(4) The second decile point is the 20th percentile point, that is, the point corresponding to $30 \times \dfrac{20}{100} = 6$ which falls in the interval 10.5 to 13.5. Hence the required point is,

$$= 10.5 + (6-4) \times \tfrac{3}{8} = 11.25.$$

In general the formula for a pth percentile point D_p can be given as follows.

$$D_p = L_p + \left(N \times \frac{p}{100} - f_{cp} \right) \frac{c}{f_p}$$

where L_p is the true lower limit of the pth percentile class (the class where the percentile point lies), f_{cp} is the cumulative frequency up to L_p, c is the class interval of the percentile class and f_p is the frequency in the percentile class.

Exercises

1.41 Calculate the median from the following observations.
  (a) $-1, 0, 2, -3, 5, 7$; (b) $-2, 0, 1, -5, 7, 6, 5, -1, 4, 1, 0.$

1.42 Calculate (1) The Median, (2) The third quartile Q_3, (3) The first decile point from the following data.

(a)

Class	Frequency
2—4	1
5—7	2
8—10	4
11—13	3
14—16	2

(b)

Class	Frequency
0—4	2
5—9	5
10—18	10
19—27	7

1.43 Calculate (a) the median from the following data. 1, 3, 4, 5, 6, 7, 8, 9, 11, 9, 10, 10, 12, 13, 12; (b) classify the data into 5 classes of equal widths taking 1 and 13 as the smallest lower limit and the largest upper limit respectively and calculate the median.

1.44 Calculate the third decile point from the ungrouped and grouped data in problem 1.43.

1.45 Calculate the median from the following data.

Class	Frequency
less than 10	5
11—16	4
17—22	4
over 23	11

1.64 *The Mode*

Sometimes the mean value or the median may not be of much value for a data analyst. A businessman interested in setting up a temporary shop for tourists at a particular place is not interested in the average number or the median number of tourists visiting that place but he will be interested in the peak period or the time when the maximum number of tourists visit the place. While scheduling flights an airline is not interested in the average number per day of people travelling by plane but the airline will be more interested in the peak seasons. There may be two peak seasons, one in summer when people take a vacation and one in the winter when people go to summer resorts.

Definition. In statistical language the mode in an ungrouped data is defined as the value corresponding to a peak when the observations are arranged according to their order of magnitudes and in a grouped data it is defined as the value corresponding to a maximum frequency.

Example 1.64.1 Obtain the mode, if it exists, for the following data (a) 1, 2, 2, 7; (b) 4, 5, 6; (c) −1, 0, 0, 1, 2, 4, 5, 6, 8; (d) 2, 7, 7, 8, 9, 10, 10, 10, 11, 12.

Solution. (a) Since 2 is the most frequently occurring item the mode is 2. (b) Since 4, 5, and 6 occur with frequency one each there is no mode. (c) Here 0 occurs with frequency 2 and hence the mode is zero. (d) Here 7 occurs with frequency 2 and 10 occurs with frequency 3 and others occur with frequency one. If these are represented by a histogram then there will be two peak points corresponding to the frequencies 2 and 3. Hence there are two modes, namely, 7 and 10. This can be called a *bimodal* data.

Comment. According to the definition a given data may or may not have a mode and if it has a mode the mode may not be unique, that is, there can be a number of modes.

Example 1.64.2 Calculate the mode, if it exists, for the following data.

Class	0—2	3—5	6—8	9—11
Frequency	2	4	5	3

Solution. Here the largest frequency is 5 and which corresponds to the interval 5.5—8.5. So any point in this interval can be the mode. Since there is only one maximum frequency there is only one mode. In order to assign a single value for the mode, in the interval 5.5—8.5, the following formula can be used.

$$\text{Mode} = L_1 + \frac{(f - f_1)\, c}{[(f-f_1) + (f-f_2)]},$$

where f is the frequency in the modal class (the class corresponding to a maximum frequency), L_1 is the lower class limit of the modal class, f_1 is the frequency in the preceding class and f_2 is the frequency in the succeeding class and c is the class interval of the modal class. Hence the mode in this example is,

$$\text{Mode} = 5.5 + \left[\frac{5-4}{(5-4) + (5-3)}\right] 3 = 6.5$$

Exercises

1.46 Write down the advantages and disadvantages of the measures the mean, the median and the mode as measures of central tendency in a data.

1.47 Calculate the mode or modes in the following data.
(a) 4, 7, 8, 9, 25; (b) $-1, 1, 0, 1, 1, 2, 4, -2, 0, 2, 2, 8, 2$;
(c) 1, 2, 3, 4, 4, 4, 5.

1.48 Calculate the mode from the following table.

(a)

Class	Frequency
10—12	2
13—15	10
16—18	5

(b)

Class	Frequency
less than 20	10
21—25	12
26—34	8
35 and over	4

1.49 Calculate the mean, the median and the mode for the data in the problem 1.48(a) and check for the approximate relation among the measures, that is,

$$\text{Mean} - \text{Mode} \approx 3 \, (\text{Mean} - \text{Median})$$

which is an approximate empirical relation for unimodal data.

1.50 Calculate the mean, median and the mode for the following data and check for the empirical relationship among them. $-1, 1, 0, 2, 3, 5, 5, 6, 8, 10, 11.$

1.7 MEASURES OF DISPERSION

A measure of central tendency is not good enough for a good description of a numerical data. Suppose that there are two townships where the average annual income of the people is $10,000 per family unit, for both the townships. This information is not enough for a welfare agency which wants to open an office in one of the townships which has a more concentration of poor people. In one township the people may have annual income around $ 10,000 and there may not be much disparity among the incomes. But in the other township there may be a few millionaires and a lot of people may be poor and thus a great disparity among the incomes. So a measure of disparity or scatter or dispersion of the elements from some value, namely $ 10,000 in the above example, is of some importance in data analysis. A house builder will be more concerned about the extreme climatic conditions rather than the average temperature or the average humidity.

A data analyst will be interested in three types of scatter in a given data. (1) the scatter or range between the extreme values; (2) the scatter or dispersion of the elements from some central value; (3) the scatter among individual elements in a given data. The following are some of the measures designed to measure these different types of scatter.

1.71 The Standard Deviation

Definition. Let $x_1, x_2, \ldots, x_n$ be a set of n numbers. The standard deviation S.D. of these numbers $x_1, \ldots, x_n$ is defined as

$$\text{S.D.} = \left\{ \frac{\sum\limits_{i=1}^{n} (x_i - \bar{x})^2}{n} \right\}^{\frac{1}{2}} \tag{1.71.1}$$

and $(S.D.)^2$ is known as the variance of $x_1, x_2, \ldots, x_n$.

The standard deviation measures the scatter of the elements x_1, ..., x_n from the mean value $\bar{x}$. The individual differences $x_1 - \bar{x}$, $x_2 - \bar{x}, ..., x_n - \bar{x}$ can be called deviations of $x_1, ..., x_n$ from $\bar{x}$. Since the standard deviation is obtained by taking the *positive square root* of the sum of squares of the deviations it is often called the *root mean square deviation* where the deviations are taken from the mean value $\bar{x}$. In general we can define a root mean square deviation from some arbitrary value d as follows:

Definition. The root mean square deviation of $x_1, ..., x_n$ from d is defined as

$$\text{R.M.S.D.} = \left\{ \frac{1}{n} \sum_{i=1}^{n} (x_i - d)^2 \right\}^{\frac{1}{2}} \qquad (1.71.2)$$

That is, when $d = \bar{x}$ we get the standard deviation of $x_1, x_2, ..., x_n$. For computational convenience we can express the standard deviation in different forms.

$$(\text{S.D.})^2 = \frac{\sum\limits_{i=1}^{n} (x_i - \bar{x})^2}{n} = \frac{1}{n} \left\{ (x_1 - \bar{x})^2 + (x_2 - \bar{x})^2 + ... + (x_n - \bar{x})^2 \right\}$$

$$= \frac{1}{n} \left\{ \sum_{i=1}^{n} (x_i^2 + \bar{x}^2 - 2x_i \bar{x}) \right\}$$

$$= \frac{1}{n} \left\{ \sum_{i=1}^{n} x_i^2 + \sum_{i=1}^{n} \bar{x}^2 - \sum_{i=1}^{n} 2x_i \bar{x} \right\}$$

$$= \frac{1}{n} \left\{ \sum_{i=1}^{n} x_i^2 + n\bar{x}^2 - 2\bar{x} \sum_{i=1}^{n} x_i \right\}$$

$$= \frac{1}{n} \left\{ \sum_{i=1}^{n} x_i^2 + n\bar{x}^2 - 2n\bar{x}^2 \right\}$$

$$\left(\text{Since } \sum_{i=1}^{n} x_i = n\bar{x} \right)$$

$$= \frac{\sum\limits_{i=1}^{n} x_i^2}{n} - \bar{x}^2$$

$$= \frac{\sum\limits_{i=1}^{n} x_i^2}{n} - \left(\frac{\sum\limits_{i=1}^{n} x_i}{n^2} \right)^2$$

For computational purposes the formula,

$$(\text{S.D.})^2 = \frac{\left(\sum\limits_{i=1}^{n} x_i^2 \right)}{n} - \frac{\left(\sum\limits_{i=1}^{n} x_i \right)^2}{n} \tag{1.71.3}$$

is often used. For the details of the above derivation see the technical note in section 1.5.

Example 1.71.1 Calculate the standard deviation of the following set of numbers. $-3, 0, 1, 2, 2, 3, 5, 6, 2$.

Solution. $n=9$ since there are 9 numbers. For illustration we will compute the S.D. by using formula 1.71.1 and 1.71.3. In order to use formula 1.71.3 we need $\sum\limits_{i=1}^{n} x_i$ and $\sum\limits_{i=1}^{n} x_i^2$.

$$\sum_{i=1}^{n} x_i = x_1 + x_2 + \ldots + x_n = -3 + 0 + 1 + 2 + 2 + 2 + 3 + 5 + 6 = 18$$

$$\sum_{i=1}^{n} x_i^2 = x_1^2 + x_2^2 + \ldots + x_n^2 = (-3)^2 + 0^2 + 1^2 + 2^2 + 2^2 + 2^2 + 3^2$$
$$+ 5^2 + 6^2 = 92$$

Hence,

$$(\text{S. D.})^2 = \frac{92}{9} - \frac{18^2}{9^2} = 10.22 - 4 = 6.22$$

Therefore the S.D. is obtained by taking the positive square root of 6.22. That is, S.D. $= \sqrt{6.22} = 2.49$.

Method 2. In order to apply formula 1.71.1 we need $\bar{x}$ and the deviations of the numbers from $\bar{x}$. It is seen that $\bar{x}=2$. The deviations are,

$$x_1 - \bar{x} = -3 - 2 = -5, \, x_2 - \bar{x} = 0 - 2 = -2, \, x_3 - \bar{x} = 1 - 2 = -1,$$
$$x_4 - \bar{x} = 2 - 2 = 0 = x_5 - \bar{x} = x_6 - \bar{x}, \, x_7 - \bar{x} = 3 - 2 = 1,$$

$x_8 - \bar{x} = 5 - 2 = 3$, and $x_9 - \bar{x} = 6 - 2 = 4$. The sum of squares of the deviations is,

$$\sum_{i=1}^{n} (x_i - \bar{x})^2 = (-5)^2 + (-2)^2 + (-1)^2 + 0^2 + 0^2 + 0^2 + 1^2 + 3^2 + 4^2$$

$$= 56$$

Therefore,

$$\text{S.D.} = \sqrt{56/9} = \sqrt{6.22} = 2.49.$$

Comment. When some of the values are repeated, that is, if $x_1, x_2, \ldots, x_n$ occur with frequencies $f_1, f_2, \ldots, f_n$ respectively then the formula 1.71.1 and 1.71.3 can be written as,

$$(\text{S. D.})^2 = \frac{\sum\limits_{i=1}^{n} f_i (x_i - \bar{x})^2}{\sum\limits_{i=1}^{n} f_i} \tag{1.71.4}$$

$$(\text{S. D.})^2 = \frac{\sum\limits_{i=1}^{n} f_i x_i^2}{\sum\limits_{i=1}^{n} f_i} - \frac{\left(\sum\limits_{i=1}^{n} f_i x_i\right)^2}{\left(\sum\limits_{i=1}^{n} f_i\right)^2} \tag{1.71.5}$$

When we want to calculate the standard deviation from a frequency table the formula 1.71.4 or 1.71.5 can be used. These formulae can again be simplified by using the following theorems.

Theorem 1.71.1   Let $x_i = a + bu_i$ for $i = 1, 2, \ldots, n$. That is, $x_1 = a + bu_1$, $x_2 = a + bu_2, \ldots, x_n = a + bu_n$, where a and b are some constants. Then the standard deviation of $x_1, \ldots, x_n$ is $|b|$ times the standard deviation of $u_1, \ldots, u_n$.

That is,

$$S_x = |b| S_u$$

where S_x and S_u denote the standard deviations of $x_1, \ldots, x_n$ and $u_1, \ldots, u_n$ respectively, and $|b|$ denotes the magnitude of b.

Proof. By definition,

$$S_x^2 = \sum_{i=1}^{n} \frac{(x_i - \bar{x})^2}{n}$$

and

$$S_u^2 = \sum_{i=1}^{n} \frac{(u_i - \bar{u})^2}{n}$$

Since $x_i = a + bu_i$,

$$\bar{x} = a + b\bar{u}.$$

Hence,

$$S_x^2 = \sum_{i=1}^{n} \frac{(x_i - \bar{x})^2}{n} = \frac{1}{n} \sum_{i=1}^{n} \{(a + bu_i) - (a + b\bar{u})\}^2$$

$$= \frac{1}{n} \sum_{i=1}^{n} \{b(u_i - \bar{u})\}^2$$

$$= \sum_{i=1}^{n} b^2 \frac{(u_i - \bar{u})^2}{n}$$

$$= b^2 \sum_{i=1}^{n} \frac{(u_i - \bar{u})^2}{n} = b^2 S_u^2$$

Example 1.71.2 Calculate the standard deviation of the following set of numbers.   20122, 20124, 20120.

Solution. If we use formula 1.71.1 or 1.71.3 it is quite difficult due to the fact that the numbers are quite large.   But the numbers can be written as,

$20122 = 20120 + 2 \times 1, 20124 = 20120 + 2 \times 2, 20120 = 20120 + 2 \times 0.$
Now by applying theorem 1.71.1 the standard deviation of the given set of numbers is the same as 2 times the standard deviation of the numbers 1, 2, 0.   The S.D. of 1, 2, 0 is

$$\text{S.D.} = \sqrt{\tfrac{2}{3}} = 0.81$$

Hence the required standard deviation is $2 \times 0.81 = 1.62$.

Example 1.71.3 Calculate the standard deviation from the following frequency table.

TABLE 1.11

Class	2—4	5—7	8—10	11—13
Frequency	4	6	7	3

Solution. Here it can be assumed that the 4 observations in the class 2—4 have the values equal to the class mark 3, the six observations in the class 5—7 have the values 6 (the class mark) each and so on. For simplification we will use the formula 1.71.4 and the theorem 1.71.1. In order to use these we will form the following table.

TABLE 1.12

Class	Frequency	Class mark	u	fu	fu^2
2—4	4	3	−1	−4	4
5—7	6	6	0	0	0
8—10	7	9	1	7	7
11—13	3	12	2	6	12
Total	20			9	23

Here u is formed by taking any one of the class marks as the starting point (here 6 is taken), subtracting the other class marks from it and then dividing by the class intervals. That is, the general formula is,

$$u = \frac{x - M}{c} \quad \text{or} \quad x = M + cu$$

where M is the class mark which is taken as the starting point from which deviations are taken and c is the class width. In our example,

$$x = 6 + 3u$$

Hence,
$$S_x = 3\, S_u$$

But,
$$S_u^2 = \frac{\Sigma f u^2}{\Sigma f} - \left(\frac{\Sigma f u}{\Sigma f}\right)^2$$
$$= \tfrac{23}{20} - (\tfrac{9}{20})^2 = 0.9475$$

That is,
$$S_u = 0.97$$

Therefore,
$$S_x = 3 \times 0.97 = 2.91$$

Comment. It is easy to use the above described method but if the class widths are different in different classes then the deviations can be divided by an appropriate number, for example, the most frequently occurring class width. The procedure remains the same and the simplification is achieved by using the theorem 1.71.1.

Exercises

1.51 Calculate the standard deviation of the following numbers:
  (a) 1, −1, 0, 2, 5, 7.
  (b) 1999, 1998, 1995, 2000, 2001, 2005.
  (c) −1, −1, −1, 2, 2, 2, 2, 5, 5, 5, 5, 7, 8.

1.52 Calculate the standard deviation from the following tables, if possible.

(a)

Observation	Frequency
−5	2
0	10
1	15
2	3

(b)

Class	Frequency
10—14	10
15—19	16
20—24	14

(c)

Class	Frequency
2—4	2
5—7	3
8—12	10
13—17	12
18—22	13

(d)

Class	Frequency
less than 10	5
10—12	4
13—15	3
greater than 15	10

1.53 Prove the following results.

  (a) In general $\{\Sigma(x_i-\bar{x})^2\}^{\frac{1}{2}} \neq \Sigma(x_i-\bar{x})$;

  (*Hint:* Give a counter example)

  (b) If $x=au+bv$ where a and b are constants and u and v take corresponding values as x then,

  $$S_x^2=a^2\,S_u^2+b^2\,S_v^2+2ab\,.\,\sum\frac{(u-\bar{u})\,(v-\bar{v})}{n}$$

  where n is the number of values assumed by x.

  (c) $S^2\left(\dfrac{x-\bar{x}}{S_x}\right)=1$ (That is the variance of the numbers

  $\dfrac{x_1-\bar{x}}{S_x},\ \dfrac{x_2-\bar{x}}{S_x},\ \ldots,\ \dfrac{x_n-\bar{x}}{S_x}$ is unity where S_x is the standard deviation of $x_1, x_2,\ldots, x_n$).

1.54 Show that the root mean square deviation is least when the deviations are taken from the mean value. That is,

$$\sqrt{\frac{\Sigma(x_i-d)^2}{n}} \quad \text{has the smallest value when } d=\bar{x}.$$

Hint. $\Sigma(x_i-d)^2=\Sigma(x_i-\bar{x}+\bar{x}-d)^2=\Sigma(x_i-\bar{x})^2+n(\bar{x}-d)^2$, since $\Sigma(x_i-\bar{x})=0$.

1.55 Obtain $S^2(x-y)$ in terms of S_x^2 and S_y^2 and show that, in general, $S(x+y)\neq S_x+S_y$, $S(x-y)\neq S_x+S_y$, and $S(x-y)\neq S_x-S_y$.

1.56 When is the variance of a set of numbers zero?

1.57 Can the standard deviation of a set of numbers be negative? Why?

1.72 *The Mean Absolute Deviation*

Another measure of scatter of the numbers $x_1, x_2,\ldots, x_n$ from some number d is the mean absolute deviation (sometimes known as the mean deviation) which is defined as,

$$\text{M.D.} = \sum_{i=1}^{n}\frac{|x_i-d|}{n}$$

where $|x_i-d|$ means the absolute value (without considering the sign) of x_i-d. Here we deal with only real numbers and, for example, $|-2|=2, |2|=2, |-8|=8, |0|=0$. When $x_1, x_2,\ldots, x_n$ occur with frequencies $f_1, f_2,\ldots, f_n$ respectively then the expression for the mean absolute deviation reduces to,

$$\frac{\sum\limits_{i=1}^{n} f_i|x_i-d|}{\sum\limits_{i=1}^{n} f_i};$$

when $d=\bar{x}$ we get the mean absolute deviation from $\bar{x}$.

Example 1.72.1 Obtain the mean absolute deviation from 3, of the following numbers: $-1, -2, 0, 1, 4, 5$.

Solution. M.D. $= \frac{1}{6} \{ |-1-3| + |-2-3| + |0-3| + |1-3|$
$$+ |4-3| + |5-3| \}$$
$$= \frac{1}{6} \{4+5+3+2+1+2\} = 2.83.$$

Comment. It may be noticed that the mean absolute deviation is a measure of scatter of the elements from a number d and is zero if and only if all the numbers coincide with d. Some other measures of scatter of the numbers $x_1, x_2, \ldots, x_n$ from d are

$$M_r = \left\{ \frac{\sum\limits_{i=1}^{n} f_i |x_i - d|^r}{\sum\limits_{i=1}^{n} f_i} \right\}^{\frac{1}{r}} \quad \text{for } r \geqslant 1.$$

When $r=1$ we get the mean absolute deviation from d and when $r=2$ and $d=\bar{x}$ we get the standard deviation.

1.73 *The Range and Interquartile Range*

The extent to which the individual elements in a data are scattered is sometimes measured by the range.

Definition. The difference between the largest and the smallest numbers, in a given set of numbers, is known as the range of the numbers.

Definition. The semi-interquartile range is defined as $(Q_3 - Q_1)/2$ where Q_3 is the third quartile point and Q_1 is the first quartile point. Sometimes for comparison of two sets of data a relative measure of scatter is used. These are defined as follows:

Coefficient of variation=(Standard deviation)/(Mean value).

Relative absolute deviation=(Mean absolute deviation from the mean)/(the mean value).

Example 1.73.1 For the following data obtain (1) the range, (2) the semi-interquartile range, (3) the coefficient of variation, 0, 1, −1, −2, 4, 5, 6, 8, 12, 10, 11.

Solution.

(1) Since the numbers vary from −2 to 12 the total range is
$12-(-2)=14$.

(2) The numbers, if arranged in the order of magnitude, are, $-2, -1, 0, 1, 4, 5, 6, 8, 10, 11, 12$. Hence $Q_1=0, Q_2=5$, $Q_3=10$ and the interquartile range is, $(Q_3-Q_1)/2=(10-0)/2 =5$.

(3) The mean value is,

$\bar{x}=(-2-1+0+1+4+5+6+8+10+11+12)/11=4\cdot9$

$(\text{S.D.})^2=[(-2)^2+(-1)^2+0^2+1^2+4^2+5^2+6^2+8^2+10^2 +11^2+12^2]/11-(4.9)^2=22.535.$

S.D.$=4.74.$

Hence the coefficient of variation,

$=(\text{S.D.})/\bar{x}=4.74/4\cdot9=0.97.$

Comment. If the data is given in a frequency table the total range can be taken as the difference between the largest true upper class limit and the smallest true lower class limit.

Exercises

1.58 Calculate the mean absolute deviation from the mean value for the following data.

(a)

Observation	Frequency
−2	3
−4	5
0	8
2	4

(b)

Class	Frequency
3—6	1
7—10	5
11—14	6
15—18	4

1.59 For the data in problem 1.58 calculate (1) the range, (2) the semi-interquartile range, (3) the coefficient of variation, (4) relative absolute deviation.

1.60 Show that, in general, $\left\{ \sum_{i=1}^{n} \frac{(x_i-\bar{x})^2}{n} \right\}^{\frac{1}{2}} \neq \sum_{i=1}^{n} \frac{|x_i-\bar{x}|}{n}$

1.61 Show that the mean absolute deviation is least when the deviations are taken from the median. That is, the least

value of $\sum_{i=1}^{n} \dfrac{|x_i - d|}{n}$ is attained when d is the median of

$x_1, x_2, ..., x_n$.

Hint. Let for $i = 1, 2, ..., m$ the x's be greater than d and the remaining x's be less than d then,

$$\sum_{i=1}^{n} |x_i - d| = \sum_{i=1}^{m} (x_i - d) + \sum_{i=m+1}^{n} (d - x_i).$$

1.62 There are some approximate relationships among the different measures of scatter for a data which when represented gives a moderately symmetric looking histogram. The relationships are given below. Check for these relationships by using the data in problem 1.58.

Mean deviation $= \frac{4}{5}$ (Standard deviation)

Semi-interquartile range $= \frac{2}{3}$ (Standard deviation).

1.63 *Moments.* The rth moment of $x_1, ..., x_n$ about the origin is defined as,

$$m_r' = \dfrac{\sum\limits_{i=1}^{n} x_i^r}{n} = (x_1^r + x_2^r + ... + x_n^r)/n \text{ for } r \geqslant 1$$

The rth central moment of $x_1, ..., x_n$ is defined as

$$m_r = \sum_{i=1}^{n} \dfrac{(x_i - \bar{x})^r}{n} \text{ for } r \geqslant 1$$

The rth absolute moment of $x_1, ..., x_n$ about the mean value is defined as

$$M_r' = \sum_{i=1}^{n} \dfrac{|x_i - \bar{x}|^r}{n} \text{ for } r \geqslant 1$$

Calculate m_3', m_4 and M_2' for the following data.

Observation	0	1	−1
Frequency	2	3	5

1.64 Show that if $x_1,...,$ x_n are real numbers then $M_2's=m_2s$ for $s=0, 1, 2,....$

1.65 If $x=a+bu$ where a and b are some constants then show that the rth absolute moment of x's equals $|b|^r$ times the rth absolute moment of u's, about the mean values.

Additional Exercises

1.1 Can the following be considered to be (a) population; (b) a sample, if so, what is the corresponding population.
   (1) The marks obtained by 10 students in a class.
   (2) The income tax returns of the residents of a township.
   (3) The prices of round steak in 5 different stores.
   (4) The prices of all different items in a shop.
   (5) The numbers of people visiting a psychiatrist on 10 consecutive days.
   (6) The demand for a particular make of car for 15 years.

1.2 Can the following be a simple random sample, if so, specify the populations.
   (1) The sale of a commodity for 5 consecutive years.
   (2) The eggs found in a bird's nest.
   (3) The rabbits shot by a hunter.
   (4) The prices of all the different items in a shop.
   (5) The diameters of the first 10 iron rods produced by a machine.
   (6) The set of numbers obtained when a die is rolled 5 times. (A die is a cube with the faces marked by the numbers 1 to 6.)

1.3 By using a table of random numbers select 2 random samples of size 10 each from the following data.
   (1) −3, −4, −7, 0, 5, 2, 8, 20, 32, 45, 40, 41, 31, 30, 21, 22, 28, 29.
   (2) 0, 1, 2, 5, 2, 7, 6, −1, 2, 3, 8, −1, −2, −5, 0, 5, 6, 8, 20, 21, 25, 28, 21, 22, 20, 6, 7, 2, −1, 2, 5, 8, 12, 11, 13, 14, 15, 18, 20.

1.4 The following data gives the numbers of times 30 women could lift a weight at one stretch in a weight lifting contest.

20, 21, 22, 20, 21, 25, 28, 26, 29, 27, 28, 29, 26, 27, 24,
23, 22, 21, 25, 26, 27, 25, 26, 24, 25, 26, 24, 25, 27, 22.

(1) Classify the data by selecting 5 classes of equal widths.
(2) What are the class intervals?
(3) What are the true class limits?
(4) What type of diagrammatic representation is appropriate for this data?
(5) Is it advisable to fit a frequency curve to this data?
(6) How many people are there who could lift the weights more than 23 times?

1.5 The following data gives the result of an endurance test where 20 people were subject to heat and the maximum endurable temperature reading (in °F) is taken. 100, 101, 110, 120, 125, 130, 135, 128, 124, 126, 120, 110, 112, 115, 116, 115, 117 118, 120, 125. Group this data into 4 classes of equal width and answer all the questions from (2) to (5) in the above problem 1.4.

1.6 Calculate the following for the data in problem 1.4 before and after classification. (1) The mean; (2) The median; (3) The mode; (4) The standard deviation; (5) The mean absolute deviation from the mean; (6) The 24th percentile point; (7) The semi-interquartile range; (8) The third absolute moment about the mean value.

1.7 Calculate the same measures asked for in problem 1.6, for the data in problem 1.5.

1.8 The following data gives the marks in a certain subject obtained by the first year students in a college;

40, 42, 41, 45, 48, 50, 55, 58, 60, 62, 65, 68, 70, 71, 72, 73, 74, 75, 75, 76, 75, 77, 78, 80, 81, 82, 85, 90, 92, 96, 94, 90, 85, 82, 83, 84, 65, 68, 67, 60, 62, 65, 70, 72, 75, 77, 75, 70, 68, 75, 64, 65, 66, 69, 52, 55, 50, 65, 66, 68, 70, 71, 75, 76, 72, 70, 68, 80, 81, 68, 75, 70, 81, 82, 83, 84, 80, 69, 70, 71, 75, 60, 58, 50, 62, 81, 75, 68, 60, 62.

(a) By using a table of random numbers select 15 samples of size 10 each and calculate the following.

(1) The sample means.
(2) The mean value of the sample means and compare it with the population mean.

(3) The standard deviation of the sample means and compare it with the population standard deviation.

(4) The sample medians and compare it with the population median.

(5) The standard deviation of the sample medians and compare it with the population standard deviation as well as with the standard deviation of the sample means.

(6) Group the sample means into an appropriate number of classes.

(7) Draw a histogram for the sample means and compare it with a histogram for the population.

(8) Smooth the histogram for the sample means by a frequency curve and compare the curve with a similar curve for the population.

(b) By using a table of random numbers select 20 samples of size 30 each and calculate and represent as in (1) to (8) of (a).

(c) Draw the two curves for the sample means in (a) and (b) in a single graph.

(d) For the given data form the standardized scores, that is, $(x - \bar{x}) / (\text{S.D.})$. Classify and draw a frequency curve by smoothing a histogram.

(e) For the sample means in (a) and (b) also draw the frequency curves after standardizing the data.

(f) Compare the three 'standardized' curves. Which one of (e) is closer to the one in (d)?

(g) By using the results in (a) (3) and (b) (3) can you guess the behaviour of the standard deviation of the sample means when the sample size increases?

(h) By using the results in (a) (5) and (b) (5) can you guess the behaviour of the standard deviation of the sample medians when the sample size increases?

1.9 Calculate the standard deviation and the mean absolute deviation from the mean for the following data which gives the classification of the visitors of a historical monument.

Number of people in a party	1	2	3	4	5	6
Number of parties	4	20	18	12	4	2

1.10 Calculate (1) the range, (2) the standard deviation for the following data which gives the age group of people in a sit-in.

Age	14—16	17—19	20—24	25—30	31—40
Number of people	40	60	55	30	15

CHAPTER TWO

An Outcome Set

In this chapter we will develop the necessary tools for assigning a numerical measure for what is known in conversational language as ' chance ' such as the chances of winning a lottery, the chances of getting rain on a particular day and so on. There are situations where the results are (1) deterministic (sure), (2) predictable but not deterministic or in other words the results depend upon some chance variation. If a coin is thrown it is not certain whether it will fall on one side or on the other side because the various factors governing the result of this experiment of throwing a coin are not completely known to us and hence these factors induce a certain element of uncertainty in determining the result. But we know for sure that when a gold coin is thrown in water it sinks because the factors governing the sinking of the coin in water are known to us so that the result of this experiment is deterministic. In statistical theory we are interested in experiments whose results are known but not deterministic.

2.1 RANDOM EXPERIMENTS

Definition. A random experiment is a procedure which results in some non-deterministic outcomes in a particular situation. An outcome is a single realization of a phenomenon under consideration. The outcomes need not always be numbers or quantities which are representable in terms of numbers.

A random experiment, thus, has the meaning that randomness is associated with the outcomes rather than in performing the experiment. If a person selects a day at random and then puts a gold coin in water to see whether it sinks or not, the experiment is not a random experiment even though the day was selected at random. But if he tosses a coin and if he is interested in the outcomes whether it falls head or tail (one side is called head and the other side is called tail) it is a random experiment due to non-deterministic nature of the outcome whether he selects a day at random or not to perform the experiment.

2.11 *The Outcome Set or The Sure Event*

Definition. The set of all possible outcomes of a random experiment is called an outcome set. It is also called *the possibility set, the sure event, the sample space,* etc. Hereafter an experiment in our discussion means a random experiment.

Example 2.11.1 A coin is tossed twice. What is the outcome set?

Answer. Denoting one side by H and the other side by T and assuming that the coin does not stand on its edge when thrown, the possible outcomes are H for the first tossing and H for the second tossing, and so on. That is $(H, H), (H, T), (T, H), (T, T)$, where the first letter stands for the result in the first tossing and the second letter stands for the result in the second tossing. Hence the outcome set,

$$S = \{(H, H), (H, T), (T, H), (T, T)\}.$$

Comment. Here the experiment consists of two throws and hence an outcome consists of a couplet of letters. In a particular experiment suppose that one got a head in the first throw and a tail in the second throw. But the outcome set is not $\{H, T\}$. It is not the set of outcomes in individual acts in an experiment but it is the set of *possible* outcomes of the whole experiment (which may consist of a number of individual acts).

Example 2.11.2 Two girls are selected at random from a set of 24 cheerleaders. What is the outcome set?

Answer. It is the set of all such samples of size 2 that can be obtained from the set of 24 girls. Here, a single outcome of the experiment is a sample of size 2. Hence all possible outcomes are all possible such samples of size 2 each. In an experiment of sampling the outcome set is more appropriately called the *sample space*.

Example 2.11.3 Is the following correct? In an experiment of throwing a coin twice the outcome set is (2 heads, zero tail), (one head, one tail), (zero head, two tails).

Answer. No. Even though $(2H, 0T)$, $(1H, 1T)$, $(0H, 2T)$ are the only possibilities the outcome (one head, one tail) consists of two outcomes (H, T) and (T, H). Hence we do not take this representation as the outcome set. *We take that representation which does not allow any further subdivision of its elements as the outcome set.* Hence the outcome set here is,

$$S = \{(H, H), (H, T), (T, H), (T, T)\}.$$

2.12 *An Event*

Definition. *Event is a subset of the outcome set (sample space), and elementary event is an event consisting of only one element.* Thus an event consists of one or more elementary events.

Example 2.12.1 In an experiment of rolling a die once obtain the events (1) of getting an even number, (2) of getting a number less than 3, (3) of getting a number 8. (A die is a cube with the faces marked 1 to 6).

Solution. The outcome set is the set of numbers 1, 2,..., 6. That is, $S = \{1, 2,..., 6\}$ since one and only one of the numbers 1, ..., 6 appears in a single roll of the die. There are 6 elementary events in S, that is, $\{1\}$, $\{2\}$, ..., $\{6\}$. The event A of getting an even number consists of the elements 2, 4, 6 and hence

$$A = \{2, 4, 6\}$$

The event B of getting a number less than 3 is,

$$B = \{1, 2\}$$

The event C of getting a number 8 is impossible because there are only six numbers 1, 2, ..., 6 involved. Hence C is a null set. That is,

$$C = \phi$$

Comment. The event of getting 1 or 2 or 3 or 4 or 5 or 6, that is, the whole outcome set, is sure to happen and hence an outcome set is appropriately called *a sure event.* The event of getting a number 8 is impossible in this case but there are events which are *almost surely sure* and *almost surely impossible.* For example, if a monkey is given a typewriter to play with, then it is not logically impossible that it will not type this book word by word but it is practically impossible and we call such events almost surely impossible events. If a religious leader makes a statement that ' everyman who is born will die one day ' it is not logically quite correct. Even though it is verified that the first one million people died it does not logically follow that ' therefore the next one will also die one day '. But we know from verifications that it is almost sure that everybody will die one day. These events are also sometimes called ' statistically impossible ' and ' statistically sure ' events.

Example 2.12.2 If a coin is tossed twice write down the events of getting (1) at least one head, (2) exactly 2 tails, (3) at the most one tail.

Solution. The outcome set is,
$$S=\{(1, 1), (1, 0), (0, 1), (0, 0)\}$$
where 1 denotes a head and 0 denotes a tail.

(1) At least one head here means, one or two heads and hence the event consists of the elementary events
$$\{(1, 1), (1, 0), (0, 1)\}.$$

(2) There is only one element in this event and it is $\{(0, 0)\}$.

(3) This is the event of getting 0 or 1 tail and hence it is,
$$\{(1, 1), (1, 0), (0, 1)\}.$$

Exercises

2.1   Are the following experiments random experiments?

(1) A chemist mixes two known compounds to study the reaction.

(2) A fisherman catching 5 fish from a lake on a Sunday.

(3) Prescribing treatments for 10 of his patients, on a full moon by a psychiatrist.

(4) Throwing 2 dice 3 times, to look for a double 5.

(5) Drawing 2 cards from a deck of cards if the aim is (a) to get any two cards, (b) to see whether the cards are aces or not.

2.2   Write down the outcome set and enumerate the number of elementary events.

(1) A die is rolled twice.

    (2) Two birds are taken at random from a set of 5 birds.

    (3) One card is taken at random from a deck of 52 cards.

    (4) A spinner, with the dial marked 0 to 10, is rotated once.

    (5) The spinner in (4) is rotated twice.

2.3   In an experiment of rolling a die twice write down the events of

    (1) rolling 10 (sum of the face numbers is 10).

    (2) rolling at least 11.

    (3) rolling 13.

2.4   From the following numbers 2, 3, 5, 7 a random sample of size 2 is taken with replacements. Write down the event of,

    (1) The sample sum is at least 10.

    (2) The sample difference (first minus the second) is less than 0.

2.2 ALGEBRA OF SETS

Analogous to the mathematical operations of addition and multiplication on numbers, we can define operations on sets and develop a theory of sets. Here we are mainly interested in the algebra of events which may help us in assigning numerical values to the chances of the occurrence of these events in an experiment. Mainly, we define two operations 'union' and 'intersection' of sets.

2.21 *Union of Sets*

Definition.   *The union of two sets A and B, denoted by $A \cup B$ is the set of elements which belong to A or B or both.* From the definition itself it is evident that $A \cup B$ is the same as $B \cup A$.

    Example 2.21.1   Consider the following sets,

$$A = \{0, 1, -1\} \text{ and } B = \{1, 2, 5\}$$

then, $A \cup B = \{0, 1, -1, 2, 5\}$

    Comment. Here 0, 1, and -1 belong to A and hence they will be in $A \cup B$. 1, 2, 5 are in B and hence they will be in $A \cup B$. A and B contain 1 but which will not appear twice in $A \cup B$ because we are interested only in the elements and not in the number of times they occur in A, B and both A and B together. From the definition itself it is apparent that for any set A,

$$A \cup A = A$$
$$\text{and} \quad A \cup \phi = A$$

Further if $C \subset A$ then $C \cup A = A$.

Example 2.21.2 In an experiment of rolling a die twice obtain the event of getting a sum less than 8 and greater than 5 or exactly 10.

Solution. Here the outcome set,
$$S = \{(6, 1), (6, 2),..., (6, 6), (5, 1), ..., (1, 6)\}.$$

The event A of getting a sum less than 8 and greater than 5 is,
$$A = \{(6, 1), (5, 2), (5, 1), (4, 3), (4, 2), (3, 4), (3, 3),$$
$$(2, 5), (2, 4), (1, 6), (1, 5)\}$$

The event B of getting a sum exactly equal to 10 is,
$$B = \{(6, 4), (5, 5), (4, 6)\}$$

Hence the event of getting either a sum between 5 and 8 or a sum exactly equal to 10 is the union of A and B. That is,
$$A \cup B = \{(6, 1), (6, 4), (5, 1), (5, 2), (5, 5), (4, 2), (4, 3),$$
$$(4, 6), (3, 3), (3, 4), (2, 4), (2, 5), (1, 5), (1,6)\}$$

Comment. It is easy to notice that the opertaion '$\cup$' corresponds to the statements '*either or*' or '*at least one of*'. That is $A \cup B$ can be interpreted as the occurrence of either A or B, or as the occurrence of at least one of the events A and B. It should be noticed that, for example, if $A = \{2\}$ and $B = \{3\}$ then $A \cup B \neq \{2+3\} = 5$ but $A \cup B = \{2, 3\}$.

2.22 *Intersection of Sets*

Definition. *The intersection of two sets A and B, denoted by $A \cap B$, is defined as the set of elements which belong to A as well as to B.* From the definition itself it follows that $A \cap B = B \cap A$.

Example 2.22.1 Let $A = \{2, 5, -1\}$, $B = \{2, 3, 4\}$, $C = \{1,3,4,8\}$. Then, $A \cap B = \{2\}$, $B \cap C = \{3, 4\}$ but $A \cap C = \phi$ since there is no element common to the sets A and C. For two sets P and Q if $P \cap Q = \phi$ then P and Q are called *disjoint sets*.

Example 2.22.2 Let $A = \{x \mid 0 \leqslant x \leqslant 5\}$ and $B = \{x \mid 2 \leqslant x \leqslant 7\}$ then $A \cup B = \{a \mid 0 \leqslant a \leqslant 7\}$ and $A \cap B = \{b \mid 2 \leqslant b \leqslant 5\}$, (all sets here are assumed to be of numbers on the same line.)

Example 2.22.3 In an experiment of rolling a die once obtain the event of getting an even number which is less than 4.

Solution. The required event $C=\{2\}$. This can be considered to be the intersection of the event A of getting an even number and the event B of getting a number less than 4. That is,

$$A=\{2, 4, 6\} \text{ and } B=\{1, 2, 3\}$$

Hence,

$$A \cap B=\{2\}=C$$

Comment.   It is easy to see that for any set A,

$$A \cap A=A \text{ and } A \cap \phi=\phi$$

and if $B \subset A$ then $A \cap B=B$. The operation ' $\cap$ ' corresponds to the statement ' and ' or ' as well as '. That is, $A \cap B$ means the simultaneous occurrence of A and B or the set of elements belonging to A as well as to B.

The definitions of ' union ' and ' intersection ' of sets can be extended to a number of sets A, B, C and so on, in a similar fashion.

Exercises

2.5   A number of persons are weighed and let x denote the weight. Interpret the following events.

$A=\{x \mid 20 \leqslant x \leqslant 80\}$, $B=\{x \mid 60 \leqslant x \leqslant 140\}$, $C=\{x \mid 130 \leqslant x \leqslant 200\}$ $A \cup B$, $B \cup C$, $A \cup B \cup C$, $A \cap B$, $A \cap B \cap C$.

2.6   In an experiment of rolling a die once interpret the events $A=\{1, 3, 5\}$, $B=\{2, 4, 6\}$, $A \cup B$, $A \cap B$.

2.7   For any three events, A, B and C verify the results
(1) $A \cup (B \cup C) = (A \cup B) \cup C$.
(2) $A \cap (B \cap C) = (A \cap B) \cap C$.
(3) $A \cup (B \cap C) = (A \cup B) \cap (A \cup C)$.
(4) $A \cap (B \cup C) = (A \cap B) \cup (A \cap C)$.

2.8   Enumerate the number of elementary events when,
(a) a coin is tossed n times.
(b) a die is rolled n times.

2.9   Let A denote the event that a person driving through a national park will see a deer and B be the event that he will see a bear.   Interpret $A \cup B$ and $A \cap B$.

2.10   A box contains 2 dimes and 2 nickels.   A person picks up 2 coins.   Let A be the event that both are dimes and B be the event that one is a dime and the other is nickel. Interpret $A \cup B$ and $A \cap B$.

2.3 MUTUALLY EXCLUSIVE AND IMPOSSIBLE EVENTS

Definition. An event is said to be impossible if it is a null set, that is, it has no element of the outcome set. Two events are said to be mutually exclusive if the events are disjoint, that is, their intersection is null. In such a case one can say that the occurrence of one excludes the occurrence of the other event.

The event of getting a sum of 15 when a die is rolled twice is an impossible event because the total can be only at the most 12. The event of getting a head and the event of getting a tail when a coin is tossed once are mutually exclusive since they cannot occur simultaneously or the occurrence of one excludes the occurrence of the other. On the other hand, when a card is drawn from a deck of playing cards the event of getting an ace and the event of getting a club are not mutually exclusive since the ace can be ace of clubs.

2.31 *Complements and Non-Occurrence of Events*

Definition. If S is an outcome set and A is an event then the non-occurrence of the event A, that is, the complement $\overline{A}$ of A, consists of the set of elements in S which are not in A. Thus complement of a set is a relative concept relative to a reference set. In the case of events, the outcome set is the reference set so that the complement of an event is taken with respect to S.

Example 2.31.1 Let $A = \{0, 1, 2, -1\}$ and $B = \{0, 2\}$ then the complement of B with respect to A is,

$$\overline{B} = \{1, -1\}$$

so that $B \cup \overline{B} = A$.

Example 2.31.2 Let A be the event of getting a student whose I.Q. is greater than 120, when a student is selected at random from a class. Then $\overline{A}$ is the event of getting a student from the same class whose I.Q. is not greater than 120 or whose I.Q. is less than or equal to 120. Here the reference set is the set of students in a class under consideration.

Comment. In our definition, the complement of a set A or the non-occurrence of an event A is defined with respect to a reference set S such that $A \subset S$ and further $A \cup \bar{A} = S$.

Naturally we have the result,

$$\bar{\phi} = S \text{ and } \bar{S} = \phi.$$

But if A and B are two sets where $A \not\subset B$ or $B \not\subset A$ then the complement of B with respect to A, denoted by $A-B$, is defined as the set of elements in A which are not in B.

Example 2.31.3 Let $A=\{1, 2, 3, 5, 7\}$ and $B=\{5, 7, 8, 10\}$ then $A-B=\{1, 2, 3\}$. In general if $P \subset Q$ then $Q-P = \bar{P}$ where $\bar{P}$ is the complement of P with respect to Q.

2.32. *A Correspondence*

Here we will list a set of correspondence between an outcome set and events.

Events	*Outcome set S*
Elementary events	Points belonging to an outcome set
Events	Subsets of an outcome set
Sure event	Whole of the outcome set S
Impossible event	Null set which is evidently a subset of S
Non-occurrence of an event A	Complement of a subset A of S
Mutually exclusive events A and B	Disjoint subsets A and B of S. That is $A \cap B = \phi$
Occurrence of at least one of the events A and B	$A \cup B$ where $A \subset S$ and $B \subset S$
Simultaneous occurrence of the events A and B	$A \cap B$ where $A \subset S$ and $B \subset S$

2.33 *Venn Diagrams*

A convenient diagrammatic representation of sets is a symbolic representation in terms of venn diagrams. In a venn diagram sets are represented by regions enclosed by closed curves, such as, circles, ellipses, rectangles and so on. The representation is purely symbolic in the sense that, for example, the representation in Figure 2.1 may be a set of numbers, a set of pairs of numbers, a set of animate or inanimate objects, or in short, any set under consideration. If the elements in a set are to be represented they are simply represented as points in the diagram.

$$A = \{\phi, 2, \text{Miss Chick}, -1\}.$$

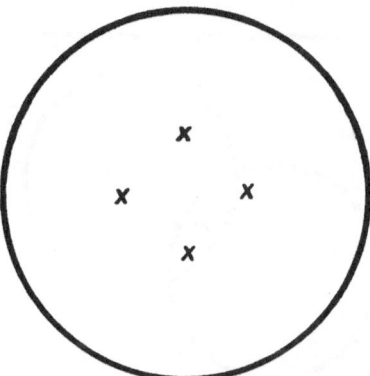

FIGURE 2.1 A Venn Diagram

In Figure 2.1 a set with 4 elements is represented and the set may be,
Evidently, none of the elements in A is representable as a geometrical point in a two dimensional space as it is done in figure 2.1 and in this sense the representation is symbolic. This representa-

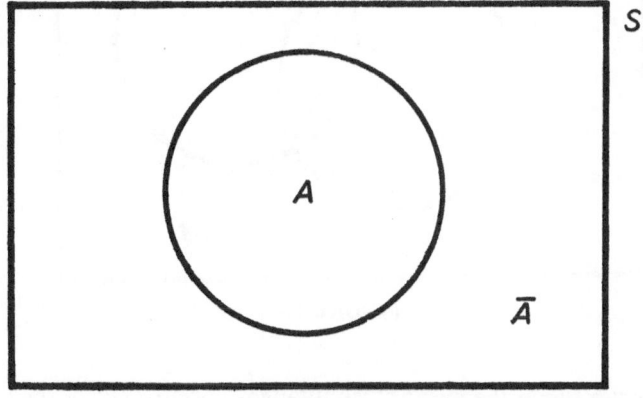

FIGURE 2.2

tion is helpful in understanding the various operations defined on sets. In Figure 2.2 we have the outcome set S, an event A and the complement or non-occurrence of the event A. Figure 2.3 shows the union of two sets A and B. Figure 2.4 gives the intersection or the simultaneous occurrence of two events A and B.

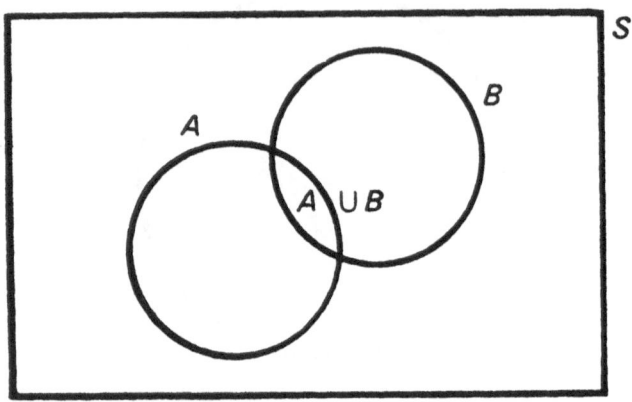

FIGURE 2.3

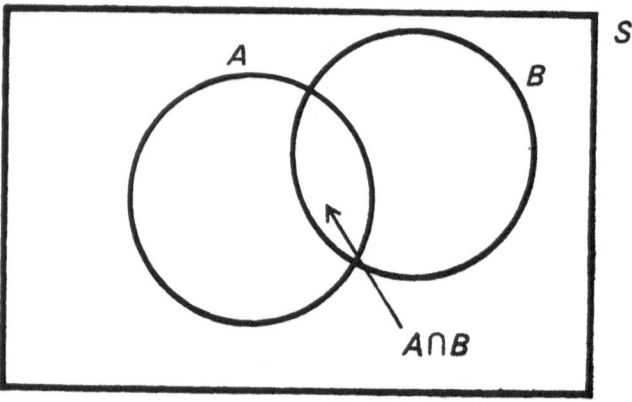

FIGURE 2.4

Figure 2.5 shows the event of simultaneous occurrence of the event A and non-occurrence of the event B. Figure 2.7 gives the simultaneous occurrence of the events A, B and C.

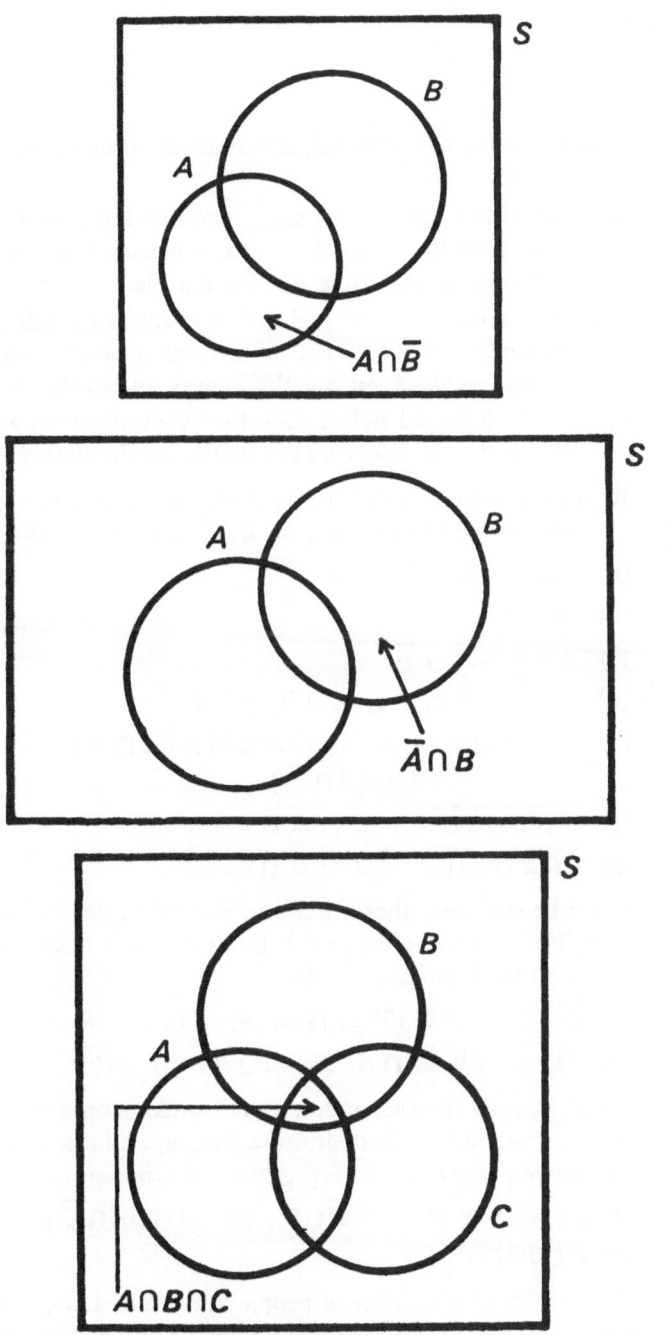

FIGURES 2.5, 2.6 and 2.7

Exercises

2.11 Check whether the following are mutually exclusive events or not.

    (a) A marble is taken from an urn containing some red and white marbles. A—the event of getting a red marble, B—the event of getting a white marble.

    (b) From a class of boys and girls one student is selected at random. A—the event of getting a female student. B—the event of getting the beauty queen in the class.

    (c) A coin is tossed twice. A—the event of getting at least one head. B—the event of getting at the most one tail.

2.12 By using a venn diagram verify the following results, where S denotes the outcome set, $A, B, C, \ldots$ denote events.

    (a) $A \cup \bar{A} = S,\ A \cap \bar{A} = \phi$

    (b) $\overline{A \cup B} = \bar{A} \cap \bar{B}$

    (c) $\overline{A \cap B} = \bar{A} \cup \bar{B}$

    (d) $A \cup B = (\bar{A} \cap B) \cup (A \cap B) \cup (A \cap \bar{B})$

    (e) $(\bar{A} \cap B) \cap (A \cap B) = \phi = (A \cap B) \cap (A \cap \bar{B})$
          $= (\bar{A} \cap B) \cap (A \cap \bar{B})$

    (f) $\overline{A \cup B \cup C \cup} \ldots = \bar{A} \cap \bar{B} \cap \ldots$

    (g) $\overline{A \cap B \cap C \cap} \ldots = \bar{A} \cup \bar{B} \cup \ldots$

2.13 Let A be the event that a business executive, selected at random, has stomach ulcer and B be that he has heart disease. Interpret the following events.

    (1) $\bar{A}$; (2) $\bar{A} \cup A$; (3) $\bar{A} \cap A$; (4) $A \cup B$; (5) $A \cap B$;

    (6) $A \cup \bar{B}$; (7) $\bar{A} \cap B$; (8) $\overline{A \cup B}$; (9) $\overline{A \cap B}$.

2.14 Let A, B, C be the events that a man walking on a street in a city will see, a new immigrant, a hippie, and a tourist from France respectively. Interpret the following events.

    (1) $A \cap B \cap C$; (2) $A \cap \overline{B \cup C}$; (3) $(A \cup B) \cap \bar{C}$;

    (4) $A \cup B \cup C$.

2.15 The heights of a number of people are taken. Let A denote the event that a height measurement falls in the interval

4—6 and B be the event that it falls in the interval 5—7. (all the 4 numbers inclusive). Interpret (1) $A \cup B$; (2) $A \cap B$; (3) $A \cup \overline{B}$.

2.4 TECHNICAL NOTES (PERMUTATION AND COMBINATION)

This section gives a note on permutations, combinations and binomial coefficients. These are not of much use in defining a numerical measure for the chances of occurrence of various events, but these will be of help in evaluating probabilities of certain events. Consider the word PET. Let us examine the set of 3 letter words that can be formed, with all the different letters, P, E and T. These are, PET, PTE, EPT, ETP, TEP and TPE. There are 6 different words possible. If the order in which the letters appear has no importance then there is only one arrangement since all the words have the same letters. Consider another word CUTE. All possible 2 letter words are given below.

PET	CUTE	
3 letter words	2 letter words	
PET	CU	UT
PTE	UC	TU
EPT	CT	UE
ETP	TC	EU
TEP	CE	TE
TPE	EC	ET
6	12	

The process of forming 3 letter words from the letters P, E, T may be considered to be a problem of filling three boxes with three different objects. The first box can be filled with one of the three objects and thus this can be done in 3 ways. After having filled the first box the second can be filled in 2 different ways because there are only 2 objects left. The third can be filled in one way and hence the total number of ways is $3 \times 2 \times 1 = 6$. This problem can also be considered to be a problem of selecting an ordered set of 3 distinct elements from the set of elements P, E and T.

2.41 *Permutations*

Definition. The number of permutations of n distinct objects, taking r at a time is defined as the number of different ordered sets of r distinct objects that can be taken from a set of n distinct objects, and is denoted by $(n)_r$, $_nP_r$, $P(n, r)$ etc. We will use the notation $(n)_r$.

Theorem 2.1 $(n)_r = n(n-1)...(n-r+1)$

Proof. Let the distinct elements be denoted by $a_1, a_2, ..., a_n$. The first element taken can be any one of them and hence this can be done in n ways. There will be $n-1$ of them left and so at the same time the second can be taken in $n-1$ ways (that is, any one of the $n-1$ remaining objects). Hence the 2 objects (first and the second) can be taken in $n(n-1)$ different ways. Extending the same argument we see that,

$$(n)_r = n(n-1)...[n-(r-1)]$$
$$= n(n-1)...(n-r+1)$$

This can be written in a convenient form.

$$(n)_r = n(n-1)...(n-r+1)$$
$$= \frac{n(n-1)...(n-r+1)\ (n-r)...2.1}{(n-r)...2.1} = \frac{n!}{(n-r)!}$$

where $n!$ (n factorial or factorial n)$=1.2...n=n(n-1)...2.1$.

Example 2.41.1 Let $S = \{1, 2, 3, 4\}$. Obtain the number of permutations, taking 2 at a time and also write down the ordered sets.

Solution. The required number of permutations is

$$(4)_2 = 4 \times 3 = 12 = \frac{4!}{2!}$$

The ordered sets are,

(1, 2), (1, 3), (1, 4), (2, 3), (2, 4), (3, 4),
(2, 1), (3, 1), (4, 1), (3, 2), (4, 2), (4, 3).

Comment. In general, $(n)_n = n(n-1)...2.1 = n!$

$$= \frac{n!}{(n-n)!} = \frac{n!}{0!} \ (0! = 1 \text{ is assumed}).$$

Example 2.41.2 Find the number of 4 letter words that can be formed by using all the letters of the word ' DOLL '.

Solution. Here there are 4 letters D, O, L, L out of which L appears twice. If the two L's are taken to be distinct such as L_1, L_2 then the total number of words is evidently $(4)_4 = 4!$. But of which L_1L_2 and L_2L_1 represent the same letters LL. Hence the number of distinct words possible is $\frac{4!}{2!} = 12$.

Comment. Even if some elements are not distinct by using the argument in this example, the number of distinct ordered sets can be found out without much difficulty.

Example 2.41.3 Find the number of distinct words that can be formed by using all letters of the word ' Panamanian '.

Solution. Here there are 10 letters and if all letters were different then the total number of words would be 10!. But *a* appears 4 times, *n* appears 3 times, *P*, *m* and *i* appear once. Hence the total number of distinct words is,

$$\frac{10!}{4!\,3!\,1!\,1!\,1!} = \frac{10!}{4!\,3!} = 25200.$$

2.42 *Combinations*

Consider the set $S = \{a, b, c\}$. The number of subsets of 2 elements each that can be taken from S is evidently 3. The subsets are $\{a, b\}$, $\{a, c\}$ and $\{b, c\}$. In a subset we are concerned only with the elements but not with the order in which the elements occur.

Definition. The number of combinations of n distinct objects, taken r at a time, is the number of subsets of r distinct elements each, that can be formed from a set of n distinct elements, and is denoted by $\binom{n}{r}$, $_nC_r$, $C(n, r)$ *and so on. We will use the notation* $\binom{n}{r}$.

Theorem 2.2 $\quad \binom{n}{r} = \dfrac{n(n-1)\ldots(n-r+1)}{r!} = \dfrac{n!}{r!\,(n-r!)}$

Proof. The number of permutations of *n* objects taken *r* at a time is given by $n! \,/\, (n-r)!$. Here the order is important. If the order is not important then the *r*! permutations of a set of *r* given objects represent the same set because the same set of elements are there in all these *r*! permutations. Hence the total number of combinations of *n* objects taken *r* at a time is

$$\binom{n}{r} = \frac{(n)_r}{r!} = \frac{n(n-1)\ldots(n-r+1)}{r!} = \frac{n!}{r!\,(n-r)!}$$

Example 2.42.1 In how many ways a set of 4 flower girls can be selected from a set of 10 girls?

Answer. Here we are interested only in a set of 4 girls and not in the order in which they are selected. Hence it is a problem of combination and the answer is,

$$\binom{10}{4} = \frac{10.9.8.7}{1.2.3.4} = 210 = \frac{10!}{4!\,6!}$$

Example 2.42.2 In a set of 20 administrators 15 are graduates and the rest are not. In how many ways a committee of 4 can be formed so that 3 are graduates and one is not a graduate?

Answer. There are 15 graduates and 5 non-graduates. 3 graduates from 15 graduates can be selected in $\binom{15}{3}$ different ways and at the same time one non-graduate can be selected in $\binom{5}{1}$ different ways. Hence the required answer is,

$$\binom{15}{3}\binom{5}{1} = \frac{15.14.13}{1.2.3}\,\frac{5}{1} = 2275$$

The following theorems can be proved by using the definition of 'number of combinations' itself.

Theorem 2.3 $\binom{n}{r} = \binom{n}{n-r}$ for n a positive integer and $r=0, 1, 2,$ ..., n. From this theorem or from the definition itself it is seen that,

$$\binom{n}{1} = n = \binom{n}{n-1}; \quad \binom{n}{n} = 1 = \binom{n}{0}$$

That is, for example,

$$\binom{4}{4} = \binom{4}{4-4} = \binom{4}{0} = 1$$

$$\binom{4}{3} = \binom{4}{4-3} = \binom{4}{1} = 4$$

Theorem 2.4 $\binom{n}{r} = \binom{n-1}{r} + \binom{n-1}{r-1}$ for n a positive integer and $r=0, 1, 2, ..., n-1$.

Theorem 2.5 $\binom{m+n}{r} = \sum_{k=0}^{r} \binom{m}{k}\binom{n}{r-k}$

Exercises

2.16 How many distinct words can be obtained by using all the letters of the word Mississippi?

2.17 A four letter word is to be made with the letters of the Latin alphabet. How many distinct words can be made if the first letter of the word should not be A?

2.18 If a telephone number is to be formed by 2 letters followed by 5 digits, how many such telephone numbers can be formed?

2.19 In how many ways a hand of 8 playing cards be selected so that there are 4 spades, 2 hearts and 2 clubs?

2.20 In how many different ways can 4 people be seated (1) in a row, (2) in a circle, (3) in a row of 6 seats?

2.21 A student's council has 5 members. How many different ways can both a programme committee and a reception committee, each with two members, be made, if (1) nobody is allowed to sit on both the committees, (b) with no restrictions?

2.22 In a community there are 400 catholics, 300 protestants and 100 others. In how many ways can a committee of 10 be formed such that there are 7 catholics, 2 protestants and 1 other?

2.23 A secretary puts 4 letters to 4 different people into 4 envelopes which are already addressed, without looking at the addresses. In how many ways can she put the letters so that nobody receives the letter he is supposed to get? This problem is often called a *matching problem.* Generalize the result to the case of n letters and n envelopes and show that the result is

$$\left\{ \frac{1}{2!} - \frac{1}{3!} + \cdots + \frac{(-1)^n}{n!} \right\} n!$$

2.24 In how many ways can 5 candy bars be distributed among 4 children so that (1) any child can receive any number of bars, (2) there should be at least one candy bar for every child. This is often called an *occupancy problem.* Show that the number of ways in which r indistinguishable articles can be assigned into n cells is $\binom{n+r-1}{r}$.

2.25 Write down (1) all the subsets of order 2, (2) all the subsets of order 3, from the following set $S = \{1, 0, -1, 3\}$.

2.43 *Stirling's Approximation*

According to the definition,
$$n! = 1.2.3...n$$
That is, when,
$$n = 5, \quad n! = 120;$$
$$n = 8, \quad n! = 40320;$$
$$n = 10, \, n! = 3628800.$$
So when n becomes large, $n!$ becomes quite large. Hence for computational convenience an approximation to $n!$, known as Stirling's approximation, is used, which is,
$$n! \approx (2\pi)^{\frac{1}{2}} \, n^{n+\frac{1}{2}} \, e^{-n} \qquad\qquad (2.43.1)$$
where ($\approx$) means approximately equal to, π and e have the usual meanings, that is, $\pi \approx 3.1416$ and $e \approx 2.71828$. A closer approximation for $n!$ is,
$$n! \approx (2\pi)^{\frac{1}{2}} \, n^{n+\frac{1}{2}} \, e^{-n+\frac{1}{12n}} \qquad\qquad (2.43.2)$$

Example 2.43.1 By using Stirling's approximation (formula (2.43.1)), show that
$$\binom{2n}{n} \approx 2^{2n}/(n\pi)^{\frac{1}{2}}$$

Answer. $n! \approx (2\pi)^{\frac{1}{2}} \, n^{n+\frac{1}{2}} \, e^{-n}.$

But, $\dbinom{2n}{n} = \dfrac{(2n)!}{n! \, n!} = \dfrac{(2n)!}{(n!)^2}$

and $(2n)! \approx (2\pi)^{\frac{1}{2}} \, (2n)^{2n+\frac{1}{2}} \, e^{-2n}$

and $(n!)^2 \approx [(2\pi)^{\frac{1}{2}} \, n^{n+\frac{1}{2}} \, e^{-n}]^2 = (2\pi)n^{2n+1} \, e^{-2n}.$

Therefore,
$$\frac{(2n)!}{(n!)^2} \approx \frac{(2\pi)^{\frac{1}{2}} \, (2n)^{2n+\frac{1}{2}} \, e^{-2n}}{(2\pi) \, n^{2n+1} \, e^{-2n}}$$
$$= 2^{2n}/(n\pi)^{\frac{1}{2}}$$

2.44 *Binomial Coefficients*

Consider the following expansions;
$$(a+b)^0 = 1$$
$$(a+b)^1 = a+b = \binom{1}{0} a^1 b^0 + \binom{1}{1} a^0 \, b^1.$$

$$(a+b)^2 = a^2+2ab+b^2 = \binom{2}{0} a^2b^0 + \binom{2}{1} a^1b^1 + \binom{2}{2}a^0b^2$$

$$(a+b)^3 = a^3+3a^2b+3ab^2+b^3$$

$$= \binom{3}{0} a^3b^0 + \binom{3}{1} a^2b^1 + \binom{3}{2} a^1b^2 + \binom{3}{3} a^0b^3.$$

In general it can be shown that

$$(a+b)^n = \binom{n}{0} a^nb^0 + \binom{n}{1}a^{n-1}b^1 + \ldots + \binom{n}{r}a^{n-r}b^r + \ldots + \binom{n}{n}a^0b^n.$$

These types of expansions are called Binomial (of two) expansions and the coefficients, $\binom{n}{0}$, $\binom{n}{1}$, $\binom{n}{2}$,..., $\binom{n}{r}$,..., $\binom{n}{n}$ are called the Binomial coefficients. The coefficient in $(a+b)^0$ is 1, the coefficients in $(a+b)^1$ are 1, 1, the coefficients in $(a+b)^2$ are 1, 2, 1 and so on. These coefficients can be easily written down by using an arrangement of numbers, called Pascal's triangle and which is given below

```
          1
        1 . 1
      1 . 2 . 1
    1 . 3 . 3 . 1
  1 . 4 . 6 . 4 . 1
```

A number in any row is obtained by taking the sum of two consecutive numbers in the preceding row and writing the sum in the row midway between the numbers added up. Each row starts with a unity and ends with a unity. First row has one number, the second row has 2 numbers and so on. In the Binomial expansions, considered above, n is a positive integer. However this restriction is not necessary. By the method of induction it can be shown that,

$$(1 +x)^n = 1 + \binom{n}{1} x + \binom{n}{2} x^2 + \ldots + \binom{n}{r} x^r + \ldots$$

where $\binom{n}{r} = \frac{(n)_r}{r!} = \frac{n(n-1)\ldots(n-r+1)}{r!}$, n is a positive integer, or a negative integer, or a positive fraction or a negative fraction; $|x|<1$ where $|x|$ denotes the absolute value of x. When n is a positive integer the expansion is valid for any finite x.

Example 2.44.1 Show that $\sum\limits_{r=0}^{n} (-1)^r \binom{n}{r} = 0$

Solution. Consider the expansion,

$$(1+x)^n = \binom{n}{0} + \binom{n}{1}x + \binom{n}{2}x^2 + \ldots + \binom{n}{n}x^n$$

(where n is a positive integer). But when n is a positive integer the expansion is valid for any finite x. Hence by putting $x = -1$ we get,

$$\binom{n}{0} - \binom{n}{1} + \binom{n}{2} - \ldots + (-1)^n \binom{n}{n} = (1-1)^n = 0$$

Example 2.44.2 Expand $(.99)^{-\frac{1}{2}}$ as a power series in 0.01.

Solution. $(1+x)^n = 1 + (n)_1 \dfrac{x}{1!} + (n)_2 \dfrac{x^2}{2!} + \ldots$

Put $x = -0.01$ and $n = -\frac{1}{2}$ then one gets,

$$(.99)^{-\frac{1}{2}} = (1-0.01)^{-\frac{1}{2}} = 1 + (-\tfrac{1}{2})_1 \frac{(-0.01)}{1!} + (-\tfrac{1}{2})_2 \frac{(-0.01)^2}{2!}$$

$$+ \ldots = 1 + \frac{1}{2}\frac{(0.01)}{1!} + \frac{1.3}{2.2}\frac{(0.01)^2}{2!} + \frac{1.3.5}{2.2.2}\frac{(0.01)^3}{3!} + \ldots$$

2.45 *Multinomial Coefficients*

The Binomial expansion, when n is a positive integer, can be written as,

$$(a+b)^n = \sum_{r=0}^{n}\binom{n}{r}a^{n-r}b^r = \sum_{r=0}^{n}\frac{n!}{r!(n-r)!}a^{n-r}b^r$$

$$= \sum_{\substack{r=0 \\ r+s=n}}^{n}\sum_{s=0}^{n}\frac{n!}{r!\,s!}a^s b^r$$

Here the double summation with respect to r and s is restricted by the condition, $r+s=n$ (that is, $s=n-r$). By using a similar technique a multinomial expansion (expansion involving a number of elements) can be written down as,

$$(P_1+P_2+\ldots+P_k)^n = \sum_{\substack{r_1=0 \\ r_1+\ldots+r_k=n}}^{n}\sum_{r_2=0}^{n}\ldots\sum_{r_k=0}^{n}\frac{n!}{r_1!\ldots r_k!}P_1^{r_1}\ldots P_k^{r_k}$$

That is, the n-tuple summation is such that for every term $r_1 + \ldots + r_k = n$. The coefficient $n!/r_1! \ldots r_k!$ of $P_1^{r_1} \ldots P_k^{r_k}$ is known as the multinomial coefficient.

Example 2.45.1 Expand $(P_1 + P_2 + P_3)^2$.

Solution. $(P_1 + P_2 + P_3)^2 = \sum\limits_{r_1=0}^{2} \sum\limits_{r_2=0}^{2} \sum\limits_{r_3=0}^{2} \dfrac{2!}{r_1! r_2! r_3!} P_1^{r_1} P_2^{r_2} P_3^{r_3}$

$$r_1 + r_2 + r_3 = 2$$

$$= \frac{2!}{2!} P_1^2 + \frac{2!}{2!} P_2^2 + \frac{2!}{2!} P_3^2 + \frac{2!}{1!} P_1^1 P_2^1 + \frac{2!}{1!} P_1^1 P_3^1$$

$$+ \frac{2!}{1!} P_2^1 P_3^1 = P_1^2 + P_2^2 + P_3^2 + 2P_1 P_2 + 2P_1 P_3 + 2P_2 P_3.$$

Exercises

2.26 If $\Gamma(n)$ (gamma n) is defined as $(n-1)!$, when n is a positive integer, show that,

$$\Gamma(n) \approx (2\pi)^{\frac{1}{2}} e^{-n} n^{n-\frac{1}{2}}.$$

2.27 Show that,

(1) $\displaystyle\sum_{r=0}^{n} \binom{n}{r}^2 = \binom{2n}{n}$

(2) $\displaystyle\sum_{r=0}^{2k} (-1)^r \binom{n}{r} \binom{n}{2k-r} = (-1)^k \binom{n}{k}$ for $0 \leqslant 2k \leqslant n$.

(3) $\displaystyle\binom{2n}{n} = (-1)^n 2^{2n} \binom{-\frac{1}{2}}{n}$

2.28 Approximate 8! and find the percentage error in the approximation.

2.29 Evaluate $\begin{pmatrix} -\frac{1}{2} \\ 2 \end{pmatrix}$ and $\begin{pmatrix} \frac{3}{4} \\ 4 \end{pmatrix}$.

2.30 Expand in powers of 0.02.
  (1) $(0.98)^{-\frac{1}{3}}$; (2) $(2.04)^{\frac{2}{3}}$.

Additional Exercises

2.1   Prove the following
      (a) Theorem 2.3   (b) Theorem 2.4.

2.2   Prove that, $\dbinom{m+n}{r} = \displaystyle\sum_{k=0}^{r} \dbinom{m}{k} \dbinom{n}{r-k}$

      where m, n are positive integers.

$$\text{Hint. } (1+x)^{m+n} \equiv (1+x)^m (1+x)^n.$$

2.3   By the method of induction prove that,

$$(1+x)^n = 1 + (n)_1 \frac{x}{1!} + (n)_2 \frac{x^2}{2!} + \ldots + (n)_r \frac{x^r}{r!} + \ldots$$

      where n is a rational number and $|x| < 1$.

2.4   Show that,

$$\frac{n!}{r_1! r_2! \ldots r_k!} = \dbinom{n}{r_1} \dbinom{n-r_1}{r_2} \dbinom{n-r_1-r_2}{r_3} \ldots \dbinom{r_k}{r_k}.$$

2.5   Evaluate the following

      (1) $\dbinom{-\frac{1}{2}}{5}$;   (2) $\dbinom{-2}{3}$;   (3) $\dbinom{5}{2}$.

2.6   Approximate and evaluate the percentage error in,
      (1) 10!; (2) 11!; (3) 12!.

2.7   A coin is tossed 10 times. What is the total possible number of ways in which (1) 4 heads and 6 tails; (2) 3 heads and 7 tails; (3) 10 heads and zero tails, can come up?

2.8   A die is rolled 10 times. In how many ways can one get (1) four 1's and six 2's; (2) two 2's, one 3, three 4's and four 5's; (3) one 1, two 2's, three 3's and four 4's?

2.9   In how many ways can an athletic team of 5 be selected from a set of 20 sportsmen?

2.10 In how many ways can a hand of 8 playing cards be selected such that it contains 3 spades, 3 clubs, 2 diamonds?

2.11 Miss Cute knows 4 different classical dances. In how many different ways can she give the performance for two occasions if (1) she does two dances for each performance and different dances for different performances; (2) two dances for each performance but she starts with the same dance for every performance.

2.12 In how many different ways can 10 candy bars be distributed among 7 children so that,
(1) each receives at least one,
(2) with no restriction,

2.13 Balls numbered 1 to 8 are put, one each, into 8 holes, at random. In how many ways can the balls be put so that no number on the ball matches with the number on the hole?

2.14 In some random experiments the number of outcomes can be easily counted by the help of a diagram called the *tree diagram*. A tree diagram, when a coin is tossed twice, is given below.

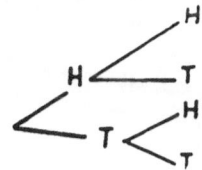

where *H* and *T* denote head and tail respectively. By using a tree diagram count the number of elementary events and write down the outcome set when
(1) a coin is tossed 4 times,
(2) a die is rolled 3 times, excluding the null set.

2.15 Write down the outcome set for the following experiment
(1) A coin is tossed till a head comes.
(2) A die is rolled until a total of 12 is obtained (face numbers added up to 12).

2.16 Give 5 examples, (from day to day events), each of
(1) two mutually exclusive events;
(2) an impossible event;
(3) a sure event;
(4) an elementary event.

2.17 $A \cup B = \{1, 2, 5\}$. Are A and B unique? If not give 3 examples each.

2.18 $A \cap B = \{0, 2\}$. Are A and B unique? If not give 4 examples each.

2.19 $\overline{A} = \{3, 7\}$. Is A unique?

2.20 If $A = \{x \mid 0 \leqslant x \leqslant 2\}$, $B = \{x \mid 0 \leqslant x \leqslant 7\}$, $C = \{x \mid 3 \leqslant x \leqslant 5\}$ evaluate   (1) $A \cup B$;   (2) $A \cap B \cap C$;   (3) $A \cup (B \cap C)$; (4) $A \cap \overline{B}$; (5) $A \cap B \cap \overline{C}$; (6) $\overline{A \cap B}$; (7) $\overline{A} \cap \overline{B} \cap \overline{C}$; where complementation is done with respect to the set of real numbers and all sets are of real numbers on the same line.

2.21 An urn contains the numbers 1, 2, 3, 4, 5, 6, 7, 8. Three numbers are selected at random with replacement. In how many ways can the numbers be selected such that the second smallest number is 6?

2.22 Two people are tossing one coin each twice. In how many different ways can both get the same number of heads?

2.23 Two men and two women are seated on a bench. In how many different ways can they be seated such that men and women alternate?

CHAPTER THREE

Probability

3.1 PROBABILITY OF AN EVENT

In day to day life we make statements such as, it is very likely that Miss Good will be elected beauty queen in the coming beauty contest, tomorrow will probably be a sunny day, drug x is more effective than drug y in curing disease A, the chances are almost nil that a man will live for ever, and so on. In all these statements there is a lack of certainty. Statistics and especially the theory of Probability play a vital role in making decisions in situations where there is a lack of certainty. There are three basic problems in the theory of probability, namely,

(1) to describe the situation or to specify the set on which probability statements are made;

(2) to define a numerical measure for a probability statement; and;

(3) to evaluate numerically the probabilities for particular events.

Even though the palmists, astrologers and fortune-tellers of ancient India might have used a record of past events to predict the future, the recorded evidence of a systematic study of present day probability is that it is developed as a theory of games of chance in the 17th century when some ardent gamblers consulted mathematicians about dividing the stake money in cases of incomplete games.

We can have a system of measurement of chance by assigning a zero for the chance of occurrence of an impossible event and a unity for the chance of occurrence of a sure event so that any event can be assigned a number between 0 and 1 as the probability of occurrence of the event. If we take a simple experiment of tossing a coin once, the events of getting a head and a tail are mutually exclusive and if they have equal chances of occurrence then we can assign numerical measure $\frac{1}{2}$ each. Further, the two events are mutually exclusive and their union is the sure event of either T or H coming in the toss. Denoting $P(A)$ the probability of any event A, we have

$$P\{H\}+P\{T\}=P(\{H\} \cup \{T\})=P\{S\}=1,$$

where S is the outcome set and evidently $\{H\}$ and $\{T\}$ are mutually exclusive. Based on these motivations, we define, $P(A)$ as follows.

Definition. The probability of occurrence of an event A, denoted by $P(A)$ is defined as a real number satisfying the following conditions:
    (1) $0 \leqslant P(A) \leqslant 1$
    (2) $P(A \cup B)=P(A)+P(B)$ whenever $A \cap B=\phi$
    (3) $P(S)=1$ where S is the sure event.

This definition does not help us in determining the probability of a particular event under consideration; for example, the probability of tomorrow being a sunny day. Since the probability of an event A is only a numerical measure of our ignorance regarding the various factors affecting this occurrence of an event there is no unique measure as $P(A)$. But, by examining the experimental conditions, we can give a few methods of assigning probabilities to particular events. This aspect will be considered in the coming sections. For a rigorous definition of probability, see *Probability Theory* by Loeve, Van Nostrand Co., New York, 1960.

Example 3.1.1 Can the following represent probability measures?
    (1) $P(A)=\ \ 0.2,\ P(B)=0.7,\ P(C)=0.1$
    (2) $P(A)=-0.5,\ P(B)=0.8,\ P(C)=0.7$
    (3) $P(A)=\ \ 0.4,\ P(B)=0.6,\ P(C)=0.2$
where $A \cup B \cup C=S$ (the sure event) and A, B, C are pairwise disjoint.

Answer. $1=P(S)=P(A \cup B \cup C)=P(A)+P(B)+P(C)$ since the events are mutually exclusive and since $S=A \cup B \cup C$.
    (1) $P(A)+P(B)+P(C)=0.2+0.7+0.1=1$
        $0 \leqslant P(A),\ P(B),\ P(C) \leqslant 1$
        Hence they represent probability measures.
    (2) $P(A)=-0.5$ which does not satisfy the condition
        $0 \leqslant P(A) \leqslant 1$ and hence is not a probability measure.
    (3) $P(A)+P(B)+P(C)=0.4+0.6+0.2=1.2 \neq 1$
        Hence they don't represent probability measures for the three events A, B, C in the problem.

Theorem 3.1 $P(\phi)=0$.

Proof. Since $\phi \subset S$ and since $\phi \cup S = S$ and $\phi \cap S = \phi$

$$1 = P(S) = P(S \cup \phi) = P(S) + P(\phi)$$

Hence,

$$P(\phi) = 1 - P(S) = 1 - 1 = 0$$

Theorem 3.2 $P(\overline{A}) = 1 - P(A)$

Proof. For any event A, $A \cup \overline{A} = S$ and $A \cap \overline{A} = \phi$

Hence, $1 = P(S) = P(A \cup \overline{A}) = P(A) + P(\overline{A})$.

That is, $P(\overline{A}) = 1 - P(A)$.

Extending the axiom that $P(A \cup B) = P(A) + P(B)$ whenever $A \cap B = \phi$, we can get the following result.

Theorem 3.3 $P(A_1 \cup A_2 \cup \ldots \cup A_n) = P(A_1) + P(A_2) + \ldots + P(A_n)$ whenever, $A_1, A_2, \ldots, A_n$ are all mutually exclusive events.

Theorem 3.4 $P(A \cup B) = P(A) + P(B) - P(A \cap B)$ where A and B are any two events.

Proof. Let S denote the sure event and A and B represent two events as shown in Figure 3.1.

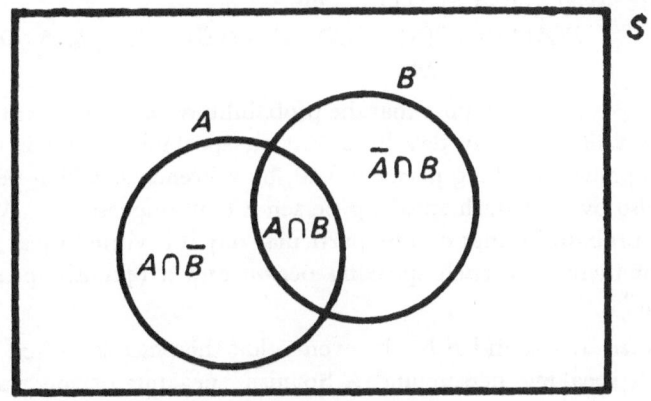

FIGURE 3.1

The events $A \cap \overline{B}$, $A \cap B$ and $\overline{A} \cap B$ are all mutually exclusive and further, $A \cup B = (A \cap \overline{B}) \cup (A \cap B) \cup (\overline{A} \cap B)$.

Therefore,

$$P(A \cup B) = P(A \cap \bar{B}) + P(A \cap B) + P(\bar{A} \cap B)$$
$$= [P(A \cap \bar{B}) + P(A \cap B)] + [P(\bar{A} \cap B) + P(A \cap B)]$$
$$- P(A \cap B)$$

by adding and subtracting $P(A \cap B)$.

But $A = (A \cap \bar{B}) \cup (A \cap B)$ where $(A \cap \bar{B}) \cap (A \cap B) = \phi$ and $B = (\bar{A} \cap B) \cup (A \cap B)$ where $(\bar{A} \cap B) \cap (A \cap B) = \phi$

(See Figure 3.1)

That is,

$$P(A) = P(A \cap \bar{B}) + P(A \cap B) \text{ and } P(B) = P(\bar{A} \cap B) + P(A \cap B).$$

Therefore,

$$P(A \cup B) = P(A) + P(B) - P(A \cap B)$$

Example 3.1.2 Suppose that the probabilities that a man entering a curio shop buys curios from the East is 0.80 that he buys curios from Europe is 0.55, that he buys curios from the East as well as the curios from Europe is 0.40. What is the probability that he buys curios of at least one of the types mentioned above?

Answer. Let A and B denote the events that he buys curios from the East and from Europe respectively. Then we are given

$$P(A) = 0.80, \ P(B) = 0.55 \text{ and } P(A \cap B) = 0.40.$$

We are asked to find $P(A \cup B)$. But,

$$P(A \cup B) = P(A) + P(B) - P(A \cap B) = 0.80 + 0.55 - 0.40$$
$$= 0.95.$$

Example 3.1.3 Suppose that the probabilities that a tourist attraction is visited on any day by a French speaking person is 0.70, by a Spanish speaking person is 0.40, by a French speaking person and also by a Spanish speaking person is 0.30 respectively. What is the probability that on any particular day it is visited by at least one of them, a French speaking person and a Spanish speaking person?

Answer. Let A and B be the events that the place is visited by a French speaking person and a Spanish speaking person respectively. Then we are given,

$$P(A) = 0.70, \ P(B) = 0.40 \text{ and } P(A \cap B) = 0.30.$$

We are asked to find out $P(A \cup B)$. But,

$$P(A \cup B) = P(A) + P(B) - P(A \cap B) = 0.70 + 0.40 - 0.30$$
$$= 0.80.$$

Exercises

3.1 Is the following correct? A—the event of seeing only French speaking persons on a particular day, B—the event of seeing only English speaking persons on the same day. Then $A \cap B$ the event of seeing a bilingual person.

3.2 Let A, B, C be a disjoint partition of an outcome set S. (That is, A, B, C are mutually exclusive and $A \cup B \cup C = S$). Can the following be probability measures?
   (1) $P(A) = 0.5, \; P(B) = 0.8, \; P(C) = 0.3$
   (2) $P(A) = -0.2, \; P(B) = 0.8, \; P(C) = 0.4$
   (3) $P(A) = 0.1, \; P(B) = 0.9, \; P(C) = 0$
   (4) $P(A) = 1.1, \; P(B) = 0.8, \; P(C) = 0.1$.

3.3 If $P(X)$ denotes the probability of any event X show that,
$$P(A \cup B \cup C) = P(A) + P(B) + P(C) - P(A \cap B) - P(B \cap C)$$
$$- P(C \cap A) + P(A \cap B \cap C).$$
Generalize the result to n events.

3.4 Let A be the event that a philosopher taking an evening walk in a Park will see a girl in her teens and let B be the event that he will see a migratory bird of a particular category. Let the probabilities of these events be, $P(A) = 0.50, \; P(B) = 0.40$ and $P(A \cap B) = 0.20$. Interpret the events (1) $A \cup B$, (2) $A \cup \bar{B}$, (3) $\bar{A} \cup B$, (4) $\overline{A \cap B}$, (5) $\bar{A} \cap B$, (6) $\overline{A \cup B}$ and evaluate the corresponding probabilities.

3.5 Can the following be probability measures?
   (1) $P(A) = 0.80, \; P(B) = 0.70$ and $P(A \cap B) = 0.40$
   (2) $P(A) = 0.3, \; P(B) = 0.5, \; P(C) = 0.2, \; P(A \cap B) = 0.1,$
   $P(A \cap C) = 0.2, \; P(B \cap C) = 0.2, \; P(A \cap B \cap C) = 0.6.$

3.6 A social worker reports that his survey of a community reveals that the chances of getting a church-going person is 90%, a person who has xenophobia is 80% and a person who is church-going and having xenophobia is only 20%. Should you question the result of this survey?

3.2 How to Assign Probabilities to Various Events

In section 3.1 we defined the probability of an event as a non-negative number between 0 and 1 and also considered some of its properties. But these properties will not help us in deciding upon the probability of any particular event under consideration. To some extent this can be achieved by considering the experimental conditions, symmetry in the experiment, past experience and so on.

3.21 *Method of Symmetry*

Consideration of symmetry in the outcomes of an experiment is a useful tool. Consider the experiment of tossing a coin once. If both sides of the coin are as symmetric and as perfect with respect to all the factors affecting the result of the experiment (in this case we call the coin unbiased) then there is some justification in assigning equal probabilities of getting a head and of getting a tail. In this case we can assign the probabilities $\frac{1}{2}$ each because the outcome set contains two 'symmetric' elementary events. That is,

$$P(H)=P(T)=\tfrac{1}{2}.$$

In an experiment of rolling a die twice, all the 36 elementary events can be considered to be 'symmetric' if all the 6 sides of the die are perfect and symmetric with respect to all the factors affecting the experiment. In this case each elementary event can be given a probability $\frac{1}{36}$. In this case the probability of getting a total 11 is obtained by considering all the mutually exclusive elementary events. The elementary events in this event are, (6, 5) and (5, 6). Since the probability of the union of mutually exclusive events is the sum of the probabilities, the probability of getting a sum 11 is $\frac{1}{36}+\frac{1}{36}=\frac{2}{36}$. From these considerations we can give the following result.

Result. When there is symmetry in the elementary events of an outcome set containing a finite number of elements the probability of any event A, is

$$P(A)=\frac{\text{Number of elementary events favourable to } A}{\text{Total number of elementary events}}$$

Example 3.21.1 What is the probability of getting at least one head when an unbiased coin is tossed twice?

Answer. Since the coin is unbiased we assume symmetry in the elementary events. The elementary events are (H, H), (H, T), (T, H) and (T, T) out of which 3 contain at least one head. Since each elementary event can be assigned a probability $\frac{1}{4}$ the required probability is $\frac{3}{4}$. In other words, there are 3 out of 4 elementary events favourable to the event of getting at least one head. Hence the probability is $\frac{3}{4}$.

Example 3.21.2 In a class of 100 students 30 are ' above average ' in intelligence, 50 are ' average ' and the rest are ' below average '. If a sample of 5 students is taken at random, what is the probability that the sample contains 2 ' above average ', and 3 ' average ' students?

Answer. The total number of elementary events equals the total number of random samples of size 5 each and that is $\binom{100}{5}$. We can assume symmetry in the elementary events. The total number of elementary events favourable to the event of getting 2 ' above average ' and 3 ' average ' is,

$$\binom{30}{2} \binom{50}{3}.$$

Hence the required probability is $\binom{30}{2}$ $\binom{50}{3}$ / $\binom{100}{5}$.

Comment. In this case the simplification of the answer or the complete evaluation of $\binom{30}{2}$ $\binom{50}{3}$ / $\binom{100}{5}$ can be achieved with the help of a table of Binomial coefficients, or by using a table of logarithms or by direct calculations. A table of Binomial coefficients is given at the end of this book.

Consideration of symmetry seems to be a very good tool in assigning probabilities to particular events. But it is important to notice that there is no symmetry in most of the day to day events. If a coin is not balanced then consideration of symmetry is of no help in finding out the chances of getting a head. Even if there is symmetry in the elementary events it is sometimes very difficult to decide upon the elementary events having symmetry. For example consider the following problem.

A businessman wants to go for a business trip. Two of his secretaries Miss Chick and Miss Cute want to go with him. He needs only one secretary for the trip. So he decided to conduct a game of chance. He would toss a balanced coin twice. If he would get at least one head he would take Miss Chick. Otherwise he would take Miss Cute. What is the probability of Miss Chick's selection?

The outcome set contains the 4 elementary events (H, H), (H, T), (T, H), (T, T) out of which 3 are favourable to Miss Chick's selection and hence the probability is $\frac{3}{4}$. Somebody may argue as follows. If head comes in the first throw then Miss Chick is already selected and the game is over. Hence the elementary events are (T, H), (T, T), (H) and the required probability is $\frac{2}{3}$.

From the above discussions it is apparent that the consideration of symmetry is not at all sufficient in all situations. But even when there is no symmetry we can assign probabilities to certain events. For example, if a monkey is given a typewriter to play with what is the probability that he will type this book word by word? Even though this event is not logically impossible but it is ' practically ' impossible and hence we are justified in assigning a probability zero to this event. Similarly we may assign a probability one that St. Lawrence River will not dry up on next Sunday morning. These are respectively, almost surely impossible and almost sure events. It should be noticed that the probability of an impossible event is zero but the converse that, if the probability of an event is zero then the event is impossible, need not be true. Similarly the probability of a sure event is unity but the converse is not true. These types of arguments have compelled some people to define probability as a *measure of the conviction of the mind based on experience*. In an experiment of throwing a coin if everything is known about the coin such as all the physical characteristics of the coin, the forces acting on it and so on then one may be able to say whether the coin is going to fall head or tail when it is tossed. So some people may argue that probability is in some sense a measure of our ignorance about the various aspects of the experiment. For an elaborate discussion of personal probability, utility, etc., see *Foundations of Statistics* by L. J. Savage.

3.22 *The Method of Relative Frequencies*

This is an empirical method of estimating the probability of an event. In order to find the probability of getting a head when a coin is tossed, if nothing is known about the unbiasedness of the coin, then we may proceed as follows. Toss the coin 100 times under similar conditions. Take the relative frequency, that is, the ratio of the number of heads to the total number of trials. Repeat the experiment 1000, 10,000, etc., times, each time taking the relative frequency. If this relative frequency is seen to approach a limit when the number of trials tends to infinity then that limit can be taken as the probability of getting a head and the relative frequency at any stage can be taken as an estimate of this probability. For example, if we get 52 heads in 100 trials then an estimate of the probability of getting a head is $\frac{52}{100}$=0.52. This technique can be used to find the probability of any event if the experiment is repeatable under similar conditions and if the relative frequency tends to a limit when the number of trials tends to infinity. In such a situation we get a fairly good estimate of the probability of an event by the relative frequency if the experiment is repeated a fairly large number of times.

Example 3.22.1 From a truck load of baskets of apples, one basket is taken at random and found that 3 out of 100 apples are spoiled. Obtain an estimate of the probability of getting a spoiled apple if one apple is taken at random from that truck load.

Solution. The relative frequency of getting a spoiled apple

$$=\frac{3}{100}=0.03.$$

Hence an estimate of the probability of getting a spoiled apple is,

$$=0.03.$$

Comment. It should be remembered that the relative frequency is different from the probability of an event. The relative frequency can be taken as an estimate of the probability of the event under consideration if the experiment is repeatable and if the relative frequency tends to a limit when the number of trials tends to infinity.

If one neglects the aspect of assigning probabilities to specific events then the mathematical theory of probability can be developed. This will then be a special branch of the mathematical theory known as Measure Theory. All the results in Measure Theory when the total measure is assumed to be unity will hold good for probabilities. But these are not of much practical use unless methods are devised to relate the theory with practical problems. Hence the aspect of assigning probabilities to specific events, building up probability models to describe random experiments and the aspect of decision making in situations having lack of certainty are the most important aspects from a practical point of view. We will be mainly concerned with these aspects in this book rather than the theory for its own sake.

Exercises

3.7　From an urn containing 10 white balls and 20 red balls, 3 balls are taken at random. What is the probability that all the 3 balls are red if (1) the balls are taken together, (2) the balls are taken one by one with replacement, (3) the balls are taken one by one without replacement?

3.8　From a well shuffled set of 52 playing cards what is the probability of getting a hand of 2 spades, 3 hearts and 3 diamonds?

3.9　What is the probability of throwing 7, 8, or 10 with 2 dice?

3.10　A line cuts the line segment AB at random into two parts. What is the probability that the ratio of the two segments (smaller to the larger) is less than $\frac{1}{3}$?

3.11　A radio station broadcasts the correct time every hour on the hour. If a listener switches on the radio at random, what is the probability that he has to wait at least 20 minutes before hearing the time broadcast?

3.12　If a secretary puts 6 letters at random into 6 envelopes which are addressed to 6 different people, what is the probability that no one receives the letter he is supposed to get?

3.13　If a boy is pelting stones at a target and if the probability of a hit is 0.5, what is the probability that he will hit exactly 3 times in 10 trials?

3.14　A balanced spinner with the dial marked 0 to 100 is rotated once. What is the probability that (1) the indicator will stop between 30 and 70, (2) the indicator will stop at 62?

3.15　What is the probability of getting 4 aces in a hand of 8 cards taken from a well shuffled deck of 52 cards?

3.3 Conditional Probability

We have defined the probability of an event A relative to the outcome set S. In other words, when an event is considered, already an outcome set S of which A is a subset is specified. Hence it is more appropriate to call the probability of A as the probability of A given S. This may be written as,

$$P(A) = P(A \mid S)$$

Every probability statement we made so far was a conditional statement of this nature. Now we will consider a different type of conditional statement such as the probability of an event A given another event B, where B need not be the sure event or the outcome set S. The usefulness of defining a conditional probability $P(A \mid B)$, that is, the probability of A, given that the event B has already occurred, can be seen from the following example. For convenience we will denote $P(A)$ for $P(A \mid S)$ and $P(A \mid B)$ if $B \neq S$.

Example 3.3.1 In a doctor's reducing laboratory people are given one of two treatments, technique T_1 and technique T_2. There are 100 women taking the treatments out of which some are divorced and looking for the next husband, some are unmarried, some weigh over 180 lbs and some weigh less than 180 lbs. The exact classification is given in the following table.

TABLE 3.1

T_1		*Over* 180 *lbs*	*Under* 180 *lbs*	
S_1	Divorced	30	10	Total 60
	Unmarried	15	5	

T_2		*Over* 180 *lbs*	*Under* 180 *lbs*	
S_2	Divorced	20	5	Total 40
	Unmarried	10	5	

A prospective patient goes to the laboratory. Assuming that every patient will be given equal chances of being called to give testimony of her achievement in weight reduction, what is the probability that she will hear from a patient who is (1) divorced, (2) undergoing treatment T_1, (3) heavier than 180 lbs, (4) heavier than 180 lbs, given that the speaker is divorced?

Solution. Here S_1 denotes the set of women undergoing treatment T_1 and S_2 is the set of women undergoing treatment T_2. The outcome set is $S = S_1 \cup S_2$. In (1), (2) and (3) we are concerned with the set S. Since there are 65 divorced women out of

100 patients the required answer in (1) is 65/100.   Let A, B, C be
the events in (1), (2) and (3) respectively.   Then,

$$P(A) = \frac{(30+10+20+5)}{100} = \frac{65}{100}$$

$$P(B) = \frac{(30+10+15+5)}{100} = \frac{60}{100}$$

$$P(C) = \frac{(30+15+20+10)}{100} = \frac{75}{100}$$

But in (4) we are concerned with the probability of getting a person heavier than 180 lbs given that the candidate is divorced.
This is a conditional statement $C \mid A$.   That is, we need,

$$P(C \mid A) = \text{Probability of } C \text{ given } A.$$

Here we are interested only in the reduced set of divorced people.
The set of divorced people S_3 is given in table 3.2.

TABLE 3.2

	Over 180 lbs	Under 180 lbs	
$S_3 \begin{cases} T_1 \\ T_2 \end{cases}$	30 20	10 5	Total 65

Hence the required probability is

$$P(C \mid A) = \frac{50}{65}$$

Comment.   Here it is interesting to notice one relationship.

$P(C \cap A) =$ probability of getting a person who is divorced as
well as over 180 lbs.   Since there are only 50 such people out of
100 people, the probability

$$P(C \cap A) = \frac{50}{100}$$

But,

$$\frac{P(C \cap A)}{P(A)} = \frac{50}{100} \times \frac{100}{65} = \frac{50}{65} = P(C \mid A)$$

Further, in the reduced outcome set S_3 the various probabilities
add up to unity.   That is, the probability of getting a person over
180 lbs and taking treatment T_1, given that she is divorced,...,
the probability of getting a person under 180 lbs and taking T_2

given that she is divorced, add up to unity. From these intuitive notions we will give the following definition for a conditional probability statement.

Definition. The probability of an event A given the event B, that is, $P(A \mid B)$ *is defined as,*

$$P(A \mid B) = \frac{P(A \cap B)}{P(B)} \quad \text{if } P(B) \neq 0.$$

That is,

$$P(A \cap B) = P(A \mid B) \, P(B) = P(B \mid A) \, P(A),$$

if $P(A) \neq 0$ and $P(B) \neq 0$.

This rule can be easily extended to a number of events. That is, for example,

$$P(A \cap B \cap C) = P(A \mid B \cap C) \, P(B \cap C)$$
$$= P(A \mid B \cap C) \, P(B \mid C) \, P(C).$$

Example 3.3.1 A consignment of 15 record players contains 4 defectives. The record players are selected at random, one by one, and examined. The ones examined are not put back. What is the probability that the 9th one examined is the last defective?

Answer. Let A be the event of getting exactly 3 defectives in 8 examinations, and let B be the event that the 9th examined is a defective. Then the required probability is $P(A \cap B)$. Since it is a problem of sampling without replacement and since there are 4 defectives out of 15 record players,

$$P(A) = \frac{\binom{4}{3} \binom{11}{5}}{\binom{15}{8}}, \quad \begin{pmatrix} 4+11=15 \\ 3+\ 5=\ 8 \end{pmatrix}.$$

It is not easy to obtain $P(B)$ but $P(B \mid A)$ is easily obtained. $P(B \mid A) =$ probability of getting the 9th examined a defective given that there were 3 defectives in the first 8 examinations. Now there is only one defective left among the remaining 7 record players. Therefore,

$$P(B \mid A) = \tfrac{1}{7}.$$

But,

$$P(A \cap B) = P(A) \, P(B \mid A) = \frac{\binom{4}{3} \binom{11}{5}}{\binom{15}{8}} \tfrac{1}{7}$$

$$= 8/195.$$

Example 3.3.2 A box contains 2 red marbles and 3 green marbles. Two marbles are taken one by one without replacement. What is the probability that (1) the two marbles are green, (2) the second marble is green given that the first marble is green?

Answer. (1) Let A be the event of getting two green marbles. It is a problem of sampling without replacement. Hence,

$$P(A) = \frac{\binom{2}{0} \binom{3}{2}}{\binom{5}{2}} = 3/10.$$

(2) If the first marble is green then there are 2 green marbles among the remaining 4 marbles. Hence the required probability is $2/4$.

Exercises

3.16 A box contains 10 red and 12 white rose flowers. Flowers are picked up at random one by one without replacement. What is the probability that (1) the first 3 flowers are red, (2) there are 2 red and 2 white flowers in the first 4 picked up, (3) the 3rd one is red given that the first two are white?

3.17 From a well shuffled deck of 52 playing cards, cards are picked up at random one by one. What is the probability that the 7th one picked up is the 5th spade?

3.18 From a band of 10 cheer leaders the girls are selected at random one by one. What is the probability that

(1) the 6th one selected is Miss Cute who is one of the 10,

(2) the 4th one selected is Miss Cute,

(3) the 3rd one selected is Miss Cute given that Miss Chick and Miss Dish are already selected?

3.19 From an urn containing 2 red and 4 green marbles, marbles are selected at random with replacement. What is the probability (1) the second one is green given that the first one is green, (2) the second one is green given that the first one is red?

3.20 Show that,

$$P(A_1 \cap A_2 \cap \ldots \cap A_n) = P(A_1 \mid A_2 \cap \ldots \cap A_n) \ldots$$
$$P(A_{n-1} \mid A_n) P(A_n).$$

3.31 *Independence of Events*

Definition. Two events A and B are said to be independent if $P(A \cap B) = P(A).P(B)$.

These may be called events where the occurrence of one does not affect the occurrence or non-occurrence of the other event. In practice it is often difficult to judge whether the occurrence of one affects the occurrence of the other or not. So it is safer to use the definition that $P(A \cap B) = P(A).P(B)$ rather than trying to find out the effect of one on the other. Exercises 3.18 and 3.19 are examples of independent events. When sampling with replacement the selection in a particular trial does not depend on what happened in the previous trial.

Example 3.31.1 What is the probability of getting heads in both trials when a balanced coin is tossed twice.

Answer. The probability of getting a head in any one trial is $\frac{1}{2}$. Since the trials are independent the required probability is $(\frac{1}{2}) (\frac{1}{2}) = \frac{1}{4}$.

Theorem. 3.3.1 If A and B are independent events and if $P(B) \neq 0$ then $$P(A \mid B) = P(A).$$

Proof. $P(A \mid B) = \dfrac{P(A \cap B)}{P(B)} = \dfrac{P(A) . P(B)}{P(B)} = P(A).$

Theorem 3.3.2 If A and B are independent events then (1) A and $\overline{B}$ are independent, (2) $\overline{A}$ and B are independent, (3) $\overline{A}$ and $\overline{B}$ are independent.

Proof. Let $P(A) \neq 0$, $P(B) \neq 0$. It is given that,
$$P(A \cap B) = P(A) . P(B).$$

(3) $P(\overline{A} \cap \overline{B}) = P(\overline{A \cup B}) = 1 - P(A \cup B)$
$= 1 - [P(A) + P(B) - P(A \cap B)]$
$= 1 - [P(A) + P(B) - P(A).P(B)]$

since $P(A \cap B) = P(A) . P(B)$.

$= [1 - P(A)] [1 - P(B)] = P(\overline{A}) . P(\overline{B})$

Hence $\overline{A}$ and $\overline{B}$ are independent if A and B are independent. (1) and (2) and the trivial cases when $P(A) = 0$ or $P(B) = 0$, are left to the reader.

3.32 *Pairwise and Mutual Independence*

Definition. A set of events A_1, A_2,...,A_n are said to be *pairwise independent if every pair of different events are independent.* That is,

$P(A_i \cap A_j) = P(A_i)P(A_j)$ for all i and j, $i \neq j$.

Definition. A set of events A_1, A_2,..., A_n are said to be *mutually independent if,*

$P(A_i \cap \ldots \cap A_k) = P(A_i)P(A_j)\ldots P(A_k)$,

for every subset $(A_i, A_j,\ldots,A_k)$ of $A_1,\ldots,A_n$.

That is, the probabilities of every two, every three,..., every n of the events are the products of the respective probabilities. It is important to notice that pairwise independence does not imply mutual independence which is evident from Figure 3.2.

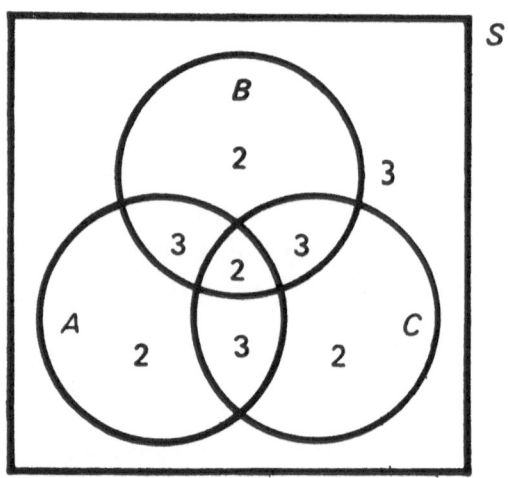

FIGURE 3.2   Pairwise independence does not imply mutual independence

In Figure 3.2 the numbers in the various subsets denote the numbers of elementary events where the elementary events are assumed to have equal probabilities. There are 10 elementary events each in A, B, and C, 5 each in $A \cap B$, $B \cap C$, and $C \cap A$, 2 in $A \cap B \cap C$ and 20 in S the outcome set. Hence,

$P(A) = P(B) = P(C) = \frac{10}{20} = \frac{1}{2}$

$P(A \cap B) = P(B \cap C) = P(C \cap A) = \frac{5}{20} = \frac{1}{4}$

and $P(A) \cdot P(B) = (\frac{1}{2})(\frac{1}{2}) = \frac{1}{4} = P(B) \cdot P(C) = P(C) \cdot P(A)$.

Therefore, the events A, B, C are pairwise independent.
But,

$$P(A \cap B \cap C) = \tfrac{2}{20} = \tfrac{1}{10} \neq P(A) \cdot P(B) \cdot P(C) = \tfrac{1}{8}.$$

Hence A, B, C are not mutually independent.

3.33 *Bayes' Rule*

This is an interesting, important and controversial rule by which we make a type of ' inverse ' reasoning in the evaluation of certain probabilities. We will develop theorem 3.3.3 and will point out Bayes' assumptions.

Lemma 3.3.3 *If* A_1, A_2, ..., A_n denote a disjoint partition of the outcome set and if $P(A_j) \neq 0$ for $j = 1, 2, ..., n$ then for any event B,

$$P(B) = \sum_{j=1}^{n} P(A_j) P(B \mid A_j).$$

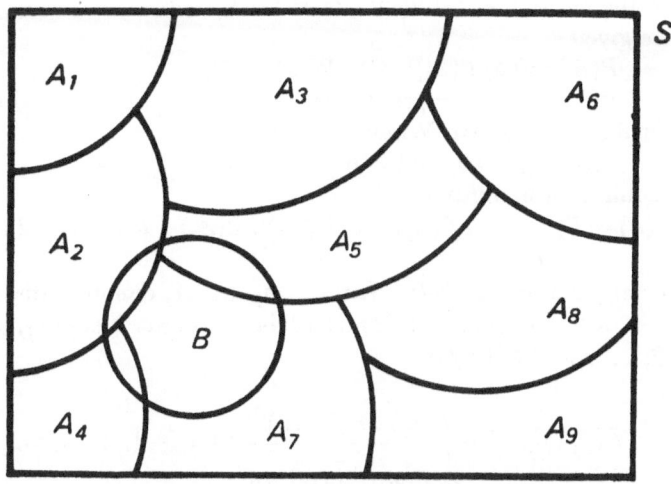

FIGURE 3.3   A partition of the outcome set S

Proof. Since A_1, A_2, ..., A_n are disjoint and $A_1 \cup A_2 \cup ... \cup A_n = S$, for any event B,

$$B = (B \cap A_1) \cup (B \cap A_2) \cup ... \cup (B \cap A_n)$$

and further, $B \cap A_1$, $B \cap A_2$, ..., $B \cap A_n$ are all disjoint. This can be verified from Figure 3.3.   Therefore,

$$P(B) = P(B \cap A_1) + P(B \cap A_2) + \ldots + P(B \cap A_n)$$
$$= \sum_{j=1}^{n} P(B \cap A_j).$$

But,

$$P(B \cap A_j) = P(B \mid A_j) \, P(A_j) \quad \text{if } P(A_j) \neq 0.$$

Therefore, $P(B) = \sum_{j=1}^{n} P(B \mid A_j) \, P(A_j).$

Example 3.33.1 Three candidates Mr. Wiseman, Miss Drink-water and Mr. Page stand for student president in a university. A public opinion poll shows their chances of winning as 0.5, 0.3, 0.2 respectively. The probabilities that they will promote 'student power' if they are elected are 0.7, 0.6, 0.9 respectively. What is the probability that 'student power' will be promoted after the election?

Solution. Let A_1, A_2, A_3 denote the events that Mr. Wiseman, Miss Drinkwater, and Mr. Page be elected respectively. Let B denote the event that 'student power' is promoted.
We are given

$P(A_1) = 0.5, \; P(A_2) = 0.3, \; P(A_3) = 0.2,$

$P(B \mid A_1) = 0.7 = $ probability that 'student power' will be promoted given that Mr. Wiseman is elected.

$P(B \mid A_2) = 0.6,$ and $P(B \mid A_3) = 0.9.$

The required probability is,

$$P(B) = P(B \mid A_1) \cdot P(A_1) + P(B \mid A_2) \cdot P(A_2) + P(B \mid A_3) \cdot P(A_3)$$
$$= 0.71.$$

Theorem 3.3.3 *Bayes' Rule.* Let A_1, A_2, ..., A_n denote a disjoint partition of the outcome set S and let B be any event. Let $P(A_j) \neq 0$, $j = 1, 2, \ldots, n$ and $P(B) \neq 0$.
Then,

$$P(A_j \mid B) = \frac{P(A_j) \cdot P(B \mid A_j)}{\sum_{j=1}^{n} P(A_j) \cdot P(B \mid A_j)} \quad \text{for } j = 1, 2, \ldots, n.$$

Proof. From the definition,

$$P(A_j \mid B) = \frac{P(A_j \cap B)}{P(B)} = \frac{P(A_j) \cdot P(B \mid A_j)}{P(B)}.$$

From Lemma 3.3.3 we have,

$$P(B) = \sum_{j=1}^{n} P(A_j) \cdot P(B \mid A_j).$$

Therefore,

$$P(A_j \,|\, B) = \frac{P(A_j) \cdot P(B \,|\, A_j)}{\displaystyle\sum_{j=1}^{n} P(A_j) \cdot P(B \,|\, A_j)} \,.$$

Here the probabilities $P(A_j \,|\, B)$ for $j = 1, 2, \ldots, n$ are the probabilities determined after observing the event B and $P(A_j)$ for $j = 1, 2, \ldots, n$ are the probabilities given beforehand. Hence $P(A_j)$ for $j = 1, 2, \ldots, n$, are called *prior* or *a priori* probabilities and $P(A_j \,|\, B)$ for $j = 1, 2, \ldots, n$ are called *posterior* or *a posteriori* probabilities. Bayes' Rule gives a relationship between $P(A_j \,|\, B)$ and $P(B \,|\, A_j)$ and thus it involves a type of inverse reasoning which can be seen from the following example.

Bayes assumed the rule of '*equal division of ignorance*', that is, he assumed that if nothing is known about the prior probabilities $P(A_1), \ldots, P(A_n)$ then they are all equal. The controversy is mainly regarding this point.

Example 3.33.2 Suppose that there is a chance for a newly constructed house to collapse whether the design is faulty or not. The chance that the design is faulty is 10%. The chance that the house collapses if the design is faulty is 95% and otherwise it is 45%. It is seen that the house collapsed. What is the probability that it is due to faulty design?

Solution. Let A_1 and A_2 denote the events that the design is faulty and the design is good respectively. Let B denote the event that the house collapses. Then we are interested in the event $(A_1 \,|\, B)$, that is, the event that the design is faulty given that the house collapsed. We are given,

$$P(A_1) = 0.1 \text{ and } P(A_2) = 0.9;$$
$$P(B \,|\, A_1) = 0.95 \text{ and } P(B \,|\, A_2) = 0.45.$$

Hence,

$$
\begin{aligned}
P(A_1 \,|\, B) &= \frac{P(A_1) \cdot P(B \,|\, A_1)}{P(A_1) \cdot P(B \,|\, A_1) + P(A_2)\, P(B \,|\, A_2)} \\
&= \frac{(0.1)\,(0.95)}{(0.1)\,(0.95) + (0.9)\,(0.45)} \\
&= 0.19.
\end{aligned}
$$

Exercises

3.21   Three machines X, Y, Z of equal capacities are producing a machine part. The probabilities that the machines produce defectives (parts which do not meet quality specifications) are 0.1, 0.2 and 0.1 respectively. An item which is produced by one of these machines is taken at random and found to be defective. What is the probability that it came from machine X?

3.22   Among three identical urns one has 2 red marbles, one has one red and one green marble and the third has 2 green marbles. One urn is selected at random and then a marble is picked up at random and is found to be red. What is the probability that the other marble in the urn is red?

3.23   Four people Mr. X, Mr. Y, Mr. Z, Mr. T compete for the presidency of Piggyland. A public opinion poll reveals their chances of winning as 0.4, 0.2, 0.3, 0.1 respectively. The probabilities that gambling will be nationalised by them if they are elected are 0.85, 0.9, 0.3 and 0.95 respectively. What is the probability that gambling will be nationalised after the presidential election?

3.24   A survey on a random sample of 200 people in a community shows that 2 out of every hundred men have stomach ulcer and one out of every hundred women has stomach ulcer. A person from that community is selected at random and found to have stomach ulcer. What is the chance that the person is a male?

3.25   The chances that doctor C will diagnose disease X correctly is 60%. The chances that a patient will die by his treatment after correct diagnosis is 40% and the chances of death by wrong diagnosis is 70%. A patient of doctor C, who had disease X, died. What is the chance that his disease was diagnosed correctly?

3.4 ENTROPY OF A FINITE SCHEME

In Information Theory and Communication Theory a concept known as ' entropy ' or ' bits of information ' is widely used. Here we will give a brief introduction and for further reading see books on Information and Communication Theories.

3.41 A Complete System of Events

Definition. *A system of events, A_1, A_2, ...,A_n in which one and only one of them occurs in each trial may be called a complete system of events.*

For example, a disjoint partition A_1, ..., A_n of an outcome set forms a complete system. That is,
  (1) A_1, ..., A_n such that $A_1 \cup ... \cup A_n = S$ (the sure event) and where A_1, ..., A_n are mutually exclusive.
  (2) The occurrence of 1 or 2 or 3 or 4 or 5 or 6 when a die is rolled once.
  (3) The occurrence of head or tail when a coin is tossed once.

3.42 A Finite Scheme

Definition. *A complete system of events, A_1, A_2, ..., A_n, where n is finite, together with their corresponding probabilities of occurrence p_1, p_2, ..., p_n is called a finite scheme, that is,*

$$A = \begin{bmatrix} A_1, ..., A_n \\ p_1, ..., p_n \end{bmatrix}$$

is a finite scheme. It should be noticed that since A_1, ..., A_n are mutually exclusive and $A_1 \cup ... \cup A_n = S$, $p_1 + p_2 + ... + p_n = 1$

3.43 Entropy

Consider the following finite schemes A, B, and C.

$$A = \begin{bmatrix} A_1, A_2 \\ 0.5, 0.5 \end{bmatrix}, \quad B = \begin{bmatrix} A_1, A_2 \\ 0.9, 0.1 \end{bmatrix}, \quad C = \begin{bmatrix} A_1, A_2 \\ 0.4, 0.6 \end{bmatrix}.$$

In A the two events A_1 and A_2 have equal chances of occurrence and hence there is a great uncertainty of A_1 or A_2's occurrence in a particular situation whereas in B there is a better chance of A_1's occurrence than that of A_2. But in C one can say that the uncertainty is somewhere between that of A and B. Therefore a measure of lack of certainty of occurrence of the events under consideration is of some use. In a finite scheme, one such measure suggested is known as 'entropy' or 'information' in a finite scheme. The entropy in a finite scheme is denoted by $H(p_1, ..., p_n)$ and is defined as,

$$H(p_1, ..., p_n) = -k \sum_{i=1}^{n} p_i \log p_i$$

where log p_i is the natural logarithm of p_i, (that is if $e^x = b$ then x is called the natural logarithm of b and is written as $x = \log b$), and k is a positive constant. For practical purposes it is enough to take k to be equal to unity and hence H is written as,

$$H(p_1, ..., p_n) = - \sum_{i=1}^{n} p_i \log p_i.$$

For various axiomatic characterizations of $H(p_1, ..., p_n)$ see Mathai and Rathie ref. 27 in Bibliography.

Example 3.4.1 In a discrete noiseless communicational system (a message is not disturbed while it travels) a coded message source produces sequences of letters chosen from among the letters a, b, c, d with probabilities 0.2, 0.3, 0.4, 0.1 respectively, where successive symbols are chosen independently. What is the entropy per symbol?

Solution. $H(p_1, p_2, ..., p_n) = - \sum_{i=1}^{n} p_i \log p_i$

$$= -[0.2 \log 0.2 + 0.3 \log 0.3 + 0.4 \log 0.4 + 0.1 \log 0.1]$$

$$= 0.5558 \times \log 10 = 1.27.$$

Other statistical concepts of 'information' will be discussed later. Some areas where the theory of probability is widely used are (1) Markov Processes, (2) Ergodic Theory, (3) Random Walk, (4) Queuing Theory, (5) Genetics, (6) Space Research —especially in predicting the operational ability of space vehicles, region of impact of rockets and so on, (7) Agricultural production, (8) Industrial production processes, (9) Sociological and other surveys. For further reading see *Introductory Probability Theory and its Applications* Vol. I and II by Feller, W., Wiley, New York.

Exercises

3.26   Find the information in the following finite schemes.

(1) $\begin{bmatrix} A_1 & A_2 & A_3 \\ \frac{1}{6} & \frac{2}{6} & \frac{3}{6} \end{bmatrix}$,   (2) $\begin{bmatrix} A_1 & A_2 & A_3 \\ \frac{1}{3} & \frac{1}{3} & \frac{1}{3} \end{bmatrix}$

3.27   Show that the entropy $H(p_1, ..., p_n)$ in a finite scheme satisfies the following conditions:

(1) H is continuous in p_i;

(2) If all the p_i's are equal that is $p_1 = p_2 = ... = p_n = 1/n$, then H is a monotonic increasing function of n. (This shows that with equally likely events there is more uncertainty in the occurrence of the events.)

(3) If a choice consists of two successive choices the original H is a weighted sum of the individual values of H. For example, if the choices are as shown below, then $H(\frac{1}{2}, \frac{1}{8}, \frac{3}{8})$ $= H(\frac{1}{2}, \frac{1}{2}) + (\frac{1}{2}) H(\frac{1}{4}, \frac{3}{4})$ where the weight $\frac{1}{2}$ is taken because the second choice occurs with probability $\frac{1}{2}$.

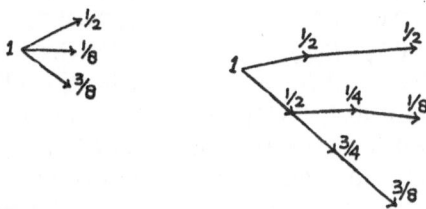

One choice   Two choices

(It can be shown that any function satisfying the conditions (1), (2) and (3) is of the form $-k \sum\limits_{i=1}^{n} p_i \log p_i$ where k is a positive constant).

(4) $H(p_1 q_1, ..., p_1 q_m, \cdots ..., p_n q_1, ..., p_n q_m)$
$= H(p_1, ..., p_n) + H(q_1 ..., q_m)$,

for $q_j \geqslant 0, \sum\limits_{j=1}^{m} q_j = 1$.

(5) $H(p_{11}, ..., p_{1n}, \cdots ..., p_{m1}, ..., p_{mn})$

$\leqslant H(\sum\limits_{j=1}^{n} p_{1j}, ..., \sum\limits_{j=1}^{n} p_{mj}) + H_n(\sum\limits_{i=1}^{m} p_{i1}, ..., \sum\limits_{i=1}^{m} p_{in})$

for $p_{ij} \geqslant 0, \sum\limits_{i=1}^{m} \sum\limits_{j=1}^{n} p_{ij} = 1$.

3.28 Show that $f(x) = H(x, 1-x)$ satisfies the functional equation

$$f(x) + (1-x) f\left(\frac{y}{1-x}\right) = f(y) + (1-y) f\left(\frac{x}{1-y}\right)$$

where $x, y \in (0, 1)$ with $x + y \in (0, 1)$.

3.29 Show that $H(p_1, ..., p_n)$ is a maximum when $p_1 = p_2 = ... = p_n$.

Additional Exercises

3.1   If A, B, C represent a disjoint partition of the outcome set can
the following be probability measures?
(1) $P(A) =$ 0.2, $P(B) = 0.7$, $P(C) = 0.1$;
(2) $P(A) =$ 0.2, $P(B) = 0.8$, $P(C) = 0.3$;
(3) $P(A) = -0.2$, $P(B) = 0.5$, $P(C) = 0.7$;
(4) $P(A) =$ 0.8, $P(B) = 0$, $P(C) = 0.2$.

3.2   Can the following be probability measures?
(1) $P(A) = 0.5$, $P(B) = 0.8$, $P(A \cap B) = 0.5$;
(2) $P(A) = 0.6$, $P(B) = 0.9$, $P(A \cap B) = 0.2$;
(3) $P(A) = 0.6$, $P(B) = 0.4$, $P(A \cap B) = 0.0$;
(4) $P(A) = 0.7$, $P(B) = 0.3$, $P(A \cap B) = 0.4$;
(5) $P(A) = 0.2$, $P(B) = 0.5$, $P(C) = 0.6$, $P(A \cap B) = 0.2$,
$P(B \cap C) = 0.4$, $P(C \cap A) = 0.4$, $P(A \cap B \cap C) = 0.8$;
(6) $P(A) = 0.5$, $P(B) = 0.6$, $P(C) = 0.5$, $P(A \cap B) = 0.2$,
$P(B \cap C) = 0.1$, $P(C \cap A) = 0.2$, $P(A \cap B \cap C) = 0.4$.

3.3   Give 3 examples, from day to day life, each of,
(1) two mutually exclusive events;
(2) four mutually exclusive events;
(3) two independent events;
(4) 4 pairwise independent events;
(5) 4 mutually independent events.

3.4   Give 2 examples each of two events which are,
(1) mutually exclusive but not independent;
(2) mutually exclusive and independent;
(3) not mutually exclusive but independent;
(4) not mutually exclusive and not independent,

3.5   From a well shuffled deck of 52 cards 5 cards are selected at
random.   What is the probability that,
(1) it contains 2 spades, 2 hearts and 1 club?
(2) it contains all hearts?
(3) it has 2 aces, and at least 2 clubs?

3.6   The probability that a man will be alive in 30 years is $\frac{1}{2}$ and
that his wife will be alive in 30 years is $\frac{2}{3}$.   What is the probability that in 30 years (1) both will be alive, (2) only the man
will be alive, (3) only his wife will be alive, (4) none of them
will be alive?

3.7 If 5 balls are placed at random into 5 cells, find the probability that (1) exactly one cell remains empty, (2) two cells remain empty.

3.8 If 5 balls numbered 1 to 5 are placed at random into 5 cells numbered 1 to 5, what is the probability that (1) none of them match, (2) 4 of them do not match?

3.9 From among the numbers, 1, 5, 6, 8 and 10 two numbers are taken at random. What is the probability that their sum is 11 if (1) the numbers are taken one by one without replacement, (2) the numbers are taken together, (3) the numbers are taken with replacement?

3.10 A club consisting of 15 married couples, chooses a president and then a secretary by random selection. What is the probability that (1) both are men, (2) one is a man and the other is a woman, (3) the president is a man and the secretary is a woman, (4) both are a married couple?

3.11 A picnic is arranged to be held on a particular day. The weather forecast says that there is 80% chance of rain on that day. If it rains the probability of a good picnic is 0.3 and if it does not the probability is 0.9. What is the probability that the picnic will be good?

3.12 A balanced die is rolled twice. What is the probability that,
(1) the sum is less than 3?
(2) the sum is an even number?
(3) the sum lies between 3 and 5?
(4) the square of the sum is less than 25?

3.13 The registration record of a university shows that out of the first 100 registrations 40 chose Science, 40 chose Arts and the remaining Professional courses. If 4 students are taken at random from this set what is the probability of getting 2 who have taken Sciences and 2 in Arts?

3.14 There are 4 boxes of one dozen eggs each. They contain 2, 3, 1, 0 spoiled eggs respectively. One box is selected at random and then an egg is taken at random. What is the probability that the egg is spoiled?

3.15 In the problem 3.14 if the egg is found to be good what is the probability that if any other egg is taken from the same box it will be good?

3.16   Mr. A is flying to a city to meet an old acquaintance. The chances that he will get immediate transportation from the airport is 60%. The probability that he will be able to meet his friend if he gets immediate transportation is 0.8 and otherwise it is 0.4. What is the probability that he will meet his friend?

3.17   A spinner is rotated twice. What is the probability that (1) the indicator stops between 0 and 10 in the first trial (2) the indicator stops between 2 and 8 in the first trial and between 20 and 30 in the second trial? (The dial is marked 0 to 100).

3.18   What is the probability that in problem 3.17 the sum of the numbers on which the indicator stopped lies between 30 and 80?

3.19   What is the probability of getting at least 2 heads when a balanced coin is tossed 3 times?

3.20   In a card game where each of 4 people has 13 cards. If one has 2 aces what is the probability that his partner has the other 2 aces?

3.21   If a balanced spinner, with dial marked 0 to 10, is rotated twice, what is the probability that the sum of the numbers on which the indicator stopped lies between 8 and 14?

3.22   For the events A_1, A_2, ..., A_n show that, $P(A_1 \cap A_2 \cap ... \cap A_n) \leqslant P(A_j) \leqslant P(A_1 \cup ... \cup A_n) \leqslant \sum\limits_{j=1}^{n} P(A_j)$.

3.23   Let A and B be two mutually exclusive events and C be any other event with $P(C) > 0$. Show that, $P(A \cup B \mid C) = P(A \mid C) + P(B \mid C)$. Generalize this result.

3.24   If A and B are two events, $P(B) > 0$, show that the following statements are not true in general. (1) $P(A \mid B) + P(\overline{A} \mid \overline{B}) = 1$, and (2) $P(A \mid B) + P(A \mid \overline{B}) = 1$.

3.25   Show that, independence need not be preserved in the conditional space. That is, if A and B are two independent events and C is any other event with $P(C) > 0$, then the following statement is not true in general:
$$P(A \cap B \mid C) = P(A \mid C) \, P(B \mid C).$$

Part II

Probability Models

CHAPTER FOUR

A Stochastic Variable

4.1 DISCRETE AND CONTINUOUS STOCHASTIC VARIABLES

The reader may be familiar with mathematical variables. Here
we will define another type of variable called a *stochastic variable*.
It is also called a *chance variable*, a *probability variable*, a *random
variable* and a *variate*. These names suggest that the variable has
something to do with probabilities.

Consider a mathematical relationship between two mathematical
variables x and y where

$$y=2x+3$$

and let x be defined on the set of numbers $A=\{1, 2, 3, 4, 5, 6\}$.
Then evidently y ranges over the set $B=\{5, 7, 9, 11, 13, 15\}$ because for a value of x in A the corresponding value of y is in B.
Now suppose that these numbers 1, 2, 3, 4, 5, 6 were the numbers
on a balanced die. If this die is rolled once then one of the numbers 1 to 6 is obtained. Suppose that x denotes the number obtained in a particular roll of the die then evidently x can take the
values 1, 2, 3, 4, 5, 6. That is, again x is defined over the set A.
But in this case there is a difference, that x takes a particular
value in A, for example 4, with a probability of $\frac{1}{6}$. In other
words x becomes a chance variable, taking the values 1, 2,..., 6
with probabilities $\frac{1}{6}, \frac{1}{6},..., \frac{1}{6}$ respectively. Consider another
example of an experiment of tossing a balanced coin twice. If y
denotes the total number of heads, then y can take the values
0, 1 and 2 with probabilities $\frac{1}{4}, \frac{2}{4}$ and $\frac{1}{4}$ respectively. Here y is a
chance variable.

*Definition. A real stochastic variable (random variable, chance variable,
variate, probability variable) is a variable defined on the outcome set of a
random experiment such that the probability statement, $P\{X \leqslant c\}$, is
defined for every real number c, where X denotes the stochastic variable
(s.v.).* (For a mathematically rigorous definition, see *Probability*

Theory by M. Loeve, Van Nostrand Co.). Stochastic variables are usually denoted by X, Y, Z, etc., and the particular values assumed by them by x, y, z, etc. Whenever there is no confusion we will use $P\{X \leqslant a\}$ or $P\{x \leqslant a\}$ to denote the probability that the s.v. X assumes values less than or equal to a.

We have seen two stochastic variables X and Y in the discussion above, where X is the number obtained when a balanced die is rolled once and Y is the number of heads when a balanced coin is tossed twice. In these two examples the outcome set is discrete (individually distinct) and thus the variables are defined on a discrete set. Such variables are called *discrete stochastic variables*. But for example, consider the life time of electric bulbs, the distance from the origin where the indicator stops when a balanced spinner is rotated once, the yield of wheat under a given set of agricultural conditions and so on, which are all defined on continuous sets and hence the chance variables associated with those are continuous variables.

Definition. If the outcome set is discrete (continuous) the chance variables on the set are called discrete (continuous).

Example 4.1.1 Consider an experiment of tossing a balanced coin twice. Let X denote the stochastic variable (s.v.) the number of heads in the outcomes and let Y denote the random variable (r.v.) the number of heads minus the number of tails in the outcomes. Evaluate the ranges and the corresponding probabilities.

Solution. The outcome set is,
$$S = \{(H, H), (H, T), (T, H), (T, T)\}.$$
Since there can be only 0, 1, 2 heads in the outcomes, X takes the values $x = 0, 1, 2$ with probabilities $\frac{1}{4}, \frac{2}{4}, \frac{1}{4}$ respectively. For the outcome (H, H) the number of heads minus the number of tails is $2 - 0 = 2$. Thus, Y takes the values $y = 2, 0, -2$, with probabilities $\frac{1}{4}, \frac{2}{4}$ and $\frac{1}{4}$ respectively. If we denote the probability associated with any particular value x of X by $f(x)$ and the probability associated with y by $g(y)$ then we have the following table.

TABLE 4.1

s.v. X		s.v. Y	
x	$f(x)$	y	$g(y)$
0	$\frac{1}{4} = f(0)$	2	$\frac{1}{4} = g(2)$
1	$\frac{2}{4} = f(1)$	0	$\frac{2}{4} = g(0)$
2	$\frac{1}{4} = f(2)$	-2	$\frac{1}{4} = g(-2)$

Here $f(x)$ and $g(y)$ are called the *probability functions* associated with the s.v.'s X and Y respectively. If the s.v.'s X and Y in this example are assumed to take all values on the real line R then we can say that X assumes the values 0, 1 and 2 with probabilities $\frac{1}{4}$, $\frac{2}{4}$ and $\frac{1}{4}$ respectively and all other values with zero probabilities. Thus, we may write the probability functions $f(x)$ and $g(y)$ as,

$f(x)$	$g(y)$
$f(0) = \frac{1}{4}$	$g(2) = \frac{1}{4}$
$f(1) = \frac{2}{4}$	$g(0) = \frac{2}{4}$
$f(2) = \frac{1}{4}$	$g(-2) = \frac{1}{4}$
and $f(x) = 0$ elsewhere;	and $g(y) = 0$ elsewhere.

Further, it should be noticed that $f(x)$ and $g(y)$ are such that

$$f(x) \geqslant 0 \text{ for all } x; \qquad g(y) \geqslant 0 \text{ for all } y.$$
$$\sum_{x} f(x) = 1 \qquad \sum_{y} g(y) = 1,$$

where, for example, $\sum_{x} f(x)$ means the sum of $f(x)$ for all values of x. In this case,

$$\sum_{x} f(x) = f(0) + f(1) + f(2) + 0 = \frac{1}{4} + \frac{2}{4} + \frac{1}{4} + 0 = 1.$$

Comment. It should be pointed out that we can define an infinite number of stochastic variables on a given outcome set. If the values assumed by a s.v., with non-zero probabilities, are all real then the s.v. is called a real s.v. and if a s.v. assumes complex values with non-zero probabilities then it is called a *complex s.v.* In example 4.1.1 we have two real discrete stochastic variables X and Y.

Example 4.1.2 In an experiment of tossing a coin twice if X denotes the s.v. the number of heads in the outcomes, evaluate the following probabilities. (1) The probability that x takes the values 0 or 1. (2) The probability that $x \leqslant 1$.

Solution. For the s.v. X in this case, the probability function $f(x)$ is obtained as,

$$f(x) = \frac{1}{4} \text{ for } x = 0$$
$$f(x) = \frac{2}{4} \text{ for } x = 1$$
$$f(x) = \frac{1}{4} \text{ for } x = 2 \text{ and } f(x) = 0 \text{ elsewhere.}$$

The probability that x takes the values 0 or 1
$$= \tfrac{1}{4} + \tfrac{2}{4} = \tfrac{3}{4}.$$
The probability that $x \leqslant 1$ is
$$= 0 + \tfrac{1}{4} + \tfrac{2}{4} = \tfrac{3}{4}.$$
These may be written as,
$$P\{x = 0 \text{ or } 1\} = \tfrac{3}{4} \text{ and } P\{x \leqslant 1\} = \tfrac{3}{4}.$$
In general if X is a s.v., then the following notations will be used.

$P\{a \leqslant x \leqslant b\}$—probability that x lies between a and b (both a and b inclusive).

$P\{x \leqslant c\}$—probability that x takes values less than or equal to c.

$P\{x \geqslant d\}$—probability that x takes values greater than or equal to d.

Example 4.1.3 If a balanced spinner, with the dial marked 0 to 100, with 0 and 100 coinciding, is rotated once and if the s.v. X is defined as the distance from the origin (in the clockwise direction) where the indicator stopped, calculate the following probabilities:

(1) $P\{0 \leqslant x \leqslant 10\}$,   (2) $P\{20 \leqslant x \leqslant 45\}$.

Solution. Since the spinner is balanced, the indicator is as likely to stop in an interval as in any other interval of the same length. Hence,

(1) $P\{0 \leqslant x \leqslant 10\} = \dfrac{10}{100}$;   (2) $P\{20 \leqslant x \leqslant 45\} = \dfrac{45 - 20}{100} = \dfrac{25}{100}$.

If x_0 denotes the distance of the indicator from 0 then,
$$P\{x \leqslant x_0\} = \frac{x_0}{100}.$$

It is intuitively apparent that the probability function associated with the s.v. X in this example is,
$$f(x) = \frac{1}{100} \quad \text{for} \quad 0 \leqslant x \leqslant 100$$
and $f(x) = 0$ elsewhere.

Thus, the probability statements in (1) and (2) can be written as,
$$P\{0 \leqslant x \leqslant 10\} = \int_0^{10} \frac{1}{100} \, dx = \frac{10}{100}$$

and
$$P\{20 \leqslant x \leqslant 45\} = \int_{20}^{45} \frac{1}{100}\, dx = \frac{25}{100},$$

since the s.v. X in this case is continuous in the sense that it can take all values in the interval 0 to 100 with non-zero probabilities, in the sense, that the probability that x falls in (a, b) is non-zero where $a \geqslant 0$, $b \leqslant 100$ and $a \neq b$. From this example it may be noticed that,

(1) $f(x) \geqslant 0$ for all x

(2) $\int_{-\infty}^{\infty} f(x)\, dx = 0 + \int_{0}^{100} \frac{1}{100}\, dx = 1.$

Similar properties were seen for a discrete probability function in example 4.1.1. The probability function when the s.v. is continuous is often called *a density function*. We will often use the term probability function when X is discrete or continuous.

4.11 *A Probability Function*

Some common properties are seen for the probability functions when the variable is discrete or continuous. Hence we will define a probability function or a density function by using the following two postulates.

Definition. $f(x)$ is a probability function if it satisfies the conditions,

(1) $f(x) \geqslant 0$ for all x

(2) $\sum_{x} f(x) = 1$ when X is discrete

$$\int_{-\infty}^{\infty} f(x)\, dx = 1 \text{ when } X \text{ is continuous.}$$

Example 4.11.1 Can the following be probability functions?

(1)
$$f(x) = \begin{cases} 0.1 \text{ for } x=-5 \\ 0.5 \text{ for } x=-1 \\ 0.2 \text{ for } x=0 \\ 0.2 \text{ for } x=1 \\ 0 \quad \text{elsewhere;} \end{cases}$$

(2)
$$g(x) = \begin{cases} \frac{1}{2} \text{ for } x=1 \\ \frac{2}{3} \text{ for } x=0 \\ \frac{1}{4} \text{ for } x=2 \\ 0 \quad \text{elsewhere;} \end{cases}$$

(3)
$$h(x) = \begin{cases} -\tfrac{1}{2} \text{ for } x=2 \\ \tfrac{1}{2} \text{ for } x=3 \\ \tfrac{1}{2} \text{ for } x=4 \\ 0 \text{ elsewhere}; \end{cases}$$

(4)
$$f_1(x) = \begin{cases} \dfrac{x}{2}, \ 0 < x < 2 \\ 0 \quad \text{elsewhere}; \end{cases}$$

(5)
$$f_2(x) = \begin{cases} \dfrac{x}{2}, \ 0 < x \leqslant 1 \\ \dfrac{1}{2}, \ 1 < x \leqslant 2 \\ \dfrac{3-x}{2}, \ 2 < x \leqslant 3 \\ 0 \quad \text{elsewhere}; \end{cases}$$

(6)
$$f_3(x) = \begin{cases} 2x, \ 0 < x \leqslant 1 \\ 4-2x, \ 1 < x < 2 \\ 0 \quad \text{elsewhere}. \end{cases}$$

Answer. (1) Evidently $f(x) \geqslant 0$ for all x

$$\sum_x f(x) = 0 + 0.1 + 0.5 + 0.2 + 0.2 = 1.$$

Hence $f(x)$ is a probability function for some s.v. X.

(2) $g(x) \geqslant 0$ for all x

$$\sum_x g(x) = \tfrac{1}{2} + \tfrac{2}{3} + \tfrac{1}{4} \neq 1.$$

Hence $g(x)$ is not a probability function.

(3) $h(2) = -\tfrac{1}{2} \ngeqslant 0.$

The first condition is violated and hence $h(x)$ is not a probability function.

(4) Since $\dfrac{x}{2}$ for $0 < x \leqslant 2$ is > 0, $f_1(x) \geqslant 0$ for all x

$$\int_{-\infty}^{\infty} f_1(x)\, dx = 0 + \int_0^2 \frac{x}{2}\, dx = \frac{x^2}{4} \Big]_0^2 = 1.$$

Hence $f_1(x)$ is a density function for some s.v. X.

(5) Since $\dfrac{x}{2} > 0$ for $0 < x \leqslant 1$, $\tfrac{1}{2} > 0$ for $1 < x \leqslant 2$

and $\dfrac{3-x}{2} \geqslant 0$ for $2 < x \leqslant 3$, $f_2(x) \geqslant 0$ for all x.

$$\int_{-\infty}^{\infty} f_2(x)\, dx = 0 + \int_0^1 \frac{x}{2}\, dx + \int_1^2 \frac{1}{2}\, dx + \int_2^3 \frac{3-x}{2}\, dx + 0$$

(Here $f_2(x)$ has different functional forms in the different intervals).

$$= \frac{x^2}{4} \Bigg]_{0}^{1} + \frac{x}{2} \Bigg]_{1}^{2} + \left(\frac{3x}{2} - \frac{x^2}{4}\right) \Bigg]_{2}^{3} + 0 = 1.$$

Hence $f_2(x)$ is a density function for some s.v. X.

(6) Evidently $f_3(x) \geqslant 0$ for all x.

$$\int_{-\infty}^{\infty} f_3(x)\, dx = 0 + \int_{0}^{1} 2x\, dx + \int_{1}^{2} (4-2x)dx + 0$$

$$= x^2 \Bigg]_{0}^{1} + (4x - x^2) \Bigg]_{1}^{2} + 0 \neq 1.$$

Hence $f_3(x)$ is not a density function.

A diagrammatic representation of the probability functions in (1) and (5) is given in Figure 4.1.

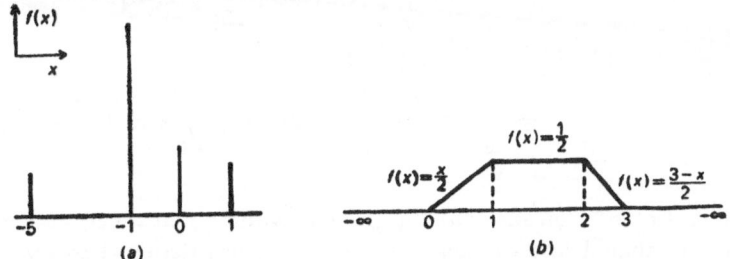

FIGURE 4.1   Discrete and continuous probability functions

Comment. When a probability function of a discrete s.v. is represented diagrammatically we can expect a set of discrete points and for a density function we can expect a continuous curve· But for a continuous s.v. we can notice the following.

$$P\{a \leqslant x \leqslant b\} = \int_{a}^{b} f(x)dx = F(b) - F(a)$$

where,

$$F(y) = \int_{-\infty}^{y} f(x)dx.$$

Therefore,

$$P\{x=c\} = P\{c \leqslant x \leqslant c\} = \int_{c}^{c} f(x)dx = 0, \text{ when } X \text{ is continuous.}$$

That is, when a s.v., is continuous the probability that it assumes a specified value is zero. Hence, a function $f(x)$ can be a density function even if $f(x)$ is not continuous provided the total probability at these discontinuity points is zero. Also it is interesting to see that if $f(x) \geqslant 0$ for all x and if $\Sigma_x f(x) = c$ or $\int_{-\infty}^{\infty} f(x)dx = c$ where c is finite (need not be unity) we can always construct a probability function or a density function by taking $g(x) = \dfrac{f(x)}{c}$ then $g(x)$ satisfies the two axioms.

Example 4.11.2 If a s.v. X has the density function
$$f(x) = \begin{cases} \frac{1}{4}, & -2 < x < 2 \\ 0 & \text{elsewhere,} \end{cases}$$
obtain (1) $P\{x < 1\}$; (2) $P\{|x| > 1\}$; (3) $P\{2x + 3 > 5\}$.

Solution. (1) $P\{x < 1\} = \int_{-\infty}^{1} f(x)dx = 0 + \int_{-2}^{1} \frac{1}{4} \, dx = \frac{3}{4}$.

(2) $P\{|x| > 1\} = P\{x > 1 \text{ or } < -1\}$
$$= \int_{-\infty}^{-1} f(x)dx + \int_{1}^{\infty} f(x)dx = 0 + \int_{-2}^{-1} \frac{1}{4} \, dx + \int_{1}^{2} \frac{1}{4} \, dx + 0$$
$$= \frac{1}{2}.$$

Here $|x| > 1$ means that the positive values that x can take are greater than 1 and the negative values are less than -1 so that in magnitude x is greater than 1.

(3) $P\{2x + 3 > 5\} = P\{2x > 2\} = P\{x > 1\}$
$$= \int_{1}^{\infty} f(x)dx = \int_{1}^{2} \frac{1}{4} \, dx + 0 = \frac{1}{4}.$$

Example 4.11.3 Check whether the following is a density function or not. If so, evaluate $P\{2 < x < 6\}$.
$$f(x) = \begin{cases} \dfrac{1}{\theta} e^{-x/\theta} & \text{for } 0 < x < \infty \\ 0 & \text{elsewhere} \end{cases}$$
where $\theta > 0$ is some constant.

Solution. Since $\theta > 0$, $\dfrac{1}{\theta} e^{-x/\theta} \geqslant 0$ for all x.

$$\int\limits_{-\infty}^{\infty} f(x)dx = 0 + \int\limits_{0}^{\infty} \frac{1}{\theta}\, e^{-x/\theta}\, dx = -\, e^{-x/\theta} \Big]_{0}^{\infty} = 1.$$

Hence $f(x)$ is a density function irrespective of the value of θ as long as $\theta > 0$ is a constant with respect to X.

$$P\{2 < x < 6\} = \int\limits_{2}^{6} \frac{1}{\theta}\, e^{-x/\theta}\, dx = -\, e^{-x/\theta} \Big]_{2}^{6}$$

$$= \left(e^{-2/\theta} - e^{-6/\theta}\right).$$

Comment. Since $P\{x{=}2\}{=}0{=}P\{x{=}6\}$ due to the continuity of X, $P\{2{<}x{<}6\}{=}P\{2{\leqslant}x{\leqslant}6\}$. Further, geometrically it is the area under the curve between the ordinates at $x{=}2$ and $x{=}6$. It is shown in Figure 4.2.

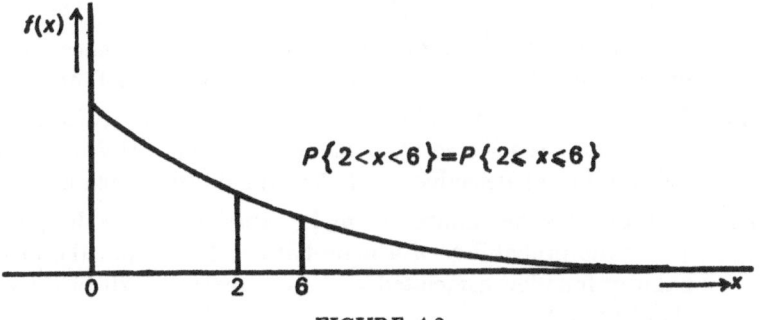

FIGURE 4.2

It is interesting to notice that, whatever may be the value of θ as long as it is a constant, finite and greater than zero, $f(x)$ is a density function. For example, for $\theta{=}1, 2, 5$ we get 3 different density functions having the same functional form. They are,

$$f(x) = \begin{cases} e^{-x}, & x > 0 \\ 0 & \text{elsewhere,} \end{cases} \qquad f(x) = \begin{cases} \frac{1}{2}\, e^{-x/2}, & x > 0 \\ 0 & \text{elsewhere.} \end{cases}$$

$$f(x) = \begin{cases} \frac{1}{5}\, e^{-x/5}, & x > 0 \\ 0 & \text{elsewhere.} \end{cases}$$

A constant, (with respect to the variate) such as θ, appearing in a probability function is known as *a parameter*. When there is a parameter in a probability function, for various possible assignable values of the parameter we get a set of probability functions called *a family of probability functions.* For example,

$$f(x) = \begin{cases} \dfrac{1}{\theta}\, e^{-x/\theta}, & x > 0,\ \theta > 0 \\ 0 & \text{elsewhere,} \end{cases}$$

represents a family of densities. An individual member in the family is obtained by taking a specific value of the parameter. It should be noticed that θ here is not treated as a variable having its own probability function.

Exercises

4.1 On the outcome set obtained by throwing a balanced coin 3 times, define 2 s.v.'s and evaluate the probability functions.

4.2 On the outcome set obtained by rolling a balanced die twice, define 2 s.v.'s and evaluate the probability functions.

4.3 A consignment of 30 radios contains 4 defectives. A random sample of 5 is selected from this consignment. If X denotes the number of defectives find the probability function of X.

4.4 If X denotes the number of male babies born in a hospital and if the probability of a male baby is $\frac{1}{2}$, find out the probability function associated with X if there are 20 births in that hospital.

4.5 Can the following be probability functions or density functions?

(1) $$f(x) = \begin{cases} \frac{3}{4} & \text{for } x = 1 \\ \frac{1}{4} & \text{for } x = 2 \\ 0 & \text{elsewhere.} \end{cases}$$

(2) $$f(x) = \begin{cases} \frac{1}{2} & \text{for } x = -1 \\ \frac{2}{3} & \text{for } x = 1 \\ 0 & \text{elsewhere.} \end{cases}$$

(3) $$f(x) = \begin{cases} \dfrac{1}{2^x} & \text{for } x = 1,2,3,\dots \\ 0 & \text{elsewhere.} \end{cases}$$

(4) $$f(x) = \begin{cases} \dfrac{1}{2x} & \text{for } x = 1,2,3. \\ 0 & \text{elsewhere.} \end{cases}$$

(5) $$f(x) = \begin{cases} x & \text{for } 0 < x \leqslant 1 \\ \dfrac{3-x}{4} & \text{for } 1 < x \leqslant 3 \\ 0 & \text{elsewhere.} \end{cases}$$

(6) $$f(x) = \begin{cases} 2(2-x), & 0 < x < 2 \\ 0 & \text{elsewhere.} \end{cases}$$

(7)
$$f(x) = \begin{cases} 4(x-1), & 0 < x < 4 \\ 0 & \text{elsewhere.} \end{cases}$$

(8)
$$f(x) = \begin{cases} \dfrac{1}{\theta}, & 0 < x < \theta \\ 0 & \text{elsewhere.} \end{cases}$$

4.6 Evaluate k if the following is the probability function or the density function of a s.v. X.

(1) $f(x) = \begin{cases} k(2-x), & 0 < x < 2, \\ 0 & \text{elsewhere.} \end{cases}$

(2)
$$f(x) = \begin{cases} \dfrac{k}{4} & \text{for } x=0 \\ \dfrac{k}{2} & \text{for } x=1 \\ 0 & \text{elsewhere.} \end{cases}$$

(3) $f(x) = \begin{cases} \dfrac{k}{\beta-a}, & a < x < \beta, \\ 0 & \text{elsewhere.} \end{cases}$

(4)
$$f(x) = \begin{cases} 2k\theta e^{-\theta x}, & x > 0, \\ & \theta > 0 \\ 0 & \text{elsewhere} \end{cases}$$
k, θ are constants.

4.7 If a s.v. X has a density function as given below evaluate and illustrate graphically,

(1) $P\{x \geqslant 3\}$,   (2) $P\{|x| < 1\}$,   (3) $P\{1 < x < 4\}$

$$f(x) = \begin{cases} \dfrac{x}{2}, & 0 < x \leqslant 1 \\ \dfrac{3-x}{4}, & 1 < x \leqslant 2 \\ \tfrac{1}{4}, & 2 < x \leqslant 3 \\ \dfrac{4-x}{4}, & 3 < x < 4 \\ 0 & \text{elsewhere.} \end{cases}$$

4.8 For the following probability function evaluate (1) $P\{|x| < 1\}$, (2) $P\{1 < x \leqslant 4\}$,   (3) $P\{x > 5\}$.

$$f(x) = \begin{cases} \tfrac{1}{4} & \text{for } x=-2 \\ \tfrac{1}{4} & \text{for } x=3 \\ \tfrac{1}{2} & \text{for } x=6 \\ 0 & \text{elsewhere.} \end{cases}$$

4.9 Can the following be a density function?
$$f(x) = \begin{cases} 5, & 0 < x < \tfrac{1}{5} \\ 0 & \text{elsewhere.} \end{cases}$$

4.10 Construct 2 examples of a density function which has different functional forms in the following intervals $-\infty < x \leqslant -1$, $-1 < x \leqslant 2$, $2 < x \leqslant 3$, $3 < x < \infty$.

4.12 *The Distribution Function*

Definition. The distribution function or the cumulative probability function is denoted by $F(x)$ and is defined as,

$$\sum_{-\infty < y \leqslant x} f(y) \quad \text{if } \varUpsilonY \text{ is discrete}$$

and $\displaystyle\int_{-\infty}^{x} f(y)\, dy$ if $\varUpsilon$ is continuous,

where $f(y)$ is the probability function of a s.v. $\varUpsilon$ and x is a fixed (given) number. Hence $F(x)$ can be called the probability that the s.v. $\varUpsilon$ will take a value less than or equal to a given value x. If $g(x)$ is the probability function then the distribution function of X, denoted by $G(x_0)$ is,

$$G(x_0) = P\{x \leqslant x_0\}.$$

Here x_0 means a given value of x and x_0 is used to avoid confusion with x (in general, any value assumed by X), x_0 is a specific value of x.   It is easy to notice the following properties, whether $\varUpsilon$ is discrete or continuous.

(1) $F(-\infty) = 0$;   (2) $F(\infty) = 1$;   (3) $F(a) \leqslant F(b)$ for $a < b$.

Hence we may use these as postulates and define a distribution function as follows.

Definition. A distribution function is a function $F(x)$ satisfying the following properties.

(1) $F(-\infty) = 0$;   (2) $F(\infty) = 1$;   (3) $F(a) \leqslant F(b)$ for all $a < b$; (4) $F(x)$ is right continuous.

Example 4.12.1 Evaluate the distribution function for the following probability function.

$$f(x) = \begin{cases} \frac{1}{8} & \text{for } x = -1 \\ \frac{2}{8} & \text{for } x = 0 \\ \frac{3}{8} & \text{for } x = 2 \\ \frac{2}{8} & \text{for } x = 3 \\ 0 & \text{elsewhere} \end{cases}$$

Solution.   By definition, the distribution function $F(x_0)$ is,

$$F(x_0) = \sum_{-\infty < x \leqslant x_0} f(x)$$

where x_0 is a specific value of x. Hence,

$$F(-1) = \sum_{-\infty < x < -1} f(x) = 0 + \tfrac{1}{8} = \tfrac{1}{8}$$

$$F(0) = \sum_{-\infty < x < 0} f(x) = 0 + \tfrac{1}{8} + \tfrac{2}{8} = \tfrac{3}{8}$$

$$F(2) = \tfrac{1}{8} + \tfrac{2}{8} + \tfrac{3}{8} = \tfrac{6}{8}$$

$$F(3) = \tfrac{1}{8} + \tfrac{2}{8} + \tfrac{3}{8} + \tfrac{2}{8} = 1$$

$$F(x) = 1 \text{ for } x \geqslant 3$$

This may be written in a better way as follows:

$$F(x) = \begin{cases} 0 & \text{for } x < -1 \\ \tfrac{1}{8} & \text{for } = x - 1 \\ \tfrac{3}{8} & \text{for } x = 0 \\ \tfrac{6}{8} & \text{for } x = 2 \\ 1 & \text{for } x \geqslant 3 \end{cases} \quad \text{or} \quad F(x) = \begin{cases} 0, & -\infty < x < -1 \\ \tfrac{1}{8}, & -1 \leqslant x < 0 \\ \tfrac{3}{8}, & 0 \leqslant x < 2 \\ \tfrac{6}{8}, & 2 \leqslant x < 3 \\ 1, & 3 \leqslant x < \infty \end{cases}$$

When $F(x)$ is represented graphically we get a diagram as in Figure 4.3 since $F(x)$ is a step function in this case. In general we can expect a step function for the distribution function when the variate is discrete.

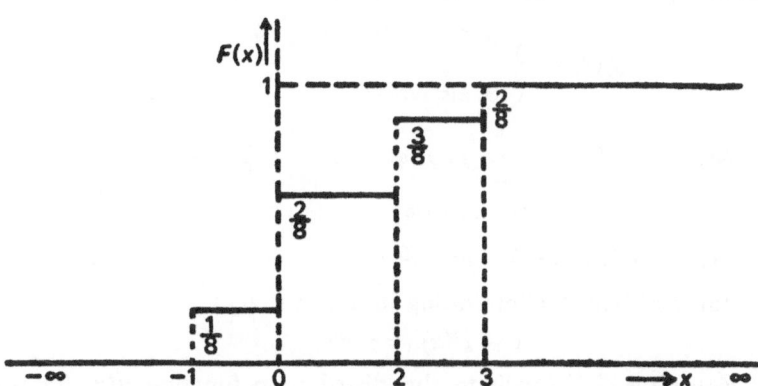

FIGURE 4.3 Distribution function of a discrete variate

That is, at -1 there is a jump of $\tfrac{1}{8}$ then at 0 there is an additional jump of $\tfrac{2}{8}$ and so on. Naturally when we plot the distribution function of a continuous s.v. we can expect a curve something similar to the one in Figure 4.4. The exact shape depends upon the exact form of the density function.

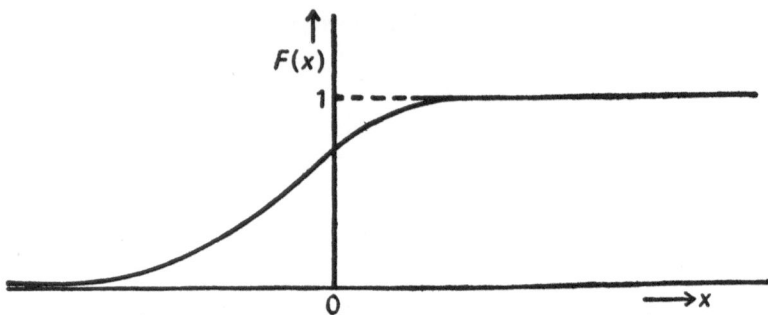

FIGURE 4.4   The distribution function of a continuous variate

It is more rigorous to define the density function, whenever it exists, as

$$f(x) = \frac{d}{dx} F(x)$$

where $F(x)$ is a differentiable function satisfying the postulates for a distribution function.

Example 4.12.2 Evaluate the distribution function $F(x)$ for the following density function and calculate $F(10)$.

$$f(x) = \begin{cases} \frac{1}{6} e^{-x/6}, & 0 < x < \infty \\ \\ 0 & \text{elsewhere.} \end{cases}$$

Solution. $F(x) = \int_{-\infty}^{x} f(y)\, dy = 0 + \int_{0}^{x} \frac{1}{6} e^{-y/6}\, dy$

$$= (1 - e^{-x/6})$$

Hence, $F(10) = (1 - e^{-10/6})$.

In this problem it is interesting to notice that,

$$1 - F(x) = e^{-x/6}.$$

Example 4.12.3 Evaluate the distribution function $F(x)$ for the following density function and calculate $F(2)$.

$$f(x) = \begin{cases} \dfrac{x}{3}, & 0 < x \leqslant 1 \\ \\ \dfrac{5}{27}(4-x), & 1 < x \leqslant 4 \\ \\ 0 & \text{elsewhere.} \end{cases}$$

Solution. By definition,

$$F(x) = \int_{-\infty}^{x} f(y)\, dy$$

Therefore, for any value of x such that $-\infty < x < 0$,

$$F(x) = \int_{-\infty}^{x} 0\, dy = 0$$

since $f(x) = 0$ in this interval. For any x in $0 < x \leqslant 1$,

$$F(x) = \int_{-\infty}^{0} 0\, dy + \int_{0}^{x} \frac{y}{3}\, dy$$

$$= 0 + \frac{x^2}{6} = \frac{x^2}{6}.$$

For any x in $1 < x \leqslant 4$,

$$F(x) = \int_{-\infty}^{0} 0\, dy + \int_{0}^{1} \frac{y}{3}\, dy + \int_{1}^{x} \frac{5}{27}(4-y)\, dy$$

$$= 0 + \frac{1}{6} + \int_{1}^{x} \frac{5}{27}(4-y)\, dy$$

$$= -\frac{13}{27} + \frac{5}{27}\left(4x - \frac{x^2}{2}\right)$$

Evidently for $x \geqslant 4$, $F(x) = 1$ and hence $F(x)$ can be written as,

$$F(x) = \begin{cases} 0, & -\infty < x < 0 \\[2mm] \dfrac{x^2}{6}, & 0 < x \leqslant 1 \\[2mm] -\dfrac{13}{27} + \dfrac{5}{27}\left(4x - \dfrac{x^2}{2}\right), & 1 < x \leqslant 4 \\[2mm] 1 & x \geqslant 4. \end{cases}$$

Hence,

$$F(2) = -\frac{13}{27} + \frac{5}{27}\left(4x - \frac{x^2}{2}\right) \text{ at } x = 2$$

$$= \frac{17}{27}.$$

Exercises

4.11 Evaluate the distribution function and represent it graphically:

$$
(1)\ f(x)=
\begin{cases}
\frac{1}{2} & \text{for } x=0 \\
\frac{1}{4} & \text{for } x=1 \\
\frac{1}{4} & \text{for } x=5 \\
0 & \text{elsewhere.}
\end{cases}
\qquad
(2)\ f(x)=
\begin{cases}
\frac{1}{5} & \text{for } x=-2 \\
\frac{1}{5} & \text{for } x=-1 \\
\frac{2}{5} & \text{for } x=1 \\
\frac{1}{5} & \text{for } x=2 \\
0 & \text{elsewhere.}
\end{cases}
$$

4.12 Evaluate the distribution function and calculate

(1) $F(2)$,   (2) $F(3)-F(2)$,   (3) $F(-1)$.

(a) $f(x)=\begin{cases}\frac{1}{5}, & 0<x<5 \\ 0 & \text{elsewhere.}\end{cases}$ (b) $f(x)=\begin{cases}\frac{1}{2}\,e^{-x/2}, & x>0 \\ 0 & \text{elsewhere.}\end{cases}$

4.13 Evaluate the distribution function and calculate

(1) $F(2)$,   (2) $F(1)$.

$$
(a)\ f(x)=
\begin{cases}
x, & 0<x\leqslant 1 \\
\dfrac{3-x}{4}, & 1<x\leqslant 3 \\
0 & \text{elsewhere.}
\end{cases}
\qquad
(b)\ f(x)=
\begin{cases}
\dfrac{x}{2}, & 0<x\leqslant 1 \\
\dfrac{3-x}{4}, & 1<x\leqslant 2 \\
\frac{1}{4}, & 2<x\leqslant 3 \\
\dfrac{4-x}{4}, & 3<x<4 \\
0 & \text{elsewhere.}
\end{cases}
$$

4.14 Evaluate the density function from the following distribution function:

$$
(1)\ F(x)=
\begin{cases}
2x^2/5, & 0<x\leqslant 1 \\
\dfrac{-3}{5}+\dfrac{2}{5}\left(3x-\dfrac{x^2}{2}\right), & 1<x\leqslant 2 \\
1, & x>2.
\end{cases}
$$

$$
(2)\ F(x)=
\begin{cases}
\dfrac{x^2}{2}, & 0<x\leqslant 1 \\
\dfrac{-1}{8}+\dfrac{1}{4}\left(3x-\dfrac{x^2}{2}\right), & 1<x\leqslant 3 \\
1, & x\geqslant 3.
\end{cases}
$$

4.15 Evaluate the probability function from the following distribution function:

$$F(x) = \begin{cases} 0, & x<1 \\ \frac{1}{4}, & 1\leqslant x<2 \\ \frac{2}{4}, & 2\leqslant x<3 \\ 1, & x\geqslant 3. \end{cases}$$

4.2 MATHEMATICAL EXPECTATION

The reader may be familiar with the integral and differential operators in Calculus. Here we will define an operator called ' Mathematical Expectation ' and which will be denoted by E.

Definition. *If X is a s.v. with probability function $f(x)$ and if $\phi(X)$ is a function of X which is again a s.v. then the mathematical expectation of $\phi(X)$ is defined as,*

$$E(\phi(X)) = \int_{-\infty}^{\infty} \phi(x)\, f(x)dx \text{ if } X \text{ is continuous}$$

$$= \sum_{-\infty<x<\infty} \phi(x)\, f(x) \text{ if } X \text{ is discrete.}$$

Example 4.2.1 Obtain (1) $E(X)$, (2) $E(|X|)$, and (3) $E(X^2)$ for the following probability function:

$$f(x) = \begin{cases} \frac{1}{4} & \text{for } x=-1 \\ \frac{1}{4} & \text{for } x=0 \\ \frac{2}{4} & \text{for } x=2 \\ 0 & \text{elsewhere.} \end{cases}$$

Solution. Since X is a discrete s.v. in this case, by definition,

(1) $E(X) = \sum_{-\infty<x<\infty} x f(x) = 0+(-1)(\frac{1}{4})+(0)(\frac{1}{4})+(2)(\frac{2}{4})$

$$= \tfrac{3}{4}.$$

(2) $E(|X|) = \sum_{-\infty<x<\infty} |x| f(x) = 0+|-1|(\frac{1}{4})+|0|(\frac{1}{4})+|2|(\frac{2}{4})$

$$= \tfrac{5}{4}.$$

(3) $E(X^2) = \sum_{-\infty<x<\infty} x^2 f(x) = 0+(-1)^2(\frac{1}{4})+(0)^2\frac{1}{4}+(2)^2(\frac{2}{4})$

$$= \tfrac{9}{4}.$$

Comment. *Whenever there is no confusion* $E(\phi(x))$ *is also written as* $E\phi(x)$.

Example 4.2.2 Obtain $E((X-EX)^2)$ for the following density function:

$$f(x) = \begin{cases} \frac{1}{7}, & 0<x<7 \\ 0 & \text{elsewhere.} \end{cases}$$

Solution. Here X is a continuous s.v. and hence, by definition,

$$E((X-EX)^2) = \int_{-\infty}^{\infty} (X-EX)^2 f(x)dx$$

$$= 0 + \int_{0}^{7} (X-EX)^2 \tfrac{1}{7} dx.$$

But,

$$EX = \int_{0}^{7} \frac{x}{7} dx = \frac{7}{2}.$$

Therefore,

$$E((X-EX)^2) = \int_{0}^{7} (x-\tfrac{7}{2})^2 \tfrac{1}{7} dx$$

$$= \int_{0}^{7} (x^2 -7x + \tfrac{49}{4}) \tfrac{1}{7} dx$$

$$= \tfrac{49}{12}.$$

Example 4.2.3 Evaluate $E(X^2)$ for the following density function,

$$f(x) = \begin{cases} x, & 0<x<1 \\ 2-x, & 1<x<2 \\ 0 & \text{elsewhere.} \end{cases}$$

Solution. By definition,

$$E(X^2)= \int_{-\infty}^{\infty} x^2 f(x)dx=0+ \int_{0}^{1} x^2(x)dx+ \int_{1}^{2} x^2(2-x)dx+0$$

(since $f(x)$ is x in $0<x<1$ and $2-x$ in $1<x<2$).

$$=\frac{x^4}{4}\Big]_{0}^{1} + 2\frac{x^3}{3} -\frac{x^4}{4}\Big]_{1}^{2} = \frac{7}{6}.$$

Example 4.2.4 A person gets dollars equal to the square of the number which comes up when a balanced die, with the faces marked 1, 2, 3, 4, 5, 6, is rolled once. If this game is repeated an indefinitely large number of times how much money can he expect in the long run, per game?

Answer. Let X be a s.v. taking the values 1^2, 2^2, 3^2, 4^2, 5^2 and 6^2. Since the probabilities of occurrence of $1, 2, 3, 4, 5$ and 6 are $\frac{1}{6}$, ...,$\frac{1}{6}$ respectively, this s.v. X takes the values 1^2, 2^2, ..., 6^2 with probabilities $\frac{1}{6}$, $\frac{1}{6}$, ..., $\frac{1}{6}$ respectively. That is, in the long run, he gets \$$1^2$, \$$2^2$, ..., \$$6^2$ with the corresponding chances $\frac{1}{6}$ each. Hence the amount that he can expect per game is,

$$= 1^2(\tfrac{1}{6}) + 2^2(\tfrac{1}{6}) + 3^2(\tfrac{1}{6}) + 4^2(\tfrac{1}{6}) + 5^2(\tfrac{1}{6}) + 6^2(\tfrac{1}{6})$$
$$= \$ \ 15.17.$$

Comment. 'In the long run' is a convenient term used in the theory of games of chance to mean that 'if the game is repeated indefinitely large number of times under similar conditions'. In this particular example, for a particular trial if 5 comes up he gets \$25 and when the game is repeated a large number of times the chance of getting 1 or 2 or ... or 6 approaches $\frac{1}{6}$.

Exercises

4.16 Evaluate $E(X)$ and $E(|X|)$ in the following problems.

(1) $f(x) = \begin{cases} \dfrac{1}{\theta} e^{-x/\theta}, \ x>0, \ \theta>0 \ \text{(constant)} \\ 0 \ \text{elsewhere.} \end{cases}$

(2) $f(x) = \begin{cases} \tfrac{1}{4}, \ -2<x<2 \\ 0 \ \text{elsewhere.} \end{cases}$

(3) $f(x) = \begin{cases} \tfrac{1}{8} \ \text{for} \ x=-4 \\ \tfrac{1}{2} \ \text{for} \ x=-1 \\ \tfrac{1}{4} \ \text{for} \ x=1 \\ 0 \ \text{elsewhere.} \end{cases}$

4.17 Calculate $E(X)$ for the following densities

(1) $f(x) = \begin{cases} \dfrac{4x}{5}, \ 0<x \leqslant 1 \\ \tfrac{2}{5}(3-x), \ 1<x \leqslant 2 \\ 0 \ \text{elsewhere.} \end{cases}$

$$(2)\ f(x) = \begin{cases} \dfrac{x}{2}, & 0 < x \leqslant 1 \\[2mm] \dfrac{3-x}{4}, & 1 < x \leqslant 2 \\[2mm] \tfrac{1}{4}, & 2 < x \leqslant 3 \\[2mm] \dfrac{4-x}{4}, & 3 < x < 4 \\[2mm] 0 & \text{elsewhere.} \end{cases}$$

4.18 If a person gets \$$(2x+5)$ where x denotes the number appearing when a balanced die is rolled once, how much money can he expect in the long run per game?

4.19 If a person gets \$2 if a head comes and loses \$ 5 when a tail comes then how much money can he expect in the long run per game, when an unbiased coin is tossed?

4.20 If a person gains or loses an amount equal to the number appearing when a balanced die is rolled once, according to whether the number is even or odd, how much money can he expect per game in the long run?

4.21 Moments

In Chapter 1 we defined the various types of moments for a set of numbers. Here we will consider the moments defined for stochastic variables. Let X be a s.v.

Definition. Moments about the origin. The rth moment about the origin $\mu_r{}'$ is defined as,

$$\mu_r{}' = E(X^r)$$

Definition. Mean value. When $r=1$, $\mu_r{}' = E(X)$ and $E(X)$ is defined as the mean value of the s.v. X.

Definition. Central moments. The rth central moment μ_r is defined as,

$$\mu_r = E((X-EX)^r)$$

Definition. The Variance. When $r=2$, $\mu_2 = E((X-EX)^2)$ and μ_2 is defined as the variance of a stochastic variable X, and $\sqrt{\mu_2}$ is known as the standard deviation of X and is usually denoted by σ (sigma).

Definition. The rth factorial moment $\mu_{[r]}$ is defined as,

$$\mu_{[r]} = E(X(X-1)\ (X-2)...(X-r+1)).$$

Definition. The absolute moments. The rth absolute moment about a point c is defined as,

$$M_r = E(|X-c|^r).$$

These different moments have uses in statistical analysis. The reader may also notice that these definitions are analogous to the definitions introduced in Chapter 1. In the above definitions E denotes 'mathematical expectation', $|X-c|$ denotes the absolute value of $X-c$.

The mean value of X can be interpreted as a measure of central tendency in the values assumed by X or as the *centre of gravity* of X. The standard deviation of X can be taken as a measure of scatter in the values assumed by X from a point of location, namely, $E(X)$. It can also be interpreted as the *moment of inertia* in certain physical set-ups. These won't be discussed in detail here.

Example 4.21.1 Let X take the values $x_1, x_2, ..., x_n$ with probabilities $p_1, p_2,...,p_n$ such that $p_1+p_2+...+p_n=1$, calculate $E(X)$ and variance of X.

Solution.
$$E(X) = \sum_{-\infty<x<\infty} xf(x) = x_1p_1 + x_2p_2 + ... + x_np_n + 0$$

$$= \sum_{i=1}^{n} x_ip_i.$$

Variance of $X = \sum_{-\infty<x<\infty} (x-E(X))^2 f(x) = \sum_{i=1}^{n} (x_i-E(X))^2 p_i$

where $E(X)$ is obtained above.

Comment. If $p_1, p_2,..., p_n$ are assumed to be forces acting at the points $x_1, ...,x_n$ then EX gives the centre of gravity of the system. The expected value of a s.v. need not always exist, which can be seen from the following example.

Example 4.21.2 Evaluate $E(2^x)$ for the following probability function,

$$f(x) = \begin{cases} \dfrac{1}{2^x}, & x=1, 2, 3,... \\ 0 \text{ elsewhere.} \end{cases}$$

Solution. By definition, since X is discrete here,

$$E(2^X) = \sum_{-\infty<x<\infty} 2^x f(x) = \sum_{x=1}^{\infty} 2^x \left(\frac{1}{2^x}\right)$$

$$= 2\left(\frac{1}{2}\right) + 2^2\left(\frac{1}{2^2}\right) + ... = 1 + 1 + ... = \infty.$$

Comment. Here $E(2^X)$ does not exist, in the sense, it is not a finite number. In general, even if a s.v. is well defined the expected value of a function of it need not exist. In our discussions we will assume that all the expected values, which are going to be used, exist unless otherwise stated.

We have defined the rth absolute moment about a point c, as,
$$M_r = E(\,|\,X-c\,|^{\,r}\,).$$

When $r=1$, we get the *absolute moment* about the point c. In Chapter 1 we have remarked that a scatter in a set of numbers can also be measured by using absolute moments. Similarly the scatter in the values assumed by X, from the point c, can be measured by the rth root of M_r. When $r=1$, this yields the mean absolute deviation and when $r=2$ we get the standard deviation for $c=E(X)$. The following are a few theorems on expected values. Wherever the proof is not given, the proof is simple and is left to the reader.

Theorem 4.21.1 Let X be a s.v. and let c be a constant with respect to X, then,
$$E(c)=c.$$

Theorem 4.21.2 Let X be a s.v. and a and b be constants then,
$$E(aX+b)=aE(X)+b.$$

Theorem 4.21.3 Let $\phi_1(X),\phi_2(X)$ be two functions of a s.v., X then,
$$E(\phi_1(X)+\phi_2(X))=E(\phi_1(X))+E(\phi_2(X)).$$

Theorem 4.21.4 Let X be a s.v. and c be a constant then,
$$E(c\phi(X))=cE(\phi(X)).$$

Theorem 4.21.5 Let X be a s.v. and a and b be constants then, variance of $aX+b$, denoted by Var $(aX+b)$ is,
$$\text{Var } (aX+b) = a^2 \text{ Var } (X).$$

Proof.
$$\begin{aligned}
\text{Var } (aX+b) &= E\{aX+b-E(aX+b)\}^2 \\
&= E\{aX+b-aE(X)-b\}^2 \\
&= E\{a(X-E(X))\}^2=a^2E(X-E(X))^2 \\
&= a^2 \text{ Var } (X).
\end{aligned}$$

Thus we have the standard deviation of $aX+b$ as $|a|$ times the standard deviation of X since *the standard deviation is defined as the positive square root of the variance.*

Example 4.21.3 Let X be a s.v. and let μ and σ denote the mean value and the standard deviation of X then evaluate the variance of $(X-\mu)/\sigma$.

Solution. $\text{Var}\left(\dfrac{X-\mu}{\sigma}\right) = E\left\{\dfrac{X-\mu}{\sigma}\right\}^2,\ \left(\text{since } E\left(\dfrac{X-\mu}{\sigma}\right) = 0\right)$

$$= \frac{1}{\sigma^2} E((X-\mu)^2) = \frac{\sigma^2}{\sigma^2} = 1.$$

Definition. **For any s.v. X the variate $\Upsilon = (X - E(X))/\sigma$ is called the standardized variable of X.** For a standardized variable Υ, it can be seen that $E(\Upsilon) = 0$ and $\text{Var}\ (\Upsilon) = 1$.

Example 4.21.4 From the following probability function obtain the standardized variable Υ,

$$f(x) = \begin{cases} \frac{1}{3} & \text{for } x=0 \\ \frac{2}{3} & \text{for } x=1 \\ 0 & \text{elsewhere.} \end{cases}$$

Solution. By definition,

$$E(X) = 0(\tfrac{1}{3}) + 1(\tfrac{2}{3}) = \tfrac{2}{3}$$
$$\text{Var}\ (X) = E(X - E(X))^2 = (0 - \tfrac{2}{3})^2\ (\tfrac{1}{3}) + (1 - \tfrac{2}{3})^2\ (\tfrac{2}{3})$$
$$= \tfrac{2}{9}.$$

$\Upsilon = (X - E(X))/(\text{S.D.})$ and here x takes the values 0 and 1. Hence y takes the values $(0 - 2/3)/\sqrt{(2/9)}$ and $(1 - 2/3)/\sqrt{(2/9)}$ with probabilities 1/3 and 2/3 respectively.

For computational purposes we may use the following formula,

$$\text{Var}\ (X) = E(X - E(X))^2 = E(X^2) - (E(X))^2,$$

which can be easily proved since,

$$E(X - E(X))^2 = E\left\{X^2 + (E(X))^2 - 2X(E(X))\right\}$$
$$= E(X^2) + (E(X))^2 - 2E(X)\ E(X)$$
$$= E(X^2) - (E(X))^2.$$

A similar formula was pointed out in Chapter 1 for the evaluation of the variance from a set of numbers.

The *percentile points, quartiles, deciles*, the median and other measures of location can also be defined for a s.v. in a similar fashion as in Chapter 1. For example *the median* of a s.v., X, denoted by M, is defined as that point on the real line such that

$$P\{x \leqslant M\} \geqslant \tfrac{1}{2} \text{ and } P\{x \geqslant M\} \geqslant \tfrac{1}{2}.$$

That is, M is that point on the real line such that

$$\int_{-\infty}^{M} f(x)dx \geqslant \tfrac{1}{2} \text{ and } \int_{M}^{\infty} f(x)dx \geqslant \tfrac{1}{2} \text{ if } X \text{ is continuous}$$

or

$$\sum_{-\infty < x \leqslant M} f(x) \geqslant \tfrac{1}{2} \text{ and } \sum_{M < x < \infty} f(x) \geqslant \tfrac{1}{2} \text{ if } X \text{ is discrete.}$$

Also a s.v. X, together with its probability function $f(x)$ is sometimes called the probability distribution (different from the distribution function), such as the Binomial distribution, Normal distribution, and so on. Individual distributions will be considered in detail later.

Example 4.21.5 Obtain the median for the following distribution.
$$X : f(x) = \begin{cases} 2x, & 0 < x < 1 \\ 0, & \text{elsewhere.} \end{cases}$$

Solution. Let M be the median. By definition,
$$P\{x \leqslant M\} \geqslant \tfrac{1}{2} \text{ and } P\{x \geqslant M\} \geqslant \tfrac{1}{2}.$$
That is, $\displaystyle\int_0^M 2x \, dx$ and $\displaystyle\int_M^1 2x \, dx \geqslant \tfrac{1}{2}$

That is, M^2 and $1 - M^2 \geqslant \tfrac{1}{2}$ and hence $M = \dfrac{1}{\sqrt{2}}$.

Comment. When a distribution is continuous, in the sense that the probability function is continuous (except for a set of points with total probability zero) then M divides the total probability into two equal parts. That is,
$$P\{x \leqslant M\} = \tfrac{1}{2} = P\{x \geqslant M\}.$$
When the distribution is not continuous sometimes a number of points in an interval may qualify for the median. In that case we take the middle point of the interval.

A *mode* of a distribution can be defined as that value on the x-axis, corresponding to a maximum point on the curve of the density function when the variate is continuous or the value assumed by the variate with maximum probability when the variate is discrete. It is interesting to notice that a distribution may have more than one mode. If there is only one mode then the distribution is called

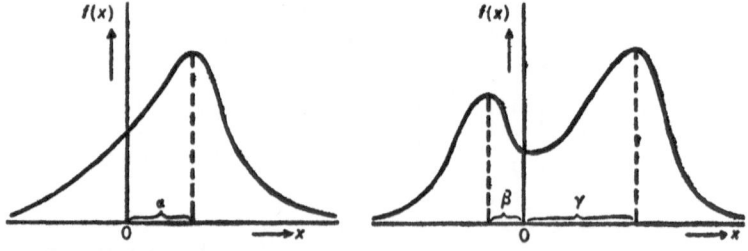

FIGURE 4.5(a) Unimodal continuous   (b) Bimodal continuous

unimodal and if there are two modes then it is called *bimodal* and sometimes there may not be any mode at all. Figure 4.5 (a) and (b) give a unimodal and a bimodal continuous distributions respectively, where α, β, γ denote the modes.

The fractile points can also be defined in a similar fashion as in Chapter 1. For example, the quartile points Q_1 and Q_3 can be defined as those points Q_1 and Q_3 such that,

$$P\{x \leqslant Q_1\} \geqslant 0.25 \text{ and } P\{x \geqslant Q_1\} \geqslant 0.75,$$
$$\text{and} \quad P\{x \leqslant Q_3\} \geqslant 0.75 \text{ and } P\{x \geqslant Q_3\} \geqslant 0.25,$$

and thus when X is continuous $P\{x \leqslant Q_1\} = 0.25$ and $P\{x \leqslant Q_3\} = 0.75$.

Exercises

4.21 Calculate (1) the median, (2) the mean value, (3) the mode if there is any, (4) the first quartile point, for the following distribution.

$$f(x) = \begin{cases} \dfrac{1}{\theta}, & 0 < x < \theta \\ 0, & \text{elsewhere.} \end{cases}$$

4.22 Calculate the variance for the following distribution:

$$f(x) = \begin{cases} \frac{1}{4} \text{ for } x = -1 \\ \frac{1}{4} \text{ for } x = 0 \\ \frac{2}{4} \text{ for } x = 1 \\ 0 \text{ elsewhere.} \end{cases}$$

4.23 Obtain the mode, if there is any, for the following distribution:

$$f(x) = \frac{1}{\sqrt{2\pi}} e^{-x^2/2}, \quad -\infty < x < \infty.$$

4.24 Show that $E((X-d)^2)$ is least when $d = E(X)$.

4.25 Show that $E(|X-d|)$ is least when d is the median of X.

4.22 *Some Decision Problems*

A basic statistical problem is a problem in making decisions in situations where there is an element of uncertainty. In many practical problems in the fields of business, transportation, games of chance, and so on, a decision can be made by simply maximizing or minimizing an expected value.   There are more complicated decision problems but here we will consider only some simple problems of decision making based on the maximization of some mathematical expectations.   Consider the following situation.   A contractor has to make a decision.   There are two jobs available to him.   Job 1 will give him a profit of $10,000 if there is no strike and he will get a profit of $2,000 if there is a strike.   Job 2 will give him a profit of $20,000 if there is no strike and only $500 if there is a strike.   The decision is quite difficult to make unless he has some idea about the chances of having a strike at these two places. Suppose that the chances of strike at the first job site is $\frac{1}{4}$ and that at the second site is $\frac{1}{2}$ then it is not difficult to make a decision.

	No Strike	*Strike*
Job 1	$ 10,000 $\frac{3}{4}$	$ 2,000 $\frac{1}{4}$
Job 2	$ 20,000 $\frac{1}{2}$	$ 500 $\frac{1}{2}$

His expected profit from job 1 $= (10,000) \ (\frac{3}{4}) + 2,000 \ (\frac{1}{4})$
$$= 8,000.$$

His expected profit from job 2 $= 20,000 \ (\frac{1}{2}) + 500 \ (\frac{1}{2})$
$$= 10,250.$$

So he will accept job 2 since more profit is expected from it.   Suppose that he is a born pessimist or he does not like to take chances then he would expect the worst situation.   That is, he will assume that there will be a strike and then he will take the job which gives him more profit.   In this case he will choose job 1 because it maximizes his minimum profit.   Suppose that he is an optimist or he neglects the chances of a strike then he will take job 2 because it gives $20,000 if there is no strike but may end up with $500.   Suppose that the chances of a strike is not known.   Then he will be in a dilemma.   He can choose one or the other.   So he may choose one of the jobs by conducting a game of chance but at the same time keeping in mind that his profit must be the same whether he chooses job 1 or job 2 and whether there is a strike or not.   Suppose that he chooses job 1 with probability p and job 2 with probability $1-p$, then he makes the profit,

$A = 10,000\ p+20,000\ (1-p)$ if there is no strike,

and $B = 2,000\ p+500\ (1-p)$ if there is strike.

He wants A to be equal to B. That is, he will select p such that,

$10,000\ p+20,000\ (1-p)=2,000\ p+500\ (1-p)$

That is, $p>1$. Since the maximum value of p is unity this indicates that he has to choose job 1 with probability 1 in order to maximize his profit.

Such a decision problem is called a *randomized decision* problem because the decision of choosing one of the jobs is left to a game of chance.

Exercises

4.26 At a beach resort a catering company wants to prepare lunch boxes. It costs them $0.50 to make a box and they can sell it for $1.50. If a box is not sold it is wasted which means a loss of $0.50 per box. If there is a large crowd they can sell 1000 and if the crowd is small they can sell 500. The company wants to maximize the expected profit. How many boxes should they prepare if (1) the chances of a big crowd is 0.3, (2) the chances of a big crowd is 0.6?

4.27 A contractor came to a grape farmer to buy all the grapes right away and the contractor will harvest it when it is ready. The offer gives him a profit of $10,000. The farmer can sell on his own and it will bring him a profit of $20,000, if the weather is good (that means, the grapes will be good). If the weather is bad the grapes will be spoiled and he will suffer a loss of $2,000 for the fertilizers, labour and so on. The chances of good weather is 0.6. Should the farmer accept the offer (1) if he wants to maximize his expected profit, (2) if he is a pessimist, (3) if he is an optimist, (4) if nothing is known about the chances of good weather what should be the chances of accepting the offer if he wants to have the same expected profit?

4.28 An engineer wants to appoint an assistant in order to complete a job on time. If he completes the job on time he makes a profit of $10,000 otherwise he suffers a loss of $2,000 compared to what he would have earned on his own if he had worked for some firm. If he works himself his chances of finishing is $\frac{1}{4}$. If he has a good assistant he can finish it on time. Two assistants came. The chances of them being good are 0.55 and 0.5 respectively. Should he employ an assistant, if so whom should he employ?

4.29 Three political parties contest in an election. The public opinion survey shows their chances of winning as 0.5 (party 1), 0.3 (party 2), 0.2 (party 3). A businessman is faced with a dilemma. If party 1 is supported and if it wins he makes a profit of $1 million through favours received but he can lose $\frac{1}{2}$ million and $\frac{1}{4}$ million if party 2 or 3 wins respectively because of other competing firms. If he supports party 2 and if it wins he can make $ 2 millions but will lose $\frac{1}{2}$ million or $\frac{1}{4}$ million if party 1 or 3 wins. If he supports party 3 and if it wins then he can make $ 4 millions but can lose $\frac{1}{4}$ or $\frac{1}{2}$ million if part 1 or 2 wins. (1) which party should he support if he wants to make a good profit, (2) if he does not believe the result of the public opinion poll with what probability should he support one party if he wants to support only either party 1 or party 2 and he wants his profit should be the same, whoever wins?

4.30 A shopkeeper has the facility to store a large number of a perishable item. He buys it at $ 30 per item and sells it at $ 50. If an item is unsold at the end of the day $ 30 is lost. The daily demand for that item is given in the following table.

Number of items	4	5	6
Probability	0.2	0.5	0.3

How many items should he store such that his profit is maximum?

4.3 THE MOMENT GENERATING FUNCTION

A moment generating function $M(t)$ which will be discussed in this section is of immense value in the theory of statistical distributions. A complete exposition of its uses is beyond the scope of this book.

Definition. The moment generating function $M(t)$ of a s.v. X, if it exists, is defined as,

$$M(t) = E(e^{tX}) = E\left(1 + \frac{tX}{1!} + \frac{(tX)^2}{2!} + \ldots\right)$$

$$= 1 + \mu_1'\frac{t}{1!} + \mu_2'\frac{t^2}{2!} + \ldots.$$

where μ_r' is the rth moment about the origin and it is the coefficient of $t^r/r!$ in the power series expansion of $M(t)$ and t is a real constant. The above expansion is obtained from the exponential series,

$$e^y = 1 + \frac{y}{1!} + \frac{y^2}{2!} + \ldots.$$

for any finite y.

Example 4.3.1 Obtain the moment generating function (M.G.F.) for the following distributions.

(1) $f(x) = \begin{cases} \frac{2}{3}, & x=1 \\ \frac{1}{3}, & x=2 \\ 0, & \text{elsewhere}; \end{cases}$
(2) $f(x) = \begin{cases} 1/\theta, & 0<x<\theta \\ 0, & \text{elsewhere}. \end{cases}$

Solution. (1) $M(t) = E(e^{tX}) = (\frac{2}{3})e^t + (\frac{1}{3})e^{2t}$

(since x takes the values 1 and 2 with probabilities $\frac{2}{3}$ and $\frac{1}{3}$ and other values with zero probabilities respectively).

(2) $M(t) = E(e^{tX}) = \int_0^\theta (e^{tx}/\theta)\, dx = (e^{t\theta}-1)/t\theta.$

Comment. For some distributions the M.G.F.'s do not exist. But there is another generating function called the characteristic function $\phi(t)$ which exists always and $\phi(t)$ is defined as,

$$\phi(t) = E(e^{itX}) = 1 + \mu_1'\frac{(it)}{1!} + \mu_2'\frac{(it)^2}{2!} + \ldots$$

where t is a real constant and $i=(-1)^{1/2}$. It should also be noticed that whenever $M(t)$ or $\phi(t)$ is differentiable the moments are obtained as,

$$\left[\frac{d^r}{dt^r} M(t)\right]_{t=0} = \mu_r'; \quad \left[\frac{d^r}{dt^r} \phi(t)\right]_{t=0} = (i)^r \mu_r'.$$

There is a uniqueness property enjoyed by $M(t)$ and $\phi(t)$ in the sense that corresponding to a distribution, there is one and only one $\phi(t)$ and whenever $M(t)$ exists, $M(t)$ as well as $\phi(t)$ will uniquely determine the distribution. One form of this result is given here without proof.   Some applications of this property will be considered in the next chapter.

We define the *Mellin transform* of $f(x)$ by

(1) $\quad g(s) = \int\limits_{0}^{\infty} x^{s-1} f(x) \, dx$

where s may be complex.   The function $f(x)$ is called the *inverse Mellin transform* of $g(s)$ and under certain conditions it is given by

(2) $\quad f(x) = (2\pi i)^{-1} \int\limits_{c-i\infty}^{c+i\infty} g(s) \, x^{-s} \, ds$

$\qquad$ where $i = (-1)^{1/2}$.

Clearly, $f(x)$ may be regarded as the density function and $g(s)$ as the $(s-1)$th moment of $f(x)$. Hence if the $(s-1)$th moment of the density function is given then the density function $f(x)$ can be uniquely determined by (2) using the usual techniques of contour integration.

Example 4.3.2 Find the density function whose $(s-1)$th moment is $\Gamma(a+s-1) / \Gamma(a)$, $a > 0$.

Solution. We have

$$f(x) = (2\pi i)^{-1} \int\limits_{c-i\infty}^{c+i\infty} \frac{\Gamma(a+s-1)}{\Gamma(a)} x^{-s} \, ds$$

The poles of the integrand are given by $s = -a+1, \ -a-0, \ -a-1, \dots$ Hence evaluating the contour integral by the method of residues, we have

$$f(x) = (1/\Gamma(a)) \sum_{r=0}^{\infty} \frac{(-1)^r x^{a+r-1}}{r!}$$

$$= \frac{x^{a-1}}{\Gamma(a)} e^{-x}$$

which is the density function of a Gamma variate.

Exercises

4.31 Evaluate the moment generating function, if it exists, for the following distributions.

(1) $f(x) = \begin{cases} xe^{-x}, & x>0 \\ 0 & \text{elsewhere}; \end{cases}$ (2) $f(x) = \dfrac{1}{\pi \{1 + x^2\}}$
$$-\infty < x < \infty.$$

4.32 Evaluate the M.G.F. for the distribution,

$f(x) = \dfrac{1}{b-a}$, $a < x < b$ and $f(x) = 0$ elsewhere, and obtain the

first two moments by (1) differentiating the M.G.F.; (2) by expanding the M.G.F.

4.33 If $M_x(t)$ denotes the M.G.F. of a s.v. X, show that,

(1) $M_{x+c}(t) = e^{tc} M_x(t)$;   (2) $M_{ax}(t) = M_x(at)$;

(3) $M_{ax+b}(t) = e^{tb} M_x(at)$, where a, b, c are constants.

4.34 Obtain the M.G.F. of the standardized variable $Y=(X-\mu)/\sigma$ in terms of the M.G.F. of X where $\mu=E(X)$ and $\sigma^2=\text{Var}(X)$.

4.35 *Factorial moment generating function.* Show that $E(t^X)$ generates the factorial moments in the sense that,

$$\mu_{[r]} = \left[\frac{d^r}{dt^r} E(t^X) \right]_{t=1}.$$

4.36 *Cumulants or Semi-invariants.* If $\log M(t) = k_1 \dfrac{t}{1!} + k_2 \dfrac{t^2}{2!} + \ldots$

then $k_1, k_2, \ldots$ are called cumulants or semi-invariants of the corresponding distribution. They are called semi-invariants because $k_2, k_3, \ldots$ for X are the same as those for $X+c$ where c is a constant. Show that for any s.v. X, k_1 and k_2 are the same as the mean value and the variance respectively.

4.37 By using some numerical examples show that, in general,

(1) $E(X-E(X))^2 \neq \{E(X-E(X))\}^2$;
(2) $\{E(X^2)\}^{1/2} \neq E(X)$;
(3) $[E(|X-EX|^2)]^{1/2} \neq E(|X-EX|)$;
(4) $E(1/X) \neq (1/E(X))$.

4.38 Find the density function whose $(s-1)$th moment is

$$\frac{\Gamma(a+s-1) \; \Gamma(a+\beta)}{\Gamma(a) \; \Gamma(a+\beta+s-1)}.$$

Additional Exercises

4.1 On each outcome set of the following experiments define two stochastic variables each and evaluate their probability functions.

(1) From a well shuffled deck of 52 cards one card is taken at random.

(2) From an urn containing 5 red and 10 green marbles 2 marbles are taken one by one without replacement.

(3) From the same urn 2 marbles are taken with replacement.

4.2 Can the following be probability functions?

(1) $f(x) = \begin{cases} 2 & \text{for } x=\frac{1}{2} \\ 1 & \text{for } x=\frac{1}{4} \\ -1 & \text{for } x=\frac{1}{4} \\ 0, & \text{elsewhere}; \end{cases}$

(2) $f(x) = \begin{cases} \frac{1}{8} & \text{for } x=1 \\ \frac{2}{8} & \text{for } x=2 \\ \frac{3}{8} & \text{for } x=3 \\ 0, & \text{elsewhere}; \end{cases}$

(3) $f(x) = \begin{cases} \dfrac{\theta}{2} & \text{for } x=1 \\ \dfrac{\theta}{2} & \text{for } x=-1 \\ 0, & \text{elsewhere}; \end{cases}$

(4) $f(x) = \begin{cases} 1 & \text{for } x=0 \\ 0, & \text{elsewhere}. \end{cases}$

4.3 Can the following be density functions?

(1) $f(x) = \begin{cases} x, & 0<x<2 \\ 0, & \text{elsewhere}; \end{cases}$

(2) $f(x) = \begin{cases} x-2, & 0<x<2 \\ 0, & \text{elsewhere}; \end{cases}$

(3) $f(x) = \begin{cases} 2\theta, & -\theta<x<\theta \\ 0, & \text{elsewhere}; \end{cases}$

(4) $f(x) = \begin{cases} xe^{-x}, & 0<x<\infty \\ 0, & \text{elsewhere}. \end{cases}$

4.4 Evaluate the distribution function for the following probability function.

(1) $f(x) = \begin{cases} \frac{1}{8}, & x=-1 \\ \frac{1}{8}, & x=0 \\ \frac{1}{8}, & x=2 \\ \frac{5}{8}, & x=4 \\ 0 & \text{elsewhere}; \end{cases}$

(2) $f(x) = \begin{cases} \frac{1}{2}, & x=-2 \\ \frac{1}{2}, & x=0 \\ 0, & \text{elsewhere}; \end{cases}$

(3) $f(x) = \begin{cases} \dfrac{1}{\theta} e^{-x/\theta}, & x>0, \ \theta>0 \\ 0, & \text{elsewhere}; \end{cases}$

(4) $f(x) = \begin{cases} xe^{-x}, & x>0 \\ 0, & \text{elsewhere}. \end{cases}$

4.5 Evaluate k if the following is a probability function

(1) $f(x) = \begin{cases} \dfrac{k}{2}, & x=0 \\ \dfrac{k}{2}, & x=1 \\ 0, & \text{elsewhere;} \end{cases}$
(2) $f(x) = \begin{cases} \dfrac{k}{3}, & x=1 \\ \dfrac{k}{4}, & x=2 \\ 0, & \text{elsewhere;} \end{cases}$

(3) $f(x) = \begin{cases} kx, & 0<x<5 \\ 0, & \text{elsewhere;} \end{cases}$
(4) $f(x) = \begin{cases} kx, & 0<x<1 \\ \dfrac{k}{2}, & 1<x<2 \\ 0, & \text{elsewhere.} \end{cases}$

4.6 Evaluate the probability function from the following distribution function.

(1) $F(x) = \begin{cases} \frac{1}{8} & \text{for } x=0 \\ \frac{2}{8} & \text{for } x=2 \\ \frac{5}{8} & \text{for } x=3 \\ 1 & \text{for } x=4 \end{cases}$
(2) $F(x) = \begin{cases} 0, & -\infty<x<1 \\ \frac{1}{4}, & 1 \leqslant x<4 \\ \frac{3}{4}, & 4 \leqslant x<6 \\ 1, & x \geqslant 6 \end{cases}$

(3) $F(x) = \begin{cases} 0, & -\infty<x<0 \\ x-\dfrac{x^2}{4}, & 0<x\leqslant2 \\ 1, & x\geqslant2 \end{cases}$
(4) $F(x) = \begin{cases} 0, & -\infty<x<0 \\ \dfrac{x^2}{2}, & 0<x\leqslant1 \\ \dfrac{x}{2}, & 1<x\leqslant2 \\ 1, & x\geqslant2. \end{cases}$

4.7 Evaluate (1) the second moment about the origin, (2) the third absolute moment about the origin, (3) the variance, (4) the second factorial moment, for the following distribution.

$$f(x) = \begin{cases} \frac{1}{4}, & x=-2 \\ \frac{1}{4}, & x=-1 \\ \frac{2}{4}, & x=1 \\ 0, & \text{elsewhere.} \end{cases}$$

4.8 Calculate the mean value and the variance for the following distribution.

$$f(x) = \begin{cases} xe^{-x}, & x>0 \\ 0, & \text{elsewhere.} \end{cases}$$

4.9   If μ_r, μ_r' and μ denote the rth central moment, rth moment about the origin (sometimes known as the *rth crude moment* or *rth raw moment*) and the mean value respectively, show that,

$$\mu_r = \mu_r' - \binom{r}{1} \mu_{r-1}' \mu + \binom{r}{2} \mu_{r-2}' \mu^2 - \ldots + (-1)^r \mu^r.$$

4.10   If $\mu_{[2]}$ denotes the second factorial moment, show that
$$\mu_{[2]} = \mu_2' - \mu = \mu_2 + \mu^2 - \mu.$$

4.11   Show that for any s.v. X, $\mathrm{Var}(X) = 0$ if and only if the variate X is degenerate. (A variate is said to be *degenerate* if it takes a particular value with probability one and all other values with probability zero).

4.12   Show that the rth absolute moment about EX is zero if and only if the variate is degenerate. Show the result for $r = 1, 2, 3$.

4.13   *A measure of skewness.* If a probability distribution is symmetric then it is easy to see that the odd central moments, μ_1, μ_3, μ_5,... are all zeros. Sometimes μ_3/σ^3, μ_5/σ^5,... are taken to be measures of skewness in a distribution. These are not good measures in the sense that if a central moment is zero it does not imply that the distribution is symmetric. Distributions can be *skewed to the right* or *skewed to the left* or *symmetric*. Take $a = \mu_3/\sigma^3$ as a crude measure of skewness and evaluate a for the following distributions.

$$(1) \ f(x) = \begin{cases} \frac{1}{4}, & x = -1 \\ \frac{2}{4}, & x = 0 \\ \frac{1}{4}, & x = 1 \\ 0, & \text{elsewhere;} \end{cases} \qquad (2) \ f(x) = \begin{cases} \frac{1}{\theta}, & 0 < x < \theta \\ 0, & \text{elsewhere.} \end{cases}$$

4.14   *Kurtosis.* Kurtosis or peakedness of a probability distribution is usually measured by,

$$\gamma_2 = \frac{\mu_4}{\mu_2^2} - 3$$

The distributions for which $\gamma_2 = 0$, > 0, < 0 are called *mesokurtic*, *leptokurtic* and *platykurtic* respectively. Again, γ_2 is not a good measure of peakedness in a distribution. Calculate γ_2 for the following distribution.

$$(1) \ f(x) = \begin{cases} \dfrac{1}{\sqrt{2\pi}} e^{-x^2/2}, & -\infty < x < \infty \\ 0, & \text{elsewhere;} \end{cases}$$

$$(2)\ f(x) = \begin{cases} \dfrac{1}{\theta}, & -\dfrac{\theta}{2} < x < \dfrac{\theta}{2} \\ 0, & \text{elsewhere.} \end{cases}$$

4.15 Evaluate the probabilities of the following events A and B in the following distribution:

$$f(x) = \begin{cases} \dfrac{1}{\theta} e^{-x/\theta}, & x > 0,\ \theta > 0 \\ 0, & \text{elsewhere}; \end{cases}$$

$A = \{0 < x < 2\}, \quad B = \{1 < x < 5\}.$

4.16 Evaluate $P(A)$, $P(B)$ and $P(A \cap B)$ for the following distribution:

$$f(x) = \begin{cases} \tfrac{1}{4}, & x = -5 \\ \tfrac{1}{4}, & x = -1 \\ \tfrac{1}{4}, & x = 0 \\ \tfrac{1}{4}, & x = 1 \\ 0, & \text{elsewhere} \end{cases}$$

and $A = \{-\infty < x < -2\}$, $B = \{-3 < x < \tfrac{1}{2}\}$.

4.17 When can $E\left(\dfrac{1}{X}\right)$ be equal to $\dfrac{1}{E(X)}$?

4.18 It costs \$ 150 to test a certain machine component and it costs \$1,000 to repair the machine if it has the defective component. Is it worth testing the component if it is known that the chances of that component being defective is (1) 1%, (2) 3%?

4.19 A shopkeeper dealing with a perishable item can buy an item at \$0.75 and sell it at \$1.50 per item. If an item is not sold by the evening it is a total loss. The daily demand for the item is given as follows.

Number of items	2	3	4	5	6 or more
Probability	.01	.02	.20	.60	.17

How many items should he store in order to have the maximum expected profit?

4.20 A contractor has found from experience that the low bid on a construction job can be considered to be a s.v. X having a distribution

$$f(x) = \begin{cases} \dfrac{2}{3\theta}, & \dfrac{\theta}{2} < x < 2\theta \\ 0, & \text{elsewhere,} \end{cases}$$

where θ is the estimated cost of construction. In order to maximize his profit, what percentage should he add to his cost estimate when submitting his bid?

4.21 Evaluate $P(A)$ and $P(\bar{A})$ for the following distribution:

$$f(x) = \begin{cases} 1, & 0 < x < 1 \\ 0, & \text{elsewhere, and } A = \{\tfrac{1}{2} < x < \tfrac{3}{4}\}. \end{cases}$$

4.22 Evaluate the M.G.F. for the following distribution.

$$(1) \ f(x) = \begin{cases} \dfrac{x^2}{16} \, e^{-x/2}, & x>0 \\ 0, & \text{elsewhere;} \end{cases} \qquad (2) \ f(x) = \begin{cases} \tfrac{1}{2}, & x=0 \\ \tfrac{1}{2}, & x=1 \\ 0, & \text{elsewhere.} \end{cases}$$

4.23 By using the uniqueness property of the M.G.F., identify the corresponding distributions from the following M.G.F.'s:

(1) $M(t) = (1-3t)^{-4}$;   (2) $M(t) = e^{5t}$.

4.24 Obtain the mean value and the variance of the distributions determined by the M.G.F.'s in problem 4.23.

4.25 If $\phi(t)$ is the characteristic function of a s.v. X, show that, (1) $\phi(0)=1$;   (2) $\phi(-t)=\bar{\phi}(t)$ where $\bar{\phi}(t)$ is the complex conjugate of $\phi(t)$;   (3) $|\phi(t)| \leqslant 1$ where $|\phi(t)|$ denotes the absolute value of $\phi(t)$.

4.26 Can the following be M.G.F's of some s.v.'s?

(1) $2t+5$;   (2) $3-t+t^2$;   (3) $2t\,e^{5t}$.

CHAPTER FIVE

Univariate Probability
Models—Discrete

The simple knowledge of a probability function and its properties will not help much in a practical situation. In many practical problems we can examine the experimental conditions and then select an appropriate probability model or probability function to describe the behaviour of the outcomes in the experiments under consideration. For example, the number of heads obtained, when a balanced coin is tossed a given number of times, may follow a certain probability law; we may be able to describe the life span of television picture tubes with the help of a probability distribution; the traffic accidents at a place, over time, may follow a certain pattern and so on. In this chapter we will analyse some experimental conditions and find out the most appropriate probability models. In the following table we give the most commonly used univariate discrete models. In section 5.2 we will consider an experimental situation where a binomial probability model is appropriate.

5.1 DISCRETE UNIVARIATE MODELS

TABLE 5.1

Name	Probability function $f(x)$	Parameters
1. The Binomial probability law	$\binom{N}{x} p^x (1-p)^{N-x}$, $x=0, 1, 2,...,N$ and $f(x)=0$, elsewhere	$0<p<1$
2. The Hypergeometric probability law	$\dfrac{\binom{a}{x} \binom{b}{n-x}}{\binom{a+b}{n}}$, $x=0, 1, ..., n$ or a and $f(x)=0$, elsewhere	(a, b, c) are all positive integers

TABLE 5.1 (*continued*)

Name	Probability function $f(x)$	Parameters
3. The Poisson law	$\dfrac{\lambda^x}{x!} e^{-\lambda}$, $x=0, 1,..., \infty$ and $f(x)=0$, elsewhere	$\lambda>0$
4. The Negative Binomial law ,,	$\dbinom{x+k-1}{x} p^k (1-p)^x$, $x=0,$ ..., ∞ and $f(x)=0$, elsewhere $\dbinom{x-1}{k-1} p^k (1-p)^{x-k}$, $x=k,$ $k+1, ..., \infty$ and $f(x)=0$, elsewhere	$0<p<1$, k—a positive integer ,,
5. The discrete uniform probability law	$\dfrac{1}{n}$ for x equals some $x_1, x_2, ...,$ x_n and $f(x)=0$, elsewhere	n—a positive integer
6. The Geometric probability law	$\theta(1-\theta)^{x-1}$, $x=1, 2, ..., \infty$ and $f(x)=0$, elsewhere	$0<\theta<1$
7. The logarithmic law	$\dfrac{-(1-p)^x}{x \log p}$, $x=1, 2,..., \infty$ and $f(x)=0$, elsewhere	$0<p<1$
8. Random Walk	$\dbinom{N}{\dfrac{x+N}{2}} (pq)^{\frac{N}{2}} \left(\dfrac{p}{q}\right)^{\frac{x}{2}}$, $x=... -2, -1, 1, 2, 3,...$ and $f(x)=0$, elsewhere	$0<p<1, q=1-p$ $\dbinom{N}{r}=0$ if r is not an integer

5.2 THE BINOMIAL MODEL

This is the most widely used discrete probability model and it is appropriate in the following experimental situations. Suppose that a random experiment is such that,

    (1) any trial results in a success or a failure;

    (2) there are N repeated trials which are independent;

    (3) the probability of a success at any trial is p.

Then such an experimental situation is called a Binomial probability situation and we will show that the behaviour of the outcomes is well described by the Binomial probability law. Let X denote the exact number of successes in N trials. Since there can be 0 or 1 or 2 or...or N successes, X can take the values 0, 1, 2,..., N with non-zero probabilities. If there are exactly x successes then the remaining $N-x$ are failures. Let us assume that the first x are successes and the remaining $N-x$ are failures. Since the probability of a success is p and that of a failure is $1-p$, the probability of getting the first x successes and the remaining $N-x$ failures is

$$= p \ldots p(1-p) \ldots (1-p) = p^x (1-p)^{N-x}$$

because, by assumption, the trials are all independent. Similarly if the first trial is a failure the next x are successes and the remaining ones are failures then the probability is

$$= (1-p)p^x (1-p)^{N-x-1} = p^x (1-p)^{N-x}.$$

It is easy to see that if any given x trials are successes and the remaining ones are failures the probability is again

$$= p^x(1-p)^{N-x}.$$

But any x trials can be selected in $\binom{N}{x}$ ways from N trials. Therefore, the probability of getting exactly x successes in N trials of a Binomial probability situation is

$$f(x) = \binom{N}{x} p^x(1-p)^{N-x} \text{ where, } x=0, 1, 2, \ldots, N.$$

That is,

$$f(x) = \begin{cases} \binom{N}{x} p^x(1-p)^{N-x}, & x=0, 1, \ldots, N \\ 0, & \text{elsewhere, where } 0<p<1 \text{ and } N \text{ is a positive} \\ & \text{integer.} \end{cases}$$

There are a number of practical situations where a trial results in a success or a failure, such as, the birth of a boy, getting a head in a toss of a coin, a defective item in an item produced and so on. In these cases the outcomes can be one or the other, that is, there are only two possibilities and the trials are independent with the same probability of success in every trial.

Example 5.2.1 Find the probability of getting (1) exactly one head, (2) at least 2 heads, (3) at the most 3 heads, when a balanced coin is tossed 4 times.

Solution. This is a Binomial probability situation since all the conditions for a Binomial situation are satisfied. Here $N=4$ and $p=\frac{1}{2}$ since there are only 4 trials and since the coin is unbiased.

(1) The probability of getting exactly x heads is,

$$f(x) = \binom{4}{x}\left(\frac{1}{2}\right)^x \left(1-\frac{1}{2}\right)^{4-x} = \binom{4}{x}\left(\frac{1}{2}\right)^4.$$

Hence the probability of getting exactly one head is

$$= \binom{4}{1}\left(\frac{1}{2}\right)^4 = \frac{4}{16} = \frac{1}{4}.$$

(2) ' At least 2 heads ' means, 2 or 3 or 4 heads.
Since these events are mutually exclusive the required probability

$$= P\{x=2\}+P\{x=3\}+P\{x=4\}$$
$$= \binom{4}{2}\left(\frac{1}{2}\right)^4 + \binom{4}{3}\left(\frac{1}{2}\right)^4 + \binom{4}{4}\left(\frac{1}{2}\right)^4 = \frac{11}{16}.$$

Since the total probability is unity,

$$P\{x=2\}+P\{x=3\}+P\{x=4\}= 1-[P\{x=0\}+P\{x=1\}]$$
$$= 1-\left[\binom{4}{0}\left(\frac{1}{2}\right)^4+\binom{4}{1}\left(\frac{1}{2}\right)^4\right]$$
$$= 1-\frac{5}{16} = \frac{11}{16}.$$

(3) ' At the most 3 heads' means, 0 or 1 or 2 or 3 heads. Hence the required probability is,

$$= P\{x=0\}+P\{x=1\}+P\{x=2\}+P\{x=3\}$$
$$= 1-[P\{x=4\}]=1-\binom{4}{4}\left(\frac{1}{2}\right)^4 = \frac{15}{16}.$$

Comment. (1) Here the parameters N and p are specified to be $N=4$ and $p=\frac{1}{2}$. In general $\binom{N}{x}p^x(1-p)^{N-x}$, for all the values N and p can assume, is a family of distributions called the *Binomial family of distributions.* $\binom{4}{x}\left(\frac{1}{2}\right)^4$ is a particular member in this family. (2) It must be noticed that $\binom{N}{x}p^x(1-p)^{N-x} \geqslant 0$ for all x and further

$$\sum_{x=0}^{N} \binom{N}{x} p^x (1-p)^{N-x} = \binom{N}{0} p^0 (1-p)^{N-0}$$

$$+ \binom{N}{1} p^1 (1-p)^{N-1} + \ldots + \binom{N}{N} (p)^N (1-p)^0$$

$$= [(1-p)+p]^N = 1^N = 1.$$

(3) When N is large it will be difficult to evaluate the Binomial probabilities by simple computation. Numerical tables of Binomial probabilities are available and a Binomial probability table is given at the end of this book. In a Binomial probability situation 'a success' means the outcome in which we are interested in. In an experiment of rolling a die, if we are interested in an even number then a success is the event of getting an even number; if we are looking for defective items in a lot then getting a defective item is a success; if we are concerned with death by snake-bite then a success is the event of a death by snake-bite and so on.

5.21 *The Mean Value and the Variance*

In statistical problems of testing hypotheses and related problems usually we will be interested in the mean values and variances of different distributions. So here we will evaluate the mean value and the variance for a Binomial probability model. By definition, the mean value, $E(X)$, for a discrete variate is defined as,

$$E(X) = \sum_{-\infty < x < \infty} x f(x).$$

Hence for the Binomial distribution,

$$E(X) = \sum_{x=0}^{N} x \binom{N}{x} p^x (1-p)^{N-x}$$

$$= \sum_{x=0}^{N} x \, \frac{N!}{x!(N-x)!} \, p^x (1-p)^{N-x}$$

$$= \sum_{x=1}^{N} x \, \frac{N!}{x!(N-x)!} \, p^x (1-p)^{N-x} \qquad \text{(Since when } x=0 \text{ the corresponding term is zero)}$$

$$= \sum_{x=1}^{N} \frac{N!}{(x-1)!\,(N-x)!}\, p^x(1-p)^{N-x} \quad \text{(by cancelling one } x)$$

$$= Np \sum_{x=1}^{N} \frac{(N-1)!}{(x-1)!\,(N-x)!}\, p^{x-1}(1-p)^{N-x}.$$

Now put, $y = x-1$ and $n = N-1$, then we get,

$$E(X) \;=\; Np \sum_{y=0}^{n} \frac{n!}{y!\,(n-y)!}\, p^y(1-p)^{n-y}$$

$$= Np(1-p+p)^n = Np\,1^n = Np$$

$$\text{Var }(X) \;=\; E(X^2) - (E(X))^2$$

$$= E(X(X-1)) + E(X) - [E(X)]^2$$

We write Var (X) in terms of the factorial moment $E(X(X-1))$ because, in general, we can see that in a number of discrete distributions, factorial moments are easier to evaluate.

$$E(X(X-1)) \;=\; \sum_{x=0}^{N} x(x-1)\binom{N}{x} p^x(1-p)^{N-x}$$

$$= \sum_{x=2}^{N} x(x-1)\binom{N}{x} p^x(1-p)^{N-x}$$

(Since the terms corresponding to $x=0$ and 1 are zeros).

$$= \sum_{x=2}^{N} x(x-1)\, \frac{N!}{x!(N-x)!}\, p^x(1-p)^{N-x}$$

$$= \sum_{x=2}^{N} \frac{N!}{(x-2)!\,(N-x)!}\, p^x(1-p)^{N-x}$$

$$= N(N-1)p^2 \sum_{x=2}^{N} \frac{(N-2)!}{(x-2)!\,(N-x)!}\, p^{x-2}(1-p)^{N-x}$$

Put $x-2=y$ and $N-2=n$, then we have

$$E(X(X-1)) = N(N-1)p^2 \sum_{y=0}^{n} \frac{n!}{y!(n-y)!} p^y(1-p)^{n-y}$$

$$= N(N-1)p^2$$

Hence,

$$\text{Var } (X) = E(X(X-1))+E(X)-(EX)^2$$
$$= N(N-1)p^2+Np-(Np)^2=Np(1-p)$$

Therefore, the standard deviation is

$$\sigma = \sqrt{Np(1-p)}$$

and the standardized Binomial variate is,

$$\Upsilon = \frac{X-Np}{\sqrt{Np(1-p)}}$$

In a Binomial case $E(X)=Np$. This can be interpreted as follows. In the long run we can expect Np successes in N trials. In other words if the probability of getting a head is $\frac{1}{2}$ when a coin is tossed and if the coin is tossed 100 times, even though in one batch of 100 trials there may not be $Np=100(\frac{1}{2})=50$ heads, if the experiment is repeated we can expect on the average 50 heads in such a batch of 100 trials.

Sometimes, instead of the Binomial variate X the Binomial proportion is of interest to the experimenter. The proportion of successes is

$$\Upsilon = \frac{X}{N}.$$

Hence, $E(\Upsilon) = E\left(\frac{X}{N}\right) = \frac{1}{N} E(X) = \frac{Np}{N} = p$

and $\text{Var } (\Upsilon) = \frac{1}{N^2} \text{Var } (X) = \frac{Np(1-p)}{N^2} = \frac{p(1-p)}{N}$

Therefore, the standardized proportion of successes is

$$Z = \frac{\Upsilon-E(\Upsilon)}{\sqrt{(\text{Var } \Upsilon)}} = \frac{\Upsilon-p}{\sqrt{\dfrac{p(1-p)}{N}}}$$

Due to the importance of the mean values and the variances of the distributions in statistical inference we will give the mean values and the variances for some distributions, in the following table.

TABLE 5.2

MEAN VALUES AND VARIANCES

X	Probability function $f(x)$ (Non-zero part)	$E(X)$	$Var(X)$
1. Binomial	$\binom{N}{x} p^x (1-p)^{N-x}$, $x=0$, $1, ..., N$	Np	$Np(1-p)$
2. Hyper-geometric	$\dfrac{\binom{a}{x}\binom{b}{n-x}}{\binom{a+b}{n}}$, $x=0,1,...,n$ or a	$\dfrac{na}{a+b}$	$\dfrac{nab\,(a+b-n)}{(a+b)^2\,(a+b-1)}$
3. Poisson	$\dfrac{\lambda^x}{x!} e^{-\lambda}$, $x=0, 1, ..., \infty$	λ	λ
4. Negative Binomial	$\binom{x-1}{k-1} p^k (1-p)^{x-k}$, $x=k, k+1, ..., \infty$	$\dfrac{k}{p}$	$\dfrac{k(1-p)}{p^2}$
5. Geometric	$p(1-p)^{x-1}$, $x=1, 2,..., \infty$	$\dfrac{1}{p}$	$\dfrac{(1-p)}{p^2}$

Sometimes, if tables are not available, it will be difficult to evaluate the Binomial probabilities. But a recurrence relationship between the probabilities of exactly x and $x+1$ successes can be easily obtained. Let $f(x)$ denote the probability of getting exactly x success in N trials in a Binomial situation. That is,

$$f(x) = \binom{N}{x} p^x (1-p)^{N-x}$$

The probability of getting exactly $x+1$ successes in N trials is

$$f(x+1) = \binom{N}{x+1} p^{x+1} (1-p)^{N-x-1}$$

Hence, $\dfrac{f(x+1)}{f(x)} = \dfrac{N-x}{x+1} \dfrac{p}{1-p}$

That is, $f(x+1) = \dfrac{N-x}{x+1} \dfrac{p}{1-p} f(x)$

If we know N and p then $f(0)$ is available and from $f(0)$ we get $f(1)$ and so on.   Such recurrence formulae can be obtained for some discrete distributions.   Table 5.3 gives the recurrence formulae for some of the commonly used discrete probability models.

TABLE 5.3

RECURRENCE FORMULAE

X	Probability function (Non-zero part)	Recurrence formulae
1. Binomial	$f(x) = \binom{N}{x} p^x (1-p)^{N-x}$, $x = 0, 1, ..., N$	$f(x+1) = \dfrac{N-x}{x+1} \dfrac{p}{1-p} f(x)$
2. Poisson	$f(x) = \dfrac{\lambda^x}{x!} e^{-\lambda}$, $x = 0, 1, ..., \infty$	$f(x+1) = \dfrac{\lambda}{x+1} f(x)$
3. Hyper-geometric	$f(x) = \dfrac{\binom{a}{x} \binom{b}{n-x}}{\binom{a+b}{n}}$, $x = 0, ..., n$ or a.	$f(x+1)$ $= \dfrac{(n-x)(a-x)}{(x+1)(b-n+x+1)} f(x)$
4. Negative Binomial	$f(x) = \binom{x-1}{k-1} p^k (1-p)^{x-k}$, $x = k, k+1, ..., \infty$.	$f(x+1) = \dfrac{x(1-p)}{x-k+1} f(x)$
5. Geometric	$f(x) = p(1-p)^{x-1}$, $x = 1, 2, ..., \infty$.	$f(x+1) = (1-p) f(x)$
6. Uniform	$f(x) = \dfrac{1}{n}$ for $x = x_1, x_2, ..., x_n$.	$f(x+1) = f(x)$

Exercises

5.1   Suppose, the probability that a bomb will hit its target is 0.8. Assuming a Binomial situation, what is the probability that out of 12 bombings exactly 3 are missed?

5.2   In a game of taking a chance, a contestant has to give correct answers to 4 out of 5 questions to win the contest.   Questions are given with 4 answers each, out of which one is a correct answer.　If a contestant, who does not know any of the correct answers, is answering the questions by selecting the answers at random, what is the probability that he will win the contest?

5.3　A manufacturer of radios claims that only 5% of his radios do not meet the quality specifications.　A shop keeper will accept a shipment, if one lot containing 10 radios, taken at random, contains only 0 or 1 defective ones; otherwise he will return the shipment.　What is the probability that (1) he will return the shipment when the manufacturer's claim is correct; (2) he will accept the shipment when there are actually 10% defectives?

5.4　What is the probability of getting (1) at least one even number; (2) exactly 2 even numbers, when a balanced die is rolled 3 times?

5.5　A social worker claims that 10% of the teenagers in a certain city have a certain disease.　A sample survey unit decided to test the claim by taking a random sample of 15 teenagers and reject the claim if only 0 or 1 is affected by the disease. What is the probability that (1) the claim is rejected when it is true; (2) the claim is accepted when only 5% are affected?

5.6　Show that, for $p=\frac{1}{2}$ the Binomial distribution has a maximum at $x=N/2$ when N is even and at $x=(N-1)/2$ and $(N+1)/2$ when N is odd.

5.7　For the Binomial distribution with parameters p and N, show

that, $\mu_{r+1}/p(1-p) = N r \mu_{r-1} + \dfrac{d}{dp} \mu_r$

where μ_r denotes the rth central moment.

5.8　Show that the moment generating function (M.G.F.), that is, Ee^{tX}, for a Binomial distribution is

M.G.F. $= (q+pe^t)^N$ where $q=1-p$.

5.9　Since $\left[\dfrac{d^r}{dt^r} M(t)\right]_{t=0} = \mu_r'$ where $M(t)$ denotes the M.G.F. of a distribution and μ_r' is the rth moment about the origin, by using $M(t)$ show that $EX=Np$ and $\mathrm{Var}(X)=Np(1-p)$ for a Binomial variate.　Evaluate the third moment about the origin.

5.10 *Uniqueness.* In general, whenever the M.G.F. $M(t)$ for a probability function $f(x)$ exists, there is a one-to-one correspondence between $f(x)$ and $M(t)$. That is, $M(t)$ uniquely determines $f(x)$ and vice versa. By using this uniqueness property determine the distribution if the M.G.F. is as follows:

(1) $M(t) = (\tfrac{1}{2} + \tfrac{1}{2} e^t)^5$

(2) $M(t) = \dfrac{(1+e^t)^4}{16}$

(3) $M(t) = \dfrac{(1+2e^t)^3}{27}$

5.3 THE POISSON MODEL

Here we will derive the Poisson probability law as a limiting case of a Binomial law and also as a model in an experimental situation. It is named after S. Poisson, a noted French Mathematician. First let us consider a limiting form of a Binomial probability situation. Let,

(1) the number of trials be very large,

(2) the probability of a success p be very small,

and (3) $Np = \lambda$ (lamda) a finite quantity, when $N \to \infty$, $p \to 0$.

This can also be called a situation of *rare events* since the probability of a success is very small and the number of trials is very large. That is, we are considering the case, $N \to \infty$, $p \to 0$ but $Np = \lambda$ is finite. Under these conditions the Binomial probability law will reduce in the following way. The Binomial law is,

$$f(x) = \binom{N}{x} p^x (1-p)^{N-x}$$

$$= \frac{N(N-1)\ldots(N-x+1)}{x!} p^x (1-p)^N (1-p)^{-x}$$

$$= \frac{N}{N}\frac{(N-1)}{N}\ldots\frac{(N-x+1)}{N}\frac{\lambda^x}{x!}\left(1-\frac{\lambda}{N}\right)^N\left(1-\frac{\lambda}{N}\right)^{-x}$$

(since $Np = \lambda$, $p = \lambda/N$).

$$= 1\left(1-\frac{1}{N}\right)\left(1-\frac{2}{N}\right)\ldots\left(1-\frac{x-1}{N}\right)\frac{\lambda^x}{x!}\left(1-\frac{\lambda}{N}\right)^N\left(1-\frac{\lambda}{N}\right)^{-x}$$

Since x is finite, when $N\to\infty$, $\left(1-\dfrac{1}{N}\right)\to 1$, $\left(1-\dfrac{2}{N}\right)\to 1,\ldots,$

$\left(1-\dfrac{x-1}{N}\right)\to 1$, $\left(1-\dfrac{\lambda}{N}\right)^{-x}\to 1$ and $\left(1-\dfrac{\lambda}{N}\right)^{N}\to e^{-\lambda}$

(since $\left(1+\dfrac{t}{n}\right)^{n}\to e^{t}$ when $n\to\infty$, where t is finite).

Hence the probability function $f(x)$ reduces to the form,

$$f(x) = \frac{\lambda^{x}}{x!}e^{-\lambda}, \quad x=0,\ 1,\ 2,\ \ldots,\ \infty$$

By taking the limit of a particular probability function, the resulting quantity need not be a probability function. But in this case, it is easy to check and see that,

$$g(x) = \frac{\lambda^{x}}{x!}e^{-\lambda}$$

is such that $g(x)\geqslant 0$ for all x and,

$$\sum_{-\infty<x<\infty} g(x) = \sum_{x=0}^{\infty}\frac{\lambda^{x}}{x!}e^{-\lambda}=e^{-\lambda}\left\{1+\frac{\lambda}{1!}+\frac{\lambda^{2}}{2!}+\ \ldots\right\}.$$
$$= e^{-\lambda}\,e^{\lambda}=1.$$

Hence $g(x)$ is a probability function and this function is the Poisson probability law given as,

$$g(x) = \begin{cases} \dfrac{\lambda^{x}}{x!}e^{-\lambda}, & x=0,\ 1,\ \ldots,\ \infty \\ 0, & \text{elsewhere,} \end{cases}$$

where the parameter $\lambda>0$ since $\lambda=Np$. The mean value and the variance or other moments can be obtained by using a method similar to the one used in section 5.2. The mean value and the variance are given in Table 5.2 and a recurrence relationship is given in Table 5.3. Since the Poisson law can be called a law of rare events it is appropriate to use it in problems such as, delivery of quintuplets, emission of alpha particles by a radio active source, traffic accidents and so on. Also it is useful to note the following. Since when $N\to\infty$, $p\to 0$ but $Np=\lambda$ (a finite quantity), the Binomial law approximates to the Poisson law, in practice when the number of trials is large and the probability of success or failure is very small instead of a Binomial distribution, a Poisson distribution is used. Due to this reason, Binomial probability tables for large N

and very small p or $(1-p)$ will not be available and in such problems one has to look into the Poisson probability tables to find the required probabilities. An extract of a table of Poisson probabilities is given at the end of this book.

Example 5.3.1 Alpha particles are emitted by a radio active source at an average rate of 5 in a 20 minute interval. Assuming that a Poisson model is appropriate, what is the probability that (1) there will be exactly 2 emissions in a particular 20 minute interval; (2) there will be at least 3 emissions in a particular 20 minute interval?

Solution. Consider the Poisson distribution,

$$f(x) = \frac{\lambda^x}{x!} e^{-\lambda}.$$

Here we know that $E(X) = \lambda$.

Hence, in this problem $\lambda = 5$ and thus the probability law is

$$f(x) = \frac{5^x}{x!} e^{-5}$$

  = probability of exactly x emissions in a 20 minute interval.

(1) The probability of exactly two emissions in a 20 minute interval is,

$$= \frac{5^2}{2!} e^{-5} = 0.0843 \text{ (Obtained from the table of Poisson}$$

probabilities).

(2) The probability of at least 3 emissions

$$= P\{x=3\} + P\{x=4\} + \ldots$$

$$= \sum_{x=3}^{\infty} \frac{5^x}{x!} e^{-5}$$

$$= 1 - \sum_{x=0}^{2} \frac{5^x}{x!} e^{-5} \quad \text{(since the total probability is unity)}.$$

$$= 0.8753.$$

Example 5.3.2 Suppose that the probability of an item being defective, in a mass production process of that item, is 0.01. If 20 items are selected at random what is the probability that exactly

2 will be defective? Approximate it by a Poisson probability and evaluate the error in this approximation.

Solution. An item can be defective or non-defective. The probability of an item being defective is 0.01 and it remains the same for every item. Obviously this is a case of a Binomial situation with a small probability of success.

Here $N=20$ and $p=0.01$. Therefore, the probability of getting exactly 2 defectives

$$= \binom{20}{2} (0.01)^2 (0.99)^{18}$$
$$= 0.0158.$$

If the Binomial probabilities are approximated by Poisson probabilities, we have seen that the Poisson parameter,

$$\lambda = Np.$$

In this problem,

$$Np = 20(0.01)=0.2=\lambda.$$

Hence, the probability of getting exactly 2 defectives, treating the problem as a problem of rare events,

$$= \frac{\lambda^2}{2!} e^{-\lambda} = \frac{(0.2)^2}{2!} e^{-0.2} = 0.0164.$$

That is, the error in this approximation

$$= 0.0164 - 0.0158 = 0.0006.$$

Comment. Here, even though $p=0.01$ which is quite small $N=20$ which is not large, but the error in the approximation seems to be quite small. Now the problem is to decide which cases can be considered to be rare events. A satisfactorily good approximation is available for N as small as 20 provided $Np < 5$. This can be taken as a rule to decide whether a Poisson distribution or a Binomial distribution is appropriate in a given situation. We will also give an independent derivation for the Poisson distribution by considering an experimental situation satisfying some conditions.

Consider the following situation, where,

(1) The probability of a success in a small time interval from t to $t+\Delta t$ is $a(\Delta t)$ where a is a positive constant and Δt denotes a small increment in time at t;

(2) The probability of getting more than one success in this interval is negligibly small.

(3) The probability of a success in the interval t to $t+\Delta t$ does not depend on the success or failure prior to time t.

Under these conditions it can be shown that the probability of getting exactly x successes in time t, denoted by $f(x, t)$, is given by

$$f(x, t) = \frac{(at)^x}{x!} e^{-at} \text{ for } x=0, 1, 2, ..., \infty,$$

which is evidently a Poisson distribution with parameter $\lambda = at$.

In order to derive this distribution we will consider two mutually exclusive events, namely (1) the event of getting exactly x successes in time t and then a failure in the interval t to $t+\Delta t$, (2) the event of getting exactly $x-1$ successes in time t and then a success in the interval t to $t+\Delta t$. These are the only possible ways in which we can get x successes in time $t+\Delta t$

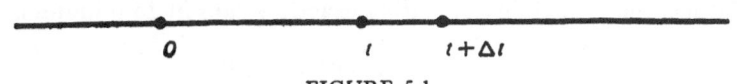

$$0 \qquad\qquad\qquad t \qquad t+\Delta t$$

FIGURE 5.1

Thus we have,

$f(x, t)$=probability of getting exactly x successes in time t.

$f(x, t+\Delta t)$=probability of getting exactly x successes in time $t+\Delta t$.

$f(x-1, t)$=probability of getting exactly $(x-1)$ successes in time t.

But, the probability of a success in time t to $t + \Delta t$ is $a(\Delta t)$ (assumed) and therefore the probability of a failure in t to $t+\Delta t$ is $1 - a(\Delta t)$ (by assumption there cannot be more than one success in t to $t+\Delta t$). Since the probability of a union of two mutually exclusive events is the sum of the probabilities, we have,

$$f(x, t+\Delta t) = f(x-1, t)\, a(\Delta t) + f(x, t)\, [1 - a(\Delta t)].$$

That is, $$\frac{f(x, t+\Delta t) - f(x, t)}{\Delta t} = a[f(x-1, t) - f(x, t)]$$

Taking the limit when $\Delta t \to 0$ we get the following difference-differential equation,

$$\frac{d}{dt} f(x, t) = a[f(x-1, t) - f(x, t)].$$

The solution of this difference-differential equation can be shown to be,

$$f(x, t) = \frac{(at)^x}{x!} e^{-at} \text{ for } x=0, 1, 2, ..., \infty.$$

(We will not discuss the proof here).

S.—12

Example 5.3.3 An office switch-board receives phone calls at the rate of 2 in every 5 minutes on the average. Assuming that this to be a Poisson situation, what is the probability of getting exactly 4 calls in 15 minutes?

Solution. Here 5 minutes is a unit of time and hence 15 minutes is 3 units. In other words $t=3$. The rate at which calls are coming is 2 per unit time. Hence $a=2$. But the probability of getting exactly x calls in time t is,

$$f(x, t) = \frac{(at)^x}{x!} e^{-at}.$$

Hence, the probability of getting exactly 4 calls in 15 minutes is,

$$= \frac{(2 \times 3)^4}{4!} e^{-2 \times 3} = \frac{6^4}{4!} e^{-6} = 0.1339.$$

Exercises

5.11   The number of traffic accidents of a city in each month is assumed to be a Poisson variate with the parameter $\lambda=3$. What is the probability that there will be (1) exactly 5 traffic accidents in a certain month, (2) at least one accident in a certain month?

5.12   A machine is known to produce 10% defective items. If a random sample of size 20 is taken what is the probability of getting exactly 3 defectives? Approximate this by a Poisson probability.

5.13   A traffic counting device is set up at a particular point on a highway. Suppose that the probability of a vehicle passing through that highway during a small interval of time Δt is $0.2(\Delta t)$, time being measured in minutes. What is the probability that the counting device will register (1) exactly 2 counts in a minute, (2) exactly 5 counts in a ten-minute interval?

5.14   A hospital switch-board receives at the rate of 2 emergency calls in every 30 minutes. What is the probability that there will be at least 3 emergency calls from 2 p.m. till 4 p.m. on a particular day?

5.15 If $\mu_r = E((X - E(X))^r)$ show that the following result is true when X is a Poisson variate with parameter λ.

$$\mu_{r+1} / \lambda = r \, \mu_{r-1} + \frac{d}{d\lambda} \, \mu_r.$$

5.16 By using the result in problem 5.15 evaluate μ_2 and μ_3.

5.17 Show that the M.G.F. $M(t) = E(e^{tX})$, for the Poisson variate X is,

$$M(t) = e^{\lambda(e^t - 1)}.$$

5.18 Since $\left[\dfrac{d^r}{dt^r} M(t) \right]_{t=0} = \mu_r'$, evaluate $E(X)$ and Var (X) by using $M(t)$ when X is a Poisson variate.

5.19 Since $M(t)$ uniquely determines the corresponding probability function, by examining the following M.G.F. find the corresponding probability function

(1) $M(t) = e^{2(e^t - 1)}$
(2) $M(t) = e^{(e^t - 1)/3}$
(3) $M(t) = (2 + e^t)^3 / 27$

5.20 Show that the M.G.F. of a Binomial distribution approaches the M.G.F. of a Poisson distribution when $N \to \infty$, $p \to 0$ but $Np = \lambda$ (a finite quantity). Hence show that the Binomial distribution approaches a Poisson distribution under these conditions.

5.4 OTHER DISCRETE MODELS

There are several other discrete models which are appropriate in problems of life testing, ecology, physical sciences, economics and several other disciplines. Some of those will be briefly discussed here. These are (1) the Hypergeometric model, (2) the Negative Binomial model and (3) the Geometric model. We will consider the experimental situation where these models are appropriate. The reader, who wishes to use these models, has to analyse the conditions of the experiment in which he is interested in and then select the best model for the particular problem.

The Hypergeometric Model. Consider the following experimental situation.

(1) A trial results in a success or a failure;

(2) There are n trials;

(3) The probability of a success in a trial does not remain the same from trial to trial.

This is a *hypergeometric situation* and a particular example is the problem of sampling without replacement. If there are $a+b$ objects out of which a are of one type and b are of the other type and n of them are selected at random without replacement, what is the probability that exactly x of them will be of the a-type? The required probability is evidently,

$$f(x) = \frac{\binom{a}{x} \binom{b}{n-x}}{\binom{a+b}{n}} \quad x = 0, 1, \ldots, n \text{ or } a.$$

This is the simple Hypergeometric distribution.

Example 5.4.1 From a well-shuffled deck of 52 cards, 2 cards are selected at random without replacement. What is the probability that they are aces?

Solution. The 52 cards can be divided into 2 sets of aces (4 of them) and non-aces (48 of them). Therefore, $a=4$, $b=48$, $n=2$, and $x=2$ and the required probability

$$= \frac{\binom{4}{2} \binom{48}{0}}{\binom{52}{2}} = \frac{4 \times 3}{52 \times 51} = \frac{1}{221}$$

The Negative Binomial Model. (Also called the *Binomial waiting time model*). Consider a Binomial probability situation. Suppose that we are interested in the number of trials needed to get exactly k successes. In other words the number of trials becomes a variable and k is a given number. This can also be called the probability of getting the kth success at the xth trial, for $x=k, k+1,\ldots,\infty$. The distribution can be easily derived by using the technique used in the independent derivation of the Poisson distribution as a Binomial waiting time distribution. Divide the event of getting the kth success at the xth trial, into two events. That is, the event of getting $k-1$ successes in $x-1$ trials and the next trial results in a success. The required probability $f(x)$ will be the product of the probabilities of these two events, and therefore,

$$f(x) = \binom{x-1}{k-1} p^{k-1} (1-p)^{(x-1)-(k-1)} p$$

$$= \begin{cases} \binom{x-1}{k-1} p^k (1-p)^{x-k} & \text{for } x = k, k+1, \ldots, \infty. \\ 0, & \text{elsewhere.} \end{cases}$$

Example 5.4.2. Mr. Feller is shooting at a target. The probability of a hit is 0.4. What is the probability that his 10th trial results in the second hit?

Solution. The number of trials required for the kth hit is 10. The number of hits (k) is 2. Evidently this is a Negative Binomial experimental situation and hence the required probability

$$= \binom{x-1}{k-1} p^k (1-p)^{x-k}$$

$$= \binom{9}{1} (0.4)^2 (0.6)^8 = 0.02418.$$

Comment. The Negative Binomial probabilities can be evaluated with the help of logarithms or by using tables of Binomial coefficients and Binomial probabilities. The different Negative Binomial probabilities can be obtained as the different terms in the expansion of $p^k(1-q)^{-k}$ where $q=1-p$ and hence the distribution is called a Negative Binomial distribution.

Example 5.4.3 An item is produced in large numbers. The machine is known to produce 2% defectives. A quality control inspector is examining the items by taking them at random. What is the probability that at least 4 items are to be examined in order to get 2 defectives?

Solution. If 2 defectives are to be obtained then it can happen in 2 or more trials. The probability of a success is 0.02 for every trial. It is a Negative Binomial situation and the required probability is,

$$= P\{x=4\}+P\{x=5\}+ \ldots$$

$$= \sum_{x=4}^{\infty} \binom{x-1}{2-1} (0.02)^2 (0.98)^{x-2}$$

$$= 1 - \sum_{x=2}^{3} \binom{x-1}{2-1} (0.02)^2 (0.98)^{x-2}$$

$$= 1 - [(0.02)^2 + 2(0.02)^2 (0.98)] = 0.998816.$$

The Geometric Model. In the Negative Binomial model when $k=1$ we get a geometric model. In other words, a Geometric distribution gives the probability of getting the first success at the xth trial or the probability that x trials are needed to get the first success. Hence the probability model is,

$$f(x) = \begin{cases} p\ q^{x-1} \text{ for } x=1, 2, ..., \infty \\ 0, \text{ elsewhere,} \end{cases}$$

where $q=1-p$ and p is the probability of success in any trial of a Binomial probability situation.

Example 5.4.4 Suppose that the probability that the prediction of a soothsayer will come true is 0.01. What is the probability that his 13th prediction is the first one to be true?

Solution. This is a Geometric distribution with $p=0.01$ and $x=13$. Hence the required probability

$$=pq^{x-1}=(0.01)\ (0.99)^{12}=0.0000203.$$

Exercises

5.21 From a box containing 20 items, out of which 3 are defectives, 2 are selected at random without replacement. What is the probability that (1) at least one is defective, (2) none of them is defective?

5.22 A number of people are swimming across a lake. Each one has a probability of 0.4 of successfully completing it. What is the probability that the 10th person is the 3rd one to complete the swimming successfully?

5.23 If a boy is pelting stones at a target and if the probability of a successful hit is 0.4, what is the probability that at least 5 trials are required to have the second successful hit?

5.24 If the probability of birth of a child with a defective heart, in a certain city, is 0.01, what is the probability that the 8th child born is the first one to have a defective heart?

5.25 Suppose that the chances of side-effects due to a particular drug is 2%. What is the probability that the 5th person who took the drug is the first person to have side-effects?

5.26 *Random walk on a line.* A point moves along a straight line in jumps of one unit each, starting from a point *o*. The point jumps to the right with a probability p and to the left with a probability $q=1-p$. Each jump is independent of the other jumps. If the distance from o after N jumps is x show that the probability of x, is

$$f(x) = \binom{N}{\dfrac{x+N}{2}} (pq)^{N/2} \left(\dfrac{p}{q}\right)^{x/2}$$

for $x=\ldots -2, -1, 0, 1, 2, \ldots$(Assume that $\binom{N}{r}=0$ if r is not an integer).

5.27 Show that the mean value and the variance, in a Negative Binomial case are k/p and kq/p^2 where $q=1-p$, respectively.

5.28 Show that the M.G.F., that is, $E(e^{tX})$, for a Negative Binomial variate is,

$$M(t)=p^k(e^{-t}-q)^{-k}, \quad q=1-p.$$

5.29 If X has a Binomial distribution with parameters N and p, show that the cumulative distribution of X can be written as,

$$(N-x) \binom{N}{x} \int_0^q y^{n-x-1} (1-y)^x dy$$

where $q=1-p$.

5.30 Identify the distributions from the following M.G.F.
(1) $M(t) = \frac{1}{2}(e^{-t}-\frac{1}{2})^{-1}$
(2) $M(t) = 4(3e^{-t}-1)^{-2}$.

5.5 ACCEPTANCE SAMPLING

In an industrial production process usually a particular item is produced in large numbers. Even if a very good machinery is used, there will be variations among the items produced. An item may be considered to be good if it satisfies certain quality specifications. In production processes involving large numbers, production engineers use quality control methods to check the quality of the outgoing items. Some quality control techniques will be discussed in a subsequent chapter. The items produced in a factory may be shipped to the consumers in lots of, say, N items. If the consumer finds some defectives, that is, items which do not meet the quality specifications, he may return the lot or the lot may not be acceptable to him. For the producer it may not be economical to inspect each and every item produced and replace the defectives by good items. Sometimes it may be impossible to perform such an inspection if the items are destroyed by the inspection or if the items are produced in very large numbers in a very short interval of time. In such cases it is good for the consumers and the producers to come to an agreement about what is meant by a ' lot being acceptable '. As a criterion one may select a lot of size N and take a random sample of size n from it and depending upon the number of defectives in that sample either the lot may be accepted or rejected. Such a decision-making process is called a single sampling plan because the decision is made based upon a single sample. If a decision is made by using a number of samples it is called a multiple sampling plan.

5.51 *The Operating Characteristic Curve*

Consider a single sampling plan where a lot of size N is accepted if the number of defectives is less than or equal to c in a random sample of size n from the lot. If θ is the probability of getting a defective item then the expected number of defective items in the lot of size N is $N\theta$. Thus the probability of getting c or less defectives is evidently

$$= \sum_{x=0}^{c} \frac{\binom{N\theta}{x} \binom{N-N\theta}{n-x}}{\binom{N}{n}}.$$

Since θ is unknown, if θ denotes the *fraction of defectives* (the number of defectives divided by the total number N) then θ may be $1/N$ or $2/N$ or... or N/N. The probability of getting c or less defectives will be a function of θ and denoting this probability by $p(\theta)$, we may write,

$$p(\theta) = \sum_{x=0}^{c} \frac{\binom{N\theta}{x} \binom{N-N\theta}{n-x}}{\binom{N}{n}}.$$

Incidentally, c is often called the *acceptance number*. If $p(\theta)$ is plotted against θ then we get what is known as an *OC*-curve. An *OC*-curve is given in Figure 5.2.

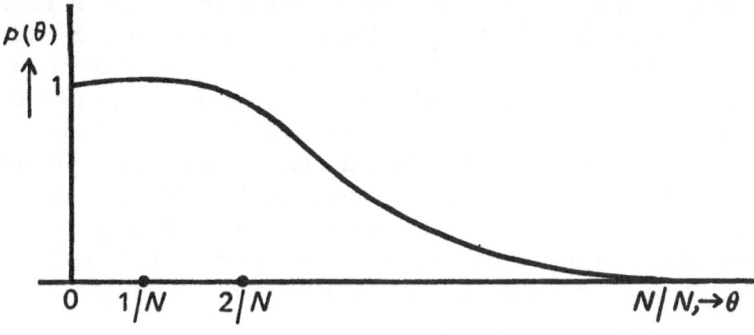

FIGURE 5.2   *OC*-Curve

When $N \to \infty$ it can be shown that the hypergeometric distribution approximates to a Binomial distribution and thus,

$$p(\theta) \to \sum_{x=0}^{c} \binom{n}{x} \theta^x (1-\theta)^{n-x} \text{ when } N \to \infty.$$

Further, when $n \to \infty$, $\theta \to 0$ but $n\theta = \lambda$ a constant, the Binomial distribution approximates to a Poisson distribution and, thus, we have,

$$p(\theta) \to \sum_{x=0}^{c} \frac{\lambda^x}{x!} e^{-\lambda},$$

under these conditions. Thus a Binomial or a Poisson approximation to $p(\theta)$ may be used according to the experimental situations.

5.52 *Producer's and Consumer's Risks*

Suppose that the producer (seller) and consumer (buyer) agree upon a particular single sampling plan, and they agree to call a lot acceptable if the fraction defective $\theta \leqslant \theta_0$ and unacceptable if $\theta \geqslant \theta_1$. Here θ_0 and θ_1 are called the *acceptable quality level* (AQL) and the *lot tolerance per cent defective* (LTPD) respectively. Thus if a lot is such that the fraction defectives is in between θ_0 and

θ_1, that is $\theta_0 < \theta < \theta_1$, then we may call it a lot of *indifferent quality*. Since a decision is taken based upon a sample, it is possible that the producer may scrap a lot when it is really acceptable, that is, when $\theta \leqslant \theta_0$ and a consumer may accept a lot when it is really unacceptable (that is, $\theta \geqslant \theta_1$). These are the two types of errors one of which can occur in a specific instance of decision making. The probability α, that a good lot $(\theta \leqslant \theta_0)$ is rejected is called the *producer's risk* and the probability β that a bad lot $(\theta \geqslant \theta_1)$ is accepted is called the *consumer's risk*. These are also known as the probabilities of type I and type II errors respectively, in the theory of testing statistical hypotheses. If the producers and consumers agree upon the values of α, β, θ_0 and θ_1 then a sampling plan can be fixed (n and c can be fixed) at least in the large lot cases (when N is large).

Example 5.5.1 Consider a simple sampling plan where the lot size is 20, the sample size is 5 and the acceptance number is one. Calculate the probability of accepting a lot when there are 10% defectives and the probability of rejecting a lot when there are only 5% defectives.

Solution. A lot is accepted if 0 or 1 are defectives in a sample of size 5 otherwise the lot is rejected. Hence the probability of accepting the lot, when there are 10% defectives (that is, 2 defectives),

$$p(\theta) = \sum_{x=0}^{1} \frac{\binom{2}{x}\binom{18}{5-x}}{\binom{20}{5}} = \frac{\binom{2}{0}\binom{18}{5}}{\binom{20}{5}} + \frac{\binom{2}{1}\binom{18}{4}}{\binom{20}{5}} = 18/19.$$

When there are only 5% defectives (that is one) the probability of rejecting a lot

$$= \sum_{x=2}^{5} \frac{\binom{1}{x}\binom{19}{5-x}}{\binom{20}{5}} = 0$$

(since $\binom{1}{x}$ is zero for $x > 1$).

Example 5.5.2 Suppose that a large lot calling for a sample of size $n = 20$ is accepted if the number of defectives is $\leqslant 1$ and is rejected if the number of defectives is $\geqslant 4$. Evaluate the producer's and consumer's risks.

Solution. Since the lot size is large we may assume a Binomial approximation. If θ is the fraction defectives then the probability of accepting the lot

$$= \sum_{x=0}^{1} \binom{20}{x} \theta^x (1-\theta)^{20-x}$$

and the probability of rejecting the lot

$$= \sum_{x=4}^{20} \binom{20}{x} \theta^x (1-\theta)^{20-x}.$$

According to the notations used,

$$\text{AQL} = \frac{1}{20} = \theta_0$$

$$\text{LTPD} = \frac{4}{20} = \theta_1$$

The consumer's risk is the probability of accepting the lot when it is to be rejected, that is, when $\theta = \frac{4}{20}$, which is,

$$= \sum_{x=0}^{1} \binom{20}{x} \left(\frac{4}{20}\right)^x \left(1-\frac{4}{20}\right)^{20-x} = 0.0692$$

The producer's risk is the probability of rejecting a lot when it is to be accepted, that is, when $\theta = \frac{1}{20}$, which is,

$$= \sum_{x=4}^{20} \binom{20}{x} \left(\frac{1}{20}\right)^x \left(1-\frac{1}{20}\right)^{20-x} = 0.0159.$$

If a lot is unacceptable then the producer can examine each and every item of the lot and replace the defectives by good ones before shipment. If θ is the fraction defectives in an incoming lot and if $p(\theta)$ is the probability of acceptance then $\theta\, p(\theta)$ can be defined as the *average outgoing quality* (*AOQ*). That is in the long-run the proportion of lots accepted is $p(\theta)$ and such lots contain a fraction defectives θ and $1-p(\theta)$ is the proportion of lots rejected which contain 0 defectives since the defectives are replaced by good ones and hence the *AOQ* is,

$$AOQ = \theta\, p(\theta) + 0\, [1 - p(\theta)] = \theta\, p(\theta).$$

Example 5.5.3 If a single sampling inspection plan calls for a sample size $n=10$ and an acceptance number 2 evaluate the average outgoing quality if the fraction defectives is $1/5$. (Assume that the lot size is large).

Solution. The probability of acceptance

$$p(\theta) = \sum_{x=0}^{2} \binom{10}{x} \left(\frac{1}{5}\right)^x \left(1-\frac{1}{5}\right)^{10-x} = 0.6778.$$

(The Binomial approximation is used since the lot size N is large). Hence the

$$AOQ = \theta\, p(\theta) = \left(\frac{1}{5}\right)(0.6778)$$
$$= 0.13556.$$

Exercises

5.31 A single sampling plan calls for $N=20$, $n=10$ and $c=2$. Draw the OC-curve. Approximate the model by a Binomial model and draw the OC-curve obtained by using this approximation. Compare the OC-curves.

5.32 For the problem 5.31 draw the AOQ curve.

5.33 A single sampling plan, where the lot size is large, calls for $c=1$ and $n=12$. Find (i) the AOQ if the producer's risk is 0.12 and (ii) the LTPD if the consumer's risk is 0.34.

5.34 A single sampling plan calls for a sample of size 20. By using a Binomial and a Poisson approximations find (i) the acceptance number c if AQL is 4.6% and the producer's risk is 0.08, (ii) by using the c in (i) obtain the consumer's risk if the LTPD is 10%, (iii) plot the OC-curve and mark the consumer's and producer's risks.

5.35 A single sampling plan calls for a sample of size 16 and $c=1$. Find the AOQ and LTPD if the producer's and consumer's risks are 0.21 and 0.14 respectively.

Additional Exercises

5.1 What is the probability of getting at least three heads when an unbiased coin is tossed 3 times?

5.2   What is the probability of getting at least one number which is less than 5 when a balanced die is rolled 3 times?

5.3   The probability of survival from the attack of a certain disease is 0.8. Twelve people are attacked by this disease. What is the probability that at least 8 of them will survive?

5.4   In a class of 50 students 20 are girls and 30 are boys. If 5 students are selected at random what is the probability that all the 5 are boys?

5.5   Approximate the probability in problem 5.4 by a Binomial distribution with parameters $N=5$ and $p=30/50$ and calculate the percentage error.

5.6   Show that when the total number N becomes large a simple hypergeometric distribution approximates to a Binomial distribution.

5.7   What are the conditions under which a hypergeometric distribution will approximate to a Poisson distribution?

5.8   *Discrete uniform distribution.* If a s.v. X takes some values $x_1, x_2, ..., x_n$ with probabilities $1/n$ each then X is called a discrete uniform variate and the distribution is called a discrete uniform distribution. For a discrete uniform distribution, obtain the mean value and the variance.

5.9   In a certain region it is noted that, on the average 2 sets of triplets are born in a five-year interval. Assuming that this is a situation of rare events obtain the probability that in 1970–1980 there will be no such births.

5.10 In a certain province there are 3 shooting accidents on the average per year. Assuming this to be a situation of rare events what is the probability that there will be (1) 2 or 3 such accidents in a particular year, (2) at least 2 accidents in a particular year?

5.11 Suppose that a particular rocket has a 70% chance of a successful hit. What is the probability that at least 10 firings are required to have 7 successful hits?

5.12 Assuming that the probability that a bomb will hit the target is 0.4, what is the probability that the 6th bomb is the 4th one that missed the target?

5.13 Obtain the M.G.F., that is, $E(e^{tX})$, of a discrete uniform distribution.

5.14 Identify the distribution corresponding to the following M.G.F.

(1) $e^t(5-4e^t)^{-1}$,   (2) $\dfrac{e^t}{2\left(1-\dfrac{e^t}{2}\right)}$.

5.15 *Logarithmic distribution.* The logarithmic distribution is given by the probability function,

$$f(x) = \frac{-(1-p)^x}{x \log p} \text{ for } x=1, 2, ..., \ 0<p<1.$$

Find $E(X)$ for this distribution.

5.16 Obtain a recurrence relation for the probability function of (i) a Poisson distribution, (ii) a Negative Binomial distribution, (iii) a Geometric distribution, (iv) a Discrete uniform distribution, (v) a Logarithmic distribution, (vi) a Random walk distribution.

5.17 For the following probability distribution obtain $P(A)$, $P(B)$, $P(A \cap B)$, $P(\bar{A})$,

(1) $f(x) = \begin{pmatrix} x-1 \\ k-1 \end{pmatrix} p^k (1-p)^{x-k}$, $x=k, k+1,...$ for $k=2$.

(2) $f(x) = -\dfrac{(1-p)^x}{x \log p}$ for $x=1, 2, ...$

where,

$$A = \{x \mid -\infty < x < 2\},$$
$$B = \{x \mid 1 < x < 5\}.$$

5.18 A certain item is shipped in lots of 20. Five of them from a lot are taken at random and inspected. Each lot is accepted if none of the 5 is defective. What is the probability that the lot is accepted when (1) there are 2 defectives, (2) there are 10 defectives, in the lot?

5.19 Suppose that the number of breakdowns of a machine, per week, can be considered to be a Poisson variate with $\lambda=0.5$. What is the probability that the machine will run for 2 weeks without any breakdown.

5.20 In a single sampling plan if $n=30$ and $c=2$, find
(1) the AQL if the producer's risk is 0.02
(2) the LTPD if the consumer's risk is 0.01.

5.21 Sketch the AOQ curve and the OC-curve for the problem 5.20.

5.22 A single sampling plan calls for a sample of size 20 and the acceptance number 2. Calculate the probability of accepting a lot of incoming quality 10% defectives and the probability of rejecting a lot of incoming quality 5 percent defectives. Assuming that the lot size is large, calculate the probabilities by (1) a Binomial approximation, (2) by a Poisson approximation.

5.23 When X has a Poisson distribution with parameter λ show that $P\{X \text{ is even}\} = \frac{1}{2}(1+e^{-2\lambda})$.

5.24 *The Relative Binomial Distribution.* Show that for the relative Binomial distribution,

$$f(x) = \binom{n}{nx} p^{nx} q^{n-nx}, \quad n = 0, \frac{1}{n}, \frac{2}{n}, ..., 1,$$

the mean value is p and the variance is pq/n.

5.25 *The Two-Point Distribution.* For the two-point distribution with the probability function,

$$f(x) = \begin{cases} q, & x = a \\ p, & x = b, \end{cases} \quad \text{where } q = 1-p, \ 0 < p < 1,$$

show that the mean value is $qa + pb$ and the variance is $pq(b-a)^2$.

5.26 *The Bernoulli Distribution.* In the two-point distribution if $a=0$ and $b=1$, we get the Bernoulli distribution. For the Bernoulli distribution show that the moment generating function is $(q + pe^t)$.

CHAPTER SIX

Univariate Probability
Models—Continuous

Among the continuous univariate probability models, the ones most commonly used are the Normal or Gaussian distribution, the Exponential distribution, the Gamma distribution and the Rectangular distribution. A list of the commonly used univariate continuous probability models is given in Table 6.1. Among these models, the Normal distribution is the most important one. There are several reasons for its importance. A good many types of data, when represented by frequency curves, the curves approximate to the Normal curve. There are also theoretical justifications for its importance. One is stated in a theorem called the Central Limit Theorem, which is given in the next chapter. It states that a certain function of the sample values has a Normal distribution when the sample size is large, whatever be the parent distribution. There are also a number of characteristic properties of the Normal distribution. That is, properties such as the statistical independence of the sample mean and the sample variance will uniquely determine the Normal distribution. In this chapter we will discuss the Normal distribution in detail and briefly mention the other distributions. Additional material is given in the problems at the end of the chapter.

6.1 Continuous Univariate Models

TABLE 6.1

Name	Density function $f(x)$	Parameters
1. The Uniform or rectangular distribution	$\dfrac{1}{\beta - a}$, $a < x < \beta$, and $f(x) = 0$, elsewhere.	$\beta - a > 0$

TABLE 6.1 (*continued*)

Name	Density function $f(x)$	Parameters
,,	$\frac{1}{\theta}$, $0<x<\theta$, and $f(x)=0$, elsewhere.	$\theta>0$
2. The Exponential distribution	$\frac{1}{\theta}e^{-x/\theta}$, $0<x<\infty$, and $f(x)=0$, elsewhere.	$\theta>0$
3. The Gamma distribution	$\frac{1}{\beta^a\Gamma(a)}x^{a-1}e^{-x/\beta}$, $0<x<\infty$ and $f(x)=0$, elsewhere.	$a, \beta>0$
,,	$\frac{1}{\Gamma(a)}x^{a-1}e^{-x}$, $0<x<\infty$,	$a>0$
4. The Beta distribution	$\frac{1}{B(a, \beta)}x^{a-1}(1-x)^{\beta-1}$, $0<x<1$, and $f(x)=0$, elsewhere.	$a, \beta>0$
5. The Cauchy distribution	$\frac{1}{\pi[1+(x-\theta)^2]}$, $-\infty<x<\infty$,	$-\infty<\theta<\infty$
,,	$\frac{1}{\pi[1+x^2]}$, $-\infty<x<\infty$,	No parameter
6. The Gaussian or Normal distribution	$\frac{1}{\beta\sqrt{2\pi}}e^{-(x-a)^2/2\beta^2}$, $-\infty<x<\infty$,	$-\infty<a<\infty, \beta>0$
,,	$\frac{1}{\beta\sqrt{2\pi}}e^{-x^2/2\beta^2}$, $-\infty<x<\infty$,	$\beta>0$
,,	$\frac{1}{\sqrt{2\pi}}e^{-(x-a)^2/2}$, $-\infty<x<\infty$,	$-\infty<a<\infty$
7. Standard Normal distribution	$\frac{1}{\sqrt{2\pi}}e^{-x^2/2}$, $-\infty<x<\infty$,	No parameter
8. The Pearson curves	$\frac{d}{dx}f(x) = \frac{d-x}{a+bx+cx^2}$	a, b, c, d such that $f(x)$ is a density function.
9. Pareto's distribution	$\frac{pa^p}{x^{p+1}}$, $x>a>0$, and $f(x)=0$, elsewhere.	$a>0, p>0$
10. Log-Normal distribution	$\frac{e^{-(\log x-\log a)^2/2\beta^2}}{\beta\sqrt{2\pi x}}$, $x>0$, and $f(x)=0$, elsewhere.	$a>0, \beta>0$

TABLE 1.6 (*continued*)

Name	Density function $f(x)$	Parameters
11. The Generalized Gamma distribution	$\dfrac{dac/d}{\Gamma(c/d)}\,x^{c-1}\,e-ax^d,\ x>0,$ and $f(x)=0$, elsewhere.	$a,\ b,\ c,\ d>0$
12. Logistic distribution	$\dfrac{\beta e-(a+\beta x)}{[1+e-(a+\beta x)]^2},\ x>0,$ and $f(x)=0$, elsewhere.	$a>0,\ \beta>0.$
13. Weibull distribution	$cax^{a-1}\,e-cx^a,\ x>0,$ and $f(x)=0$, elsewhere.	$c>0,\ a>1$
14. Laplace distribution	$\dfrac{e^{\dfrac{-\lvert x-\mu\rvert}{\lambda}}}{2\lambda},\ -\infty<x<\infty,$ and $f(x)=0$, elsewhere.	$-\infty<\mu<\infty,\ \lambda>0$
15. Rayleigh distribution	$\dfrac{x-1/2}{\sigma\sqrt{2\pi}}\,e-x/2\sigma^2,\ x>0,$ and $f(x)=0$, elsewhere.	$\sigma>0$
16. Makeham's distribution	$(\gamma+\delta e^{\mu x})\,e^{-\left[\gamma x+\dfrac{\delta}{\mu}e^{\mu x}\right]}$ for $x>0$ and $f(x)=0$, elsewhere.	$\gamma,\ \delta,\ \mu>0$
17. Waiting time distribution	$\lambda e-\lambda x,\ x>0,$ and $f(x)=0$, elsewhere.	$\lambda>0$
18. The Student-t distribution	$\dfrac{\Gamma\left(\dfrac{k+1}{2}\right)}{\sqrt{k\pi}\,\Gamma\left(\dfrac{k}{2}\right)}\left(1+\dfrac{x^2}{k}\right)^{-\dfrac{k+1}{2}},\ -\infty<x<\infty,$	k is a positive integer
19. The Chisquare distribution	$\dfrac{1}{2^{k/2}\,\Gamma\left(\dfrac{k}{2}\right)}\,x^{k/2-1}\,e-x/2,\ x>0,$ and $f(x)=0$, elsewhere.	k is a positive integer
20. The F-distribution	$\dfrac{\Gamma\left(\dfrac{m+n}{2}\right)}{\Gamma\left(\dfrac{m}{2}\right)\Gamma\left(\dfrac{n}{2}\right)}\left(\dfrac{m}{n}\right)^{m/2}\dfrac{x^{m/2-1}}{\left(1+\dfrac{m}{n}x\right)^{\dfrac{m+n}{2}}},$ for $x>0$ and $f(x)=0$, elsewhere.	$m,\ n$ are positive integers
21. Fisher's z-distribution	$\dfrac{\left(\dfrac{m}{n}\right)^{m/2}e(m/2)x}{B\left(\dfrac{m}{2},\dfrac{n}{2}\right)\left(1+\dfrac{m}{n}e^x\right)^{\dfrac{m+n}{2}}},\ x>0,$ and $f(x)=0$, elsewhere.	$m,\ n$ are positive integers

Table 1.6 (*continued*)

Name	Density function $f(x)$	Parameters
22. The Generalized F-distribution	$\dfrac{da^{c/d}\, \Gamma(a)x^{c-1}\,(1+ax^d)^{-a}}{\Gamma(c/d)\Gamma(a-c/d)}$ for $x>0$ and $f(x)=0$, elsewhere.	$a, a, b, c, d>0$ $a-c/d>0$
23. Pearson's p_λ distribution	$\dfrac{e^{-x/2}}{2}$, for $x>0$ and $f(x)=0$, elsewhere.	No parameter

From 11 one can obtain 2, 3, 13, 15, 17, 19 and 23 as special cases.

6.11 *Technical Notes*

Definition. The Gamma function, $\Gamma(a)$ (Gamma alpha) is defined as,

$$\Gamma(a) = \int_0^\infty x^{a-1}\, e^{-x}\, dx.$$

Definition. The Beta function, $B(a, \beta)$ (Beta alpha beta) is defined as,

$$B(a, \beta) = \int_0^1 x^{a-1}\,(1-x)^{\beta-1}\, dx.$$

The following results can be proved without much difficulty.

(1) $\Gamma(a) = (a-1)\,\Gamma(a-1)$

(2) $\Gamma(a) = (a-1)!$ if a is a positive integer.

(3) $\Gamma(\tfrac{1}{2}) = \sqrt{\pi}$

(4) $B(a, \beta) = \dfrac{\Gamma(a)\,\Gamma(\beta)}{\Gamma(a+\beta)}.$

Definition. If a function $\psi(x)$ is such that $\psi(x) = \psi(-x)$ then it is called an even function. Then,

$$\int_{-a}^{a} \psi(x)dx = 2 \int_{0}^{a} \psi(x)dx.$$

Definition. For a function $\varphi(x)$ if $\varphi(-x)=-\varphi(x)$ then $\varphi(x)$ is called an odd function. Then,

$$\int_{-a}^{a} \varphi(x)dx = 0.$$

Examples

(1) $x^4+2x^2+5;\quad 3x^8+7$ (Even functions)

(2) $2x^3+x;\quad 6x^5$ (Odd functions)

(3) $2x^2+3x+7;$ (Neither even nor odd)

6.2 THE NORMAL DISTRIBUTION

This is also known as Gaussian distribution and the error curve.
The name 'Normal' like many terms in statistics, is a little unfortunate.   This does not in any way mean that this particular
distribution is some sort of a standard distribution or other distributions are all abnormal.   The density function is given as,

$$f(x) = \frac{1}{\beta \sqrt{(2\pi)}} e^{-\frac{(x-a)^2}{2\beta^2}}, \quad -\infty < x < \infty$$

$$-\infty < a < \infty, \beta > 0,$$

where a and β are two parameters.   We can show that

$$E(X) = a \text{ and Var } (X) = \beta^2$$

when X is a Normal variate.   Figure 6.1 gives the Normal distribution.

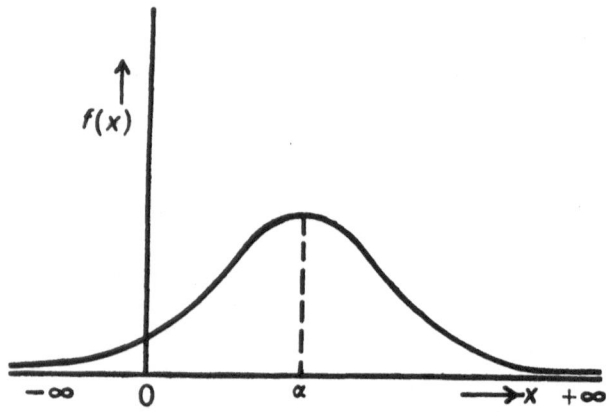

FIGURE 6.1   The Normal distribution

From the functional form of $f(x)$ itself it is clear that the distribution is *symmetric* about the ordinate at $x = a$, in the sense that the
value of $f(x)$ remains the same for $x = a + d$ and for $x = a - d$ where
d is any positive number.

6.21   *The Mean Value and the Variance*

By definition, the mean value $E(X)$, usually denoted by μ, is

$$\mu = E(X) = \int_{-\infty}^{\infty} x \, \frac{e^{[-(x-a)^2/2\beta^2]}}{\beta \sqrt{2\pi}} \, dx.$$

Put $y = \dfrac{x-a}{\beta}$ then $dx = \beta\, dy$, $x = a + y\beta$ and when $x = -\infty$,

$y = -\infty$ and when $x = \infty$, $y = \infty$, since a and β are finite.

$$\mu = E(X) = \int_{-\infty}^{\infty} \frac{(a+y\beta)}{\beta\sqrt{2\pi}}\, e^{-y^2/2}\, \beta\, dy$$

$$= a \int_{-\infty}^{\infty} \frac{e^{-y^2/2}}{\sqrt{2\pi}}\, dy + \beta \int_{-\infty}^{\infty} \frac{y e^{-y^2/2}}{\sqrt{2\pi}}\, dy$$

$$= a \int_{-\infty}^{\infty} \frac{e^{-y^2/2}}{\sqrt{2\pi}}\, dy + 0$$

(The second term is zero because $y e^{-y^2/2}$ is an odd function).

$$= 2a \int_{0}^{\infty} \frac{e^{-y^2/2}}{\sqrt{2\pi}}\, dy \text{ (since } e^{-y^2/2} \text{ is even).}$$

Put $z = \dfrac{y^2}{2}$ then $y^2 = 2z$, $dy = \dfrac{\sqrt{2}}{2} z^{-1/2}\, dz$ and when $y = 0$, $z = 0$

and when $y = \infty$, $z = \infty$. That is,

$$\mu = E(X) = a \int_{0}^{\infty} \frac{z^{-1/2}}{\sqrt{\pi}}\, e^{-z}\, dz$$

$$= \frac{a}{\sqrt{\pi}} \int_{0}^{\infty} z^{1/2-1}\, e^{-z}\, dz$$

$$= \frac{a}{\sqrt{\pi}}\, \Gamma\left(\tfrac{1}{2}\right) = a \text{ (See the technical note 6.11).}$$

$$\sigma^2 = \mathrm{Var}(X) = E(X - E(X))^2 = E((X-a)^2)$$

$$= \int_{-\infty}^{\infty} (x-a)^2\, \frac{e^{-\frac{(x-a)^2}{2\beta^2}}}{\beta\sqrt{2\pi}}\, dx$$

Put $y = \dfrac{x-a}{\beta}$ then,

$$\text{Var}\,(X) = \int_{-\infty}^{\infty} \beta^2 y^2 \frac{e^{-y^2/2}}{\beta\sqrt{2\pi}} \beta\, dy$$

$$= 2\beta^2 \int_{0}^{\infty} y^2 \frac{e^{-y^2/2}}{\sqrt{2\pi}} dy \quad \begin{array}{l}(\text{Since } y^2\, e^{-y^2/2} \text{ is an even} \\ \text{function}).\end{array}$$

Put $z = y^2/2$ then,

$$\text{Var}\,(X) = \beta^2 \int_{0}^{\infty} 2\, \frac{z^{1/2}\, e^{-z}}{\sqrt{\pi}}\, dz$$

$$= \frac{2\beta^2}{\sqrt{\pi}} \int_{0}^{\infty} z^{3/2-1}\, e^{-z}\, dz$$

$$= \frac{2\beta^2}{\sqrt{\pi}}\, \Gamma\,(\tfrac{3}{2}) = \frac{2\beta^2}{\sqrt{\pi}} \tfrac{1}{2}\; \Gamma\,(\tfrac{1}{2}) \quad \begin{array}{l}(\text{See the technical} \\ \text{note 6.11})\end{array}$$

$$= \beta^2.$$

Since for a Normal distribution, the mean value μ is a and the variance σ^2 is β^2 the parameters a and β^2 are often written as μ and σ^2 respectively and the density function is written as,

$$f(x) = \frac{1}{\sigma\sqrt{2\pi}}\, e^{-\frac{(x-\mu)^2}{2\sigma^2}}, \quad -\infty < x < \infty.$$

When $\mu = 0$ the density function becomes,

$$f(x) = \frac{1}{\sigma\sqrt{2\pi}} e^{-x^2/2\sigma^2}$$

and for a *standardized Normal variable* Y, that is,

$$Y = \frac{X-\mu}{\sigma},$$

the density function is given by,

$$f(y) = \frac{1}{\sqrt{2\pi}} e^{-y^2/2}, \quad -\infty < y < \infty.$$

In this case $\mu = E(X) = 0$ and $\sigma^2 = \mathrm{Var}(Y) = 1$. The Normal density is denoted by various notations such as,

$\mathcal{N}(\mu,\ \sigma^2)$ — a Normal distribution with mean μ and variance σ^2;

$\mathcal{N}(\mu,\ \sigma)$ — a Normal distribution with mean μ and standard deviation σ;

$X : \mathcal{N}(\mu,\ \sigma^2)$ — the s.v. X has the Normal distribution with mean μ and variance σ^2.

In this book we will follow the notation $\mathcal{N}(\mu,\ \sigma^2)$. Some Normal densities are given in Figure 6.2.

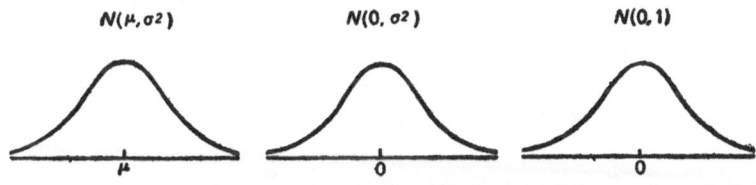

FIGURE 6.2 The Normal distribution

| Symmetric about the ordinate $x=\mu$; Maximum ordinate $= \dfrac{1}{\sigma\sqrt{2\pi}}$; Points of inflection at $x=\mu+\sigma$ and $x=\mu-\sigma$. | Symmetric about the $f(x)$-axis; Maximum ordinate $\dfrac{1}{\sigma\sqrt{2\pi}}$; Points of inflection at $x=\sigma$ and at $x=-\sigma$. | Symmetric about the $f(x)$-axis; Maximum ordinate $= \dfrac{1}{\sqrt{2\pi}}$; Points of inflection at $x=1$ and at $x=-1$. |

Example 6.2.1 For a Normal distribution with mean value 5 and variance 2, evaluate the following probabilities (1) $P\{x \geqslant 6\}$; (2) $P\{|x| < 3\}$.

Solution. Since $\mu=5$ and $\sigma^2=2$ the density function in this case is given by,

$$f(x) = \frac{1}{\sqrt{2}\ \sqrt{2\pi}}\ e^{-\frac{(x-5)^2}{4}}.$$

(1) $P\{x \geqslant 6\} = \int\limits_{6}^{\infty} f(x)\ dx = \int\limits_{6}^{\infty} \frac{1}{\sqrt{2}\ \sqrt{2\pi}}\ e^{-\frac{(x-5)^2}{4}}\ dx.$

Put $y = \dfrac{x-5}{\sqrt{2}}$ then $dx = \sqrt{2}\, dy$ and when $x=6$, $y = \dfrac{6-5}{\sqrt{2}} = \dfrac{1}{\sqrt{2}}$
and when $x = \infty$, $y = \infty$. That is,

$$P\{x \geqslant 6\} = \int_{1/\sqrt{2}}^{\infty} \frac{1}{\sqrt{2\pi}}\, e^{-y^2/2}\, dy$$

$$= 0.5 - \int_0^{1/\sqrt{2}} \frac{1}{\sqrt{2\pi}}\, e^{-y^2/2}\, dy = 0.2420 \text{ (from Normal pro-}$$

bability tables. See
also the comments).

(2) $P\{|x| < 3\} = P\{-3 < x < 3\} = \int_{-3}^{3} f(x)\, dx$

$$= \int_{-3}^{3} \frac{1}{\sqrt{2}\sqrt{2\pi}}\, e^{-\frac{(x-5)^2}{4}}\, dx$$

Put $y = \dfrac{x-5}{\sqrt{2}}$ then when $x=-3$, $y=-\dfrac{8}{\sqrt{2}}$ and when $x=3$,

$$y = -\frac{2}{\sqrt{2}}.$$

That is,

$$P\{-3 < x < 3\} = \int_{-8/\sqrt{2}}^{-2/\sqrt{2}} \frac{1}{\sqrt{2\pi}}\, e^{-y^2/2}\, dy = \int_{2/\sqrt{2}}^{8/\sqrt{2}} \frac{1}{\sqrt{2\pi}}\, e^{-y^2/2}\, dy$$

(by symmetry).

$$= \int_0^{8/\sqrt{2}} \frac{1}{\sqrt{2\pi}}\, e^{-y^2/2}\, dy - \int_0^{2/\sqrt{2}} \frac{1}{\sqrt{2\pi}}\, e^{-y^2/2}\, dy = 0.0793.$$

(from Normal probability tables. See also the comments).

Comment. The transformation used in the evaluation of the above integrals is the usual standardization of the variable. That is,

$$y = \frac{x-\mu}{\sigma} = \frac{x-5}{\sqrt{2}}.$$

Under this transformation a $\mathcal{N}(\mu, \sigma^2)$ transforms to a $\mathcal{N}(0, 1)$. The density function of $\mathcal{N}(0, 1)$ denoted by $\phi(y)$, is

$$\phi(y) = \frac{1}{\sqrt{2\pi}} e^{-y^2/2}$$

Tables are available for the integral

$$\int_0^t \phi(y)\, dy,$$

for the various values of t. Sometimes tables are given for the integrals,

$$\int_{-\infty}^t \phi(y)\, dy \text{ or } \int_t^\infty \phi(y)\, dy.$$

But,

$$\int_0^t \phi(y)\, dy$$

can be evaluated by using any one of the above integrals because,

$$\int_t^\infty \phi(y)\, dy = 0 \cdot 5 - \int_0^t \phi(y)\, dy = 1 - \int_{-\infty}^t \phi(y)\, dy.$$

Further, due to symmetry,

$$\int_0^t \phi(y)\, dy = \int_{-t}^0 \phi(y)\, dy$$

or

$$\int_{-\infty}^{-t} \phi(y)\, dy = \int_t^\infty \phi(y)\, dy.$$

A Normal probability table is given at the end of this book. Figure 6.3 gives the probabilities, $P\{x \geqslant 6\}$ and $P\{|x| < 3\}$, that is, the probabilities, $P\left\{ \dfrac{1}{\sqrt{2}} < y < \infty \right\}$ and $P\left\{ -\dfrac{8}{\sqrt{2}} < y < -\dfrac{2}{\sqrt{2}} \right\}$, respectively.   N(0, 1)

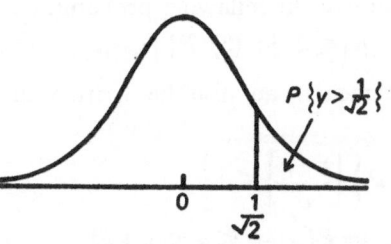
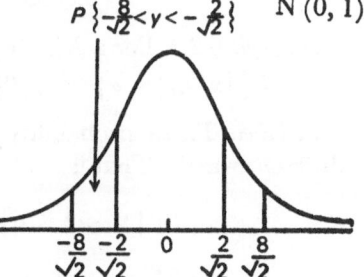

FIGURE 6.3 Certain Normal probabilities

Example 6.2.2 The marks obtained by a number of students for a certain subject are assumed to be approximately Normally distributed with mean value 65 and with a standard deviation of 5. If 3 students are taken at random from this set, what is the probability that exactly 2 of them will have marks over 70?

Solution. Let X be the marks obtained by a student of the set under consideration. Then X has an approximate Normal distribution with mean value 65 and standard deviation 5. The density function is

$$f(x) = \frac{1}{5\sqrt{2\pi}} \, e^{-\frac{(x-65)^2}{50}}.$$

Let p be the probability that a student selected at random has marks over 70. Then,

$$p = \int_{70}^{\infty} f(x)\,dx = \int_{70}^{\infty} \frac{1}{5\sqrt{2\pi}} \, e^{-\frac{(x-65)^2}{50}}\,dx$$

$$= \int_{\frac{70-65}{5}}^{\infty} \frac{1}{\sqrt{2\pi}} \, e^{-y^2/2}\,dy = \int_{1}^{\infty} \frac{1}{\sqrt{2\pi}} \, e^{-y^2/2}\,dy$$

$$= 0.1586. \quad \text{(from Normal probability tables)}$$

This probability p remains the same for every student taken at random. Now the probability of getting exactly 2 students having marks over 70 is a Binomial probability and the required probability

$$= \binom{3}{2} p^2 (1-p)^1 = \binom{3}{2} (0.1586)^2 \, (0.8414)^1$$

$$= 0.0635.$$

Example 6.2.3 For a $N(\mu, \sigma^2)$ evaluate the following probabilities
(1) $P\{|x-\mu| \leqslant \sigma\}$; (2) $P\{|x-\mu| \leqslant 2\sigma\}$; (3) $P\{|x-\mu| \leqslant 3\sigma\}$

Solution. These probability statements can also be written in different ways. That is,

$$P\{|x-\mu| \leqslant \sigma\} = P\left\{\left|\frac{x-\mu}{\sigma}\right| \leqslant 1\right\}$$

$$= P\{-\sigma \leqslant x-\mu \leqslant \sigma\} = P\{\mu-\sigma \leqslant x \leqslant \mu+\sigma\}$$

Similarly, $\quad P\{\,|\,x-\mu\,|\leqslant 2\sigma\} = P\{\mu-2\sigma\leqslant x\leqslant\mu+2\sigma\}$

and $\quad P\{\,|\,x-\mu\,|\leqslant 3\sigma\} = P\{\mu-3\sigma\leqslant x\leqslant\mu+3\sigma\}$

But, $\qquad P\{\mu-\sigma\leqslant x\leqslant\mu+\sigma\} = \int\limits_{\mu-\sigma}^{\mu+\sigma} \frac{1}{\sigma\sqrt{2\pi}}\, e^{-(x-\mu)^2/2\sigma^2}\, dx$

Standardizing the variable, that is, putting $y = \dfrac{x-\mu}{\sigma}$, we have,

$$P\{\mu-\sigma\leqslant x\leqslant\mu+\sigma\} = \int\limits_{-1}^{1} \frac{1}{\sqrt{2\pi}}\, e^{-y^2/2}\, dy$$

$$= \int\limits_{-1}^{0} \frac{1}{\sqrt{2\pi}}\, e^{-y^2/2}\, dy + \int\limits_{0}^{1} \frac{1}{\sqrt{2\pi}}\, e^{-y^2/2}\, dy$$

$$= 2\int\limits_{0}^{1} \frac{1}{\sqrt{2\pi}}\, e^{-y^2/2}\, dy$$

$$= 0.6828 \text{ (from the Normal probability tables).}$$

Similarly,

$$P\{\mu-2\sigma\leqslant x\leqslant\mu+2\sigma\} = \int\limits_{-2}^{2} \frac{1}{\sqrt{2\pi}}e^{-y^2/2}\, dy$$

$$= 2\int\limits_{0}^{2} \frac{1}{\sqrt{2\pi}}e^{-y^2/2}\, dy$$

$$= 0.9546 \text{ (from Normal probability tables),}$$

and $\qquad P\{\mu-3\sigma\leqslant x\leqslant\mu+3\sigma\} = 2\int\limits_{0}^{3} \frac{1}{\sqrt{2\pi}}e^{-y^2/2}\, dy$

$$= 0.9972.$$

Comment. It is interesting to notice that approximately 68%, 95% and 99% probabilities lie in the intervals, $(\mu-\sigma,\ \mu+\sigma)$, $(\mu-2\sigma,\ \mu+2\sigma)$ and $(\mu-3\sigma,\ \mu+3\sigma)$ respectively. These are very important results since the basic results for most of the statistical inference of hypothesis testing are these probability statements. If an obsrevation is available from a Normal distribution we can say that its value will be in the interval $(\mu-\sigma,\ \mu+\sigma)$ and the probability for such an event is 0.6828. In other words, there is a 68% chance for such a statement to be true. Also, these probability statements will give an indication about the closeness of the s.v. X to its expected value μ, measured in terms of a measure of dispersion, namely, the standard deviation. Figure 6.4 shows the areas over these intervals $(\mu-\sigma,\ \mu+\sigma)$, $(\mu-2\sigma,\ \mu+2\sigma)$ and $(\mu-3\sigma,\ \mu+3\sigma)$.

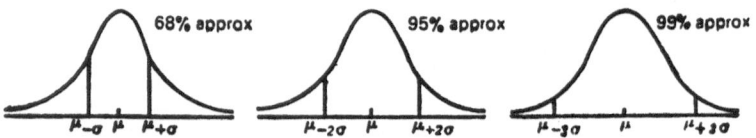

FIGURE 6.4 Certain probability statements for a $\mathcal{N}(\mu,\ \sigma^2)$

Since these probability statements occur very often in Statistical inference, the reader is advised to memorize these results. Now it is natural to ask the question: ' What is the probability that a binomial variate X will fall in the interval $(\mu-\sigma,\ \mu+\sigma)$ where μ and σ are the expected value and the standard deviation of the Binomial variate respectively?' This can be evaluated with the help of a Binomial probability table. But usually, in a practical problem, the exact underlying distribution is unknown and if we can get such a probability statement which is distribution free, in the sense, that the statement will be true whatever be the distribution, it is of great value for the experimental scientists. In the following section we consider a distribution-free statement.

6.3 A DISTRIBUTION-FREE PROPERTY

Here we consider a simple theorem known as Chebyshev's Inequality which is a probability statement which will hold good whatever be the underlying distribution.

Theorem 6.3.1 Let X be a s.v. with a finite variance σ^2 and mean value $\mu=E(X)$.

Then,
$$P\{\,|x-\mu|\geqslant k\sigma\} \leqslant \frac{1}{k^2}$$

where k is a positive real number.

Proof. $P\{\,|x-\mu|\geqslant k\sigma\} = P\{(x-\mu)\leqslant -k\sigma \text{ or } (x-\mu)\geqslant k\sigma\}$

$$= P\{x\leqslant\mu -k\sigma \text{ or } x\geqslant\mu +k\sigma\}$$

$$= \int_{-\infty}^{\mu -k\sigma} f(x)\,dx + \int_{\mu +k\sigma}^{\infty} f(x)\,dx \text{ if } X \text{ is continuous,}$$

where $f(x)$ is the density function of X. We will consider the case when X is continuous and the proof when X is discrete is left to the reader.

$$\sigma^2 = \text{Var }(X) = E(X - E(X))^2 = E((X-\mu)^2)$$

$$= \int_{-\infty}^{\infty} (x-\mu)^2 f(x)\,dx$$

For convenience, we will split this integral into three parts. That is,

$$\text{Var }(X) = \int_{-\infty}^{\mu -k\sigma} (x-\mu)^2 f(x)\,dx + \int_{\mu -k\sigma}^{\mu +k\sigma} (x-\mu)^2 f(x)\,dx$$

$$+ \int_{\mu +k\sigma}^{\infty} (x-\mu)^2 f(x)\,dx \geqslant \int_{-\infty}^{\mu -k\sigma} (x-\mu)^2 f(x)\,dx$$

$$+ \int_{\mu +k\sigma}^{\infty} (x-\mu)^2 f(x)\,dx.$$

But in the intervals $(-\infty, \mu -k\sigma)$ and $(\mu +k\sigma, \infty)$, $|x-\mu|\geqslant k\sigma$ and hence replacing $(x-\mu)^2$ by its lowest value in these intervals, we get,

$$\sigma^2 = \text{Var }(X) \geqslant \int_{-\infty}^{\mu -k\sigma} (k\sigma)^2 f(x)\,dx + \int_{\mu +k\sigma}^{\infty} (k\sigma)^2 f(x)\,dx$$

$$= (k\sigma)^2 \left[\int_{-\infty}^{\mu -k\sigma} f(x)\,dx + \int_{\mu +k\sigma}^{\infty} f(x)\,dx \right]$$

$$= (k\sigma)^2 \, P\{\,|x-\mu|\geqslant k\sigma\}.$$

Therefore, $P\{\,|x-\mu|\geqslant k\sigma\} \leqslant \dfrac{1}{k^2}.$

Figure 6.5 gives a graphical representation of the Theorem 6.1.

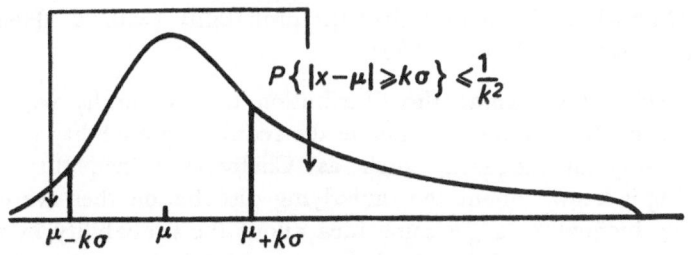

$$P\{|x-\mu|\geqslant k\sigma\} \leqslant \frac{1}{k^2}$$

$\mu -k\sigma \qquad \mu \qquad \mu +k\sigma$

FIGURE 6.5   Chebyshev's Theorem

According to the theorem, whatever be the distribution,

$$P\{|x-\mu| \geqslant 2\sigma\} \leqslant \frac{1}{2^2} = 0.25.$$

But for a Normal distribution, we know that this probability is $1-0.95=0.05$ approximately. So it is evident that the inequality given by Chebyshev's theorem is not very sharp. But the importance of the result is that the inequality is true for all the distributions with finite mean value and variance. Sharper inequalities can be obtained but we will not discuss them here.

Example 6.3.1. For a Binomial distribution with parameters $N=10$ and $p=0.4$ obtain, $P\{|x-\mu| \geqslant 2\sigma\}$ and compare it to the probability given by Chebyshev's inequality and evaluate the error.

Solution. For a Binomial distribution, $\mu=Np$ and $\sigma^2=Np(1-p)$. For this problem,

$$\mu = Np = 10 \ (0.4) = 4$$

and $\quad \sigma = \sqrt{Np(1-p)} = \sqrt{10(0.4) \ (0.6)} = 1.6$

approximately.

Hence, $\quad \mu-2\sigma = 4-3.2 = 0.8$

and $\quad \mu+2\sigma = 4+3.2 = 7.2$.

Therefore,

$$P\{|x-\mu| \geqslant 2\sigma\} = P\{x \leqslant 0.8 \text{ or } x \geqslant 7.2\}.$$

For this Binomial distribution, if $x \leqslant 0.8$ or $\geqslant 7.2$ then x can take the values 0, 8, 9 and 10. Hence the required probability

$$= \binom{10}{0} (0.4)^0 \ (0.6)^{10} + \sum_{x=8}^{10} \binom{10}{x} (0.4)^x \ (0.6)^{10-x}$$

$$= 0.0183.$$

But Chebyshev's inequality gives the probability $\frac{1}{4}=0.25$. Hence the error is $0.25-0.0183=0.2317$.

Comment. If we know the distribution then naturally we will use a probability table to evaluate the required probability rather than using an inequality such as Chebyshev's inequality. If nothing is known about the underlying distribution then we can use the inequality to get some idea about the probability in the *tail ends*, namely, in the intervals $(-\infty, \mu-k\sigma)$ and $(\mu+k\sigma, \infty)$.

Exercises

6.1 Write down the Normal densities if (1) $\mu=0$, $\sigma=0.2$, (2) $\mu=-5$, $\sigma=4$; (3) $\mu=1$, $\sigma=2$.

6.2 Evaluate the following probabilities for a $\mathcal{N}(\mu, \sigma^2)$ with $\mu=2$, $\sigma=4$.
   (1) $P\{x \geqslant 5\}$; (2) $P\{|x|>2\}$; (3) $P\{|x-1| \geqslant 2\}$;
   (4) $P\{2x+1 \geqslant 5\}$; (5) $P\{|x+2| \geqslant 3\}$.

6.3 Can the following be the parameters for a Normal distribution.
   (1) $\mu = -5$; (2) $\mu = 20$; (3) $\mu = 0$; (4) $\mu = 0.4$;
   (5) $\sigma = 0.6$; (6) $\sigma = 0$; (7) $\sigma = -4$.

6.4 For a $\mathcal{N}(\mu, \sigma^2)$ show that the maximum ordinate is $\dfrac{1}{\sqrt{2\pi\sigma^2}}$ and the points of inflection are at $x=\mu-\sigma$ and at $x=\mu+\sigma$.

6.5 For a Normal distribution $\mathcal{N}(\mu, \sigma^2)$ show that the odd central moments are all zeros.

6.6 For a Normal distribution with parameters μ (unknown) and $\sigma=1$ find t such that,
   (1) $P\{-t \leqslant x-\mu \leqslant t\} = 0.99$
   (2) $P\{x-t \leqslant \mu \leqslant x+t\} = 0.95$

6.7 For a $\mathcal{N}(\mu, \sigma^2)$ obtain t_0 and t_1 such that
$$P\left\{-t_0 \leqslant \frac{x-\mu}{\sigma} \leqslant t_1\right\} = P\{x-t_0\sigma \leqslant \mu \leqslant x+t_1\sigma\} = 0.95.$$
Are t_0 and t_1 unique?

6.8 An indicator moves from a particular point to either side. If the deviation from this point to either side is approximately a Normal variate with mean zero and $\sigma=0.2$, what is the probability of getting a deviation (1) between -0.1 and 0.1, (2) as large as 0.3?

6.9 Show that the M.G.F., that is, $E(e^{tX})$, for a Normal distribution is $M(t)=\exp\left(t\mu + \dfrac{t^2\sigma^2}{2}\right)$.

6.10 Using the uniqueness property of the M.G.F., (that is, M.G.F. will uniquely determine the distribution) identify the distribution from the following M.G.F. (1) e^{t^2}; (2) e^{-2t+4t^2}; (3) e^{3t+t^2}.

6.11 Consider a Binomial distribution with parameters N and p. When $N \to \infty$ and p remains a constant, show that the standardized Binomial distribution approaches a standardized Normal distribution. *Hint:* Show that the M.G.F. of the standardized Binomial variate, $Y = (X - Np)/\sqrt{Np(1-p)}$, approaches the M.G.F. of the standardized Normal variate.

6.12 Suppose that the profit or loss per day of a shopkeeper, dealing in a perishable item, is approximately a Normal variable with mean value \$10 and with a standard deviation \$5. If 4 days are selected at random, what is the probability that (1) on at least 2 days he will have a profit of \$12 or more, (2) on exactly one day he has a profit of \$14 or more, (3) on exactly one day he has a profit of \$10?

6.13 For any s.v. X with finite μ and σ^2 show that,

(1) $P\{ |x - \mu| \leqslant k\sigma \} \geqslant 1 - \dfrac{1}{k^2}$; (2) $P\{ |x - \mu| \geqslant k \} \leqslant \dfrac{\sigma^2}{k^2}$;

(3) $P\{ |x - \mu| \leqslant k \} \geqslant 1 - \dfrac{\sigma^2}{k^2}$.

6.14 Evaluate the following probabilities $P\{ |x - \mu| > 2\sigma \}$, $P\{ |x - \mu| \geqslant 3 \}$ for a Poisson distribution with parameter $\lambda = 4$ and compare them with the corresponding values obtained by Chebyshev's inequality.

6.15 For a $\mathcal{N}(\mu, \sigma^2)$ with $\mu = 2$, $\sigma^2 = 4$ evaluate the following probabilities and interpret them graphically.
(1) $P\{ |x - 2| \geqslant 5 \}$, (2) $P\{ x \leqslant 7 \}$.

6.16 (law of large numbers) Applying Chebyshev's inequality to the binomial distribution, prove that as the sample size increases, the probability that the observed proportion of successes differs from p by more than any arbitrary positive constant θ approaches 0.

6.17 (law of large numbers) Let $X_1, X_2, \ldots, X_n$ be a sequence of independent random variables with the same distribution having mean μ and variance σ^2. Then for any $\epsilon > 0$, prove that

$$\lim_{n \to \infty} P\left(\left| \frac{X_1 + \ldots + X_n}{n} - \mu \right| \geqslant \epsilon \right) = 0.$$

6.4 Other Continuous Models

There are a good number of other univariate continuous models used in economic problems, life testing, biological growth problems

and in other disciplines.   A list of them was given in Table 6.1.
Here we will discuss some of them briefly.

The Rectangular Distribution. This is also known as the continuous
uniform distribution and the density function is,

$$f(x) = \begin{cases} \dfrac{1}{\beta-a}, & a < x < \beta. \\ 0 & \text{elsewhere.} \end{cases}$$

Figure 6.6 gives a graphical representation of the distribution.
Due to its rectangular shape it is called a rectangular distribution.

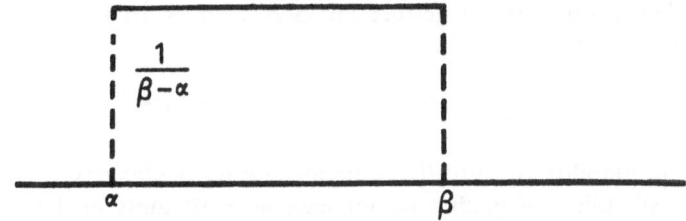

FIGURE 6.6 A Rectangular distribution

The Exponential Distribution. The density function is given by,

$$f(x) = \begin{cases} \dfrac{1}{\theta}\, e^{-x/\theta}, & 0 < x < \infty, \ \theta > 0 \\ 0, & \text{elsewhere,} \end{cases}$$

where $\theta > 0$ is a parameter.   Figure 6.7 gives a graphical representation.

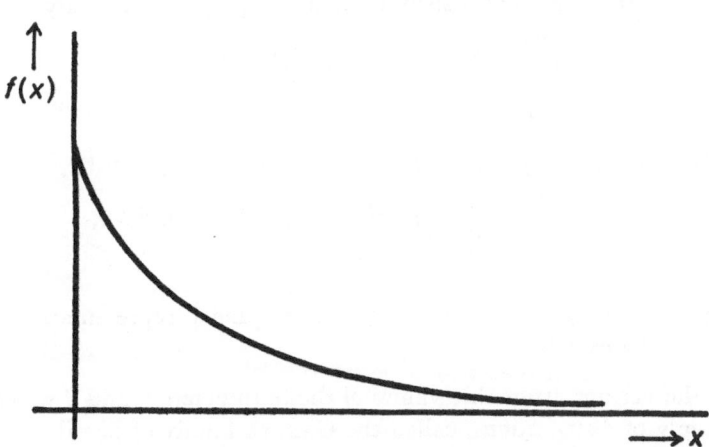

FIGURE 6.7 An Exponential distribution

Example 6.4.1 The increase in sales in a shop, per day after the appointment of a new sales-girl, is approximately exponentially distributed with the parameter $\theta = 2$. If 2 days are selected at random, what is the probability that (1) on both the days the increase is over 10 units, (2) the increase is over 8 on at least one of the days?

Solution. The distribution is

$$f(x) = \begin{cases} \frac{1}{2} e^{-x/2}, & x > 0 \\ 0, & \text{elsewhere.} \end{cases}$$

(1) The probability of getting an increase over 10 units on any one randomly selected day

$$= P\{x \geqslant 10\} = \int_{10}^{\infty} \tfrac{1}{2} e^{-x/2} \, dx = e^{-5}.$$

This probability remains the same for any particular day. Hence the probability of getting an increase over 10 units on both the days is given by the following Binomial probability and

$$= \binom{2}{2} (e^{-5})^2 (1 - e^{-5})^0 = e^{-10}.$$

(2) The probability of getting an increase over 8 units

$$= P\{x \geqslant 8\} = \int_{8}^{\infty} \tfrac{1}{2} e^{-x/2} \, dx = e^{-4}.$$

By using the Binomial distribution, the required probability

$$= \binom{2}{1} (e^{-4})^1 (1 - e^{-4}) + \binom{2}{2} (e^{-4})^2 (1 - e^{-4})^0$$

$$= 2 e^{-4} - e^{-8}.$$

The Gamma Distribution. The density function is given by,

$$f(x) = \begin{cases} \dfrac{1}{\beta^a \, \Gamma(a)} x^{\alpha-1} e^{-x/\beta}, & x > 0, \ a > 0, \ \beta > 0, \\ 0, & \text{elsewhere.} \end{cases}$$

where a and β are parameters. A graphical representation is given in figure 6.8.

For the various assignable values of the parameters a and β we get a family of distributions, called the Gamma family of distributions and three members of this family are given in figure 6.8.

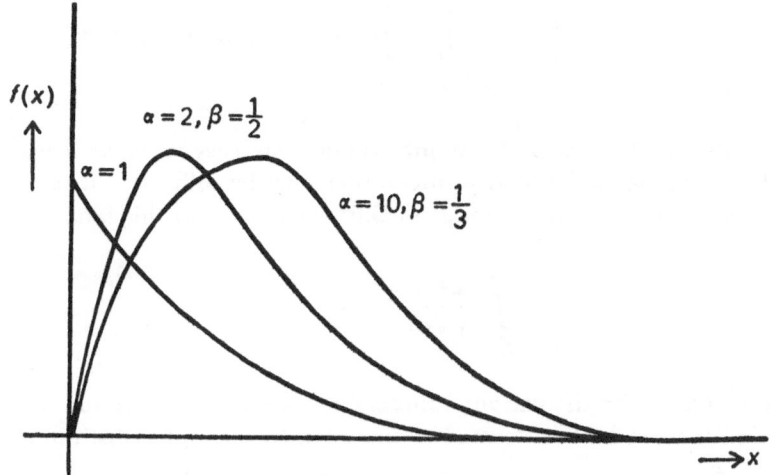

FIGURE 6.8 A Gamma distribution

Example 6.4.2   The daily consumption of milk in a city, in excess of 20,000 gallons, is approximately distributed as a Gamma distribution with the parameters $a=2$ and $\beta=10,000$.   The city has a daily stock of 30,000 gallons.   What is the probability that the stock is insufficient on a particular day?

Solution.   If Y denotes the daily consumption of milk then $X=Y-20,000$ has a Gamma distribution with $a=2$ and $\beta=10,000$. That is, the density function is,

$$f(x) = \frac{1}{(10,000)^2 \ \Gamma(2)} \ x^{2-1} \ e^{-x/10,000}$$
$$= \frac{x \ e^{-x/10,000}}{(10,000)^2}$$

If the stock on a particular day is to be insufficient then the consumption on that day must exceed 30,000.   Hence the required probability is,

$$P\{y \geqslant 30,000\} = P\{x \geqslant 10,000\}$$
$$= \int_{10,000}^{\infty} \frac{x \ e^{-x/10,000}}{(10,000)^2} \ dx = \int_{1}^{\infty} z \ e^{-z} \ dz$$

$\left(\text{by making the substitution } z = \dfrac{x}{10,000}\right).$

$$= \left[-z\, e^{-z} \right]_1^\infty - \left[e^{-z} \right]_1^\infty \text{ (Integration by parts)}$$
$$= e^{-1} + e^{-1} = 2e^{-1}.$$

Comment. In this problem integration was easy because a was 2. For a general a and β integration may be difficult. In such cases one can use an Incomplete Gamma table. In this table,

$$\int_0^t \frac{z^{a-1}}{\Gamma(a)}\, e^{-z}\, dz$$

is tabulated for the various values of t and a general Gamma distribution can be transformed to this form by a simple substitution $z = \dfrac{x}{\beta}$. For an incomplete Gamma table see, K. Pearson, *Tables of Incomplete Gamma Functions* (Cambridge University Press).

In the following table we give the mean values and the variances of some of the commonly used univariate continuous models.

TABLE 6.2

X	Density $f(x)$	Mean value	Variance
1. Normal	$\dfrac{1}{\sigma\sqrt{2\pi}} e^{\frac{-(x-\mu)^2}{2\sigma^2}},$ $-\infty < x < \infty$	μ	σ^2
2. Rectangular	$\dfrac{1}{(\beta-a)},\ a < x < \beta$	$\left(\dfrac{\beta+a}{2}\right)$	$\dfrac{(\beta-a)^2}{12}$
3. Exponential	$\dfrac{1}{\theta} e^{-x/\theta},\ x > 0$	θ	θ^2
4. Gamma	$\dfrac{1}{\beta^a \Gamma(a)} x^{a-1} e^{-x/\beta},\ x > 0$	$a\beta$	$a\beta^2$
5. Beta	$x^{a-1}(1-x)^{\beta-1},\ 0 < x < 1$	$\dfrac{a}{a+\beta}$	$\dfrac{a\beta}{(a+\beta)^2(a+\beta+1)}$

Exercises

6.16 If a balanced spinner with the dial marked 0 to 10 with 0 and 10 coinciding, is rotated twice, what is the probability that the indicator will stop in between 6 and 8 on both the trials?

6.17 Suppose that the duration of a shower in a tropical island, is approximately exponentially distributed with the parameter $\theta=5$ minutes. Out of 3 showers, what is the probability that not more than 2 will last for 10 minutes or more?

6.18 In the same problem in 6.17, what is the probability that (1) a shower will last at least 2 minutes more, given that it has already lasted for 5 minutes, (2) a shower will last not more than 6 minutes more if it has already lasted for 3 minutes?

6.19 The demand for a certain item per day is approximately distributed as a Gamma distribution with parameters $a=2$ and $\beta=4$. What is the probability that (1) there will be a demand for at least 10 units on a particular day, (2) there will be a demand for exactly 15 units on a particular day?

6.20 The sales per day in a shop is approximately exponentially distributed with mean value \$100. If sales tax is levied at the rate of 8%, what is the probability that the sales tax return from that shop will exceed \$30 on two consecutive days?

6.21 Integrating by parts show that,
$$\Gamma(a) = (a-1)\ \Gamma(a-1)$$

6.22 Show that $\Gamma\left(\dfrac{1}{2}\right) = \sqrt{\pi}$. Hint: $\left\{\ \Gamma\left(\dfrac{1}{2}\right)\right\}^2$
$$= \left(\int_0^\infty x^{1/2-1}\ e^{-x}\ dx\right)\left(\int_0^\infty y^{1/2-1}\ e^{-y}\ dy\right)$$
$$= \left(\int_0^\infty \int_0^\infty x^{-1/2}\ y^{-1/2}\ e^{-(x+y)}\ dx\ dy\right)$$

Change to polar co-ordinates.

6.23 Evaluate the probabilities of the following events, when X is an exponential variate with mean value 5.
(1) A, (2) B, (3) $\bar{A}$, (4) $A \cap B$ where $A = \{x \mid 0 < x < 1\}$, $B = \{x \mid -\infty < x < 10\}$.

6.24   Show that the M.G.F. that is $E(e^{tX})$, of a Gamma variate with parameters a and β, is,
$$M(t) = (1-\beta t)^{-a}.$$
By expanding $M(t)$ show that $E(X) = a\beta$ and $\mathrm{Var}(X) = a\beta^2$.

6.25   By using the uniqueness property, identify the distribution from the following M.G.F.

(1) $\dfrac{1}{16}(1+3e^t)^2$;   (2) $e^{2(e^t-1)}$;   (3) e^{3t^2};   (4) $(1-2t)^{-3}$;

(4) $(1-5t)^{-1}$

6.26   Show that the rth moment of a Gamma distribution is,
$$= a(a+1) \ldots (a+r-1)\beta^r$$

6.5 Change of Variables

In practical problems one often needs the distribution of a well-defined function $\psi(X)$ of a s.v. X. We may know the distribution of X and we may need the distribution of functions such as $2X$, X^2+1, and so on. The main result concerning the change of variable is contained in the following theorem:

Theorem 6.5.1   Let X be a continuous s.v. with p.d.f. $f(x)$, and let $y = \psi(x)$ be strictly monotonic with continuous non-vanishing derivative in some open interval I. Let I_1 be the image of I in the y-space. Then the s.v. Y with p.d.f. $g(y)$ exists in I_1 and is given by

$$g(y) = f(x)\frac{dx}{dy}\ \Big|$$
where $x = \psi^{-1}(y)$

Proof: Let $F(x)$ and $G(y)$ be the distribution functions of X and Y respectively. Let $y = \psi(x)$ be a monotonic increasing function as shown in figure 6.9.

Then,
$$F(a) = P\{X \leqslant a\} = P\{Y \leqslant b\} = G(b) = G(\psi(a))$$
$$f(a) = \frac{d}{da}F(a) = \frac{d}{da}G(\psi(a)) = \left[\frac{d}{d\psi}G(\psi(a))\right]\frac{d\psi}{da}$$
$$= g(\psi(a))\frac{d\psi}{da} = g(b)\frac{d\psi}{da}\ \text{for all } a.$$

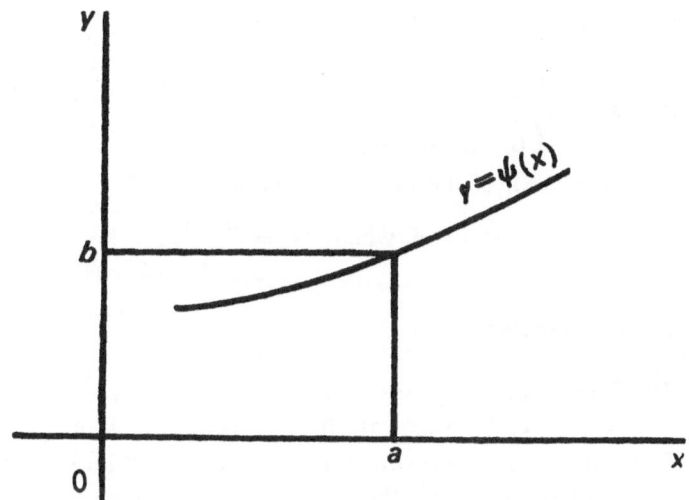

FIGURE 6.9

That is,

$$f(x) = g(y) \frac{dy}{dx}$$

But if $\psi(x)$ is a decreasing function then $\frac{d\psi}{dx}$ will be negative.

Hence in general,

$$f(x) = g(y) \left| \frac{dy}{dx} \right|$$

The preceding result can be extended to functions of several random variables (this will be discussed later on).

For example, if $f(x) = \begin{cases} 2x, & 0 < x < 1 \\ 0, & \text{elsewhere,} \end{cases}$

and if $Y = 2X$ then the density of Y, denoted by $g(y)$ is given as,

$$g(y) = 2x \left| \frac{dx}{dy} \right| = 2x \left(\frac{1}{2} \right) = x = \frac{y}{2}.$$

But when $0 < x < 1$, $0 < y < 2$ and hence,

$$g(y) = \begin{cases} \dfrac{y}{2}, & 0 < y < 2 \\ 0, & \text{elsewhere.} \end{cases}$$

Exercises

6.27 If $f(x) = \begin{cases} \dfrac{1}{a}, & 0 < x < a \\ 0, & \text{elsewhere.} \end{cases}$

Find the p.d.f. of the random variable $Y = X^n$.

6.28 Find $g(y)$, where $y = x^2$, when $f(x) = \dfrac{(x+1)}{8}$, $-1 \leqslant x \leqslant 3$.

Additional Exercises

6.1 Draw the Normal curves $\mathcal{N}(\mu, \sigma^2)$ where $\sigma^2 = 4$ and $\mu = 0$, $-1, 1, 2, -2, 3$. Draw all the 6 curves on the same graph.

6.2 Draw the Normal curves $\mathcal{N}(\mu, \sigma^2)$ where $\mu = 2$ and $\sigma^2 = 16$, 9, 3, and 1. Draw all the 4 curves on the same graph and compare them.

6.3 What happens to the Normal curves if (1) σ^2 remains the same and μ varies; (2) μ remains the same and σ^2 varies?

6.4 Calculate a measure of kurtosis, namely, $\gamma = \dfrac{\mu_4}{\mu_2^2} - 3$ for a $\mathcal{N}(\mu, \sigma^2)$ and show that $\gamma = 0$.

6.5 By using a Normal probability table evaluate the following probabilities (1) $P\{x \mid -\infty < x < 1\}$, (2) $P\{x \mid 0 < x < 5\}$, (3) $P\{x = 7\}$, (4) $P\{x \mid 1 < x < 7\}$, (5) $P\{x \mid -\infty < x < -5, 10 < x < \infty\}$, where X is a $\mathcal{N}(\mu, \sigma^2)$ with $\mu = 4$ and $\sigma^2 = 4$.

6.6 By using a table of Binomial probabilities draw a frequency polygon for a Binomial distribution with the parameters $N = 20$ and $p = \frac{1}{2}$. On the same graph draw a Normal curve with $\mu = 10$ and $\sigma^2 = 5$ and compare the two curves if the frequency polygon is smoothed by a curve.

6.7 If the I.Q. of students in a particular university is assumed to be approximately Normally distributed with mean value 100 and variance 25, what is the probability that if 2 students are selected at random from this university both will have I.Q.'s between 102 and 110?

6.8 For a $\mathcal{N}(\mu, \sigma^2)$ evaluate the following probabilities (1) $P\{\mu - \frac{2}{3}\sigma < x < \mu + \frac{2}{3}\sigma\}$, (2) $P\{\mu - \sigma < x < \mu + 2\sigma\}$, (3) $P\{x - 2\sigma < \mu < x + 3\sigma\}$.

6.9 For a rectangular distribution, $f(x) = 1/(\beta-a)$, $a < x < \beta$, evaluate the M.G.F.

6.10 By using the M.G.F. of a rectangular distribution evaluate the mean value and the variance.

6.11 Evaluate the mean values and the variances of the various probability models given in Table 6.1.

6.12 Evaluate the M.G.Fs, if they exist, for the various probability models in Table 6.1.

6.13 If X is a $\mathcal{N}(\mu, \sigma^2)$ with $\mu=5$, $\sigma^2=4$, evaluate the following probabilities (1) $P\{2x+4 \geqslant 7\}$, (2) $P\{|3x+4| < 8\}$.

6.14 If the life time of a certain type of electric bulbs is approximately distributed as an exponential variate with mean value 1000, what is the probability that a bulb will last for 1000 hours more if it has already lasted for 1000 hours?

6.15 In the problem 6.14 if two bulbs are selected at random and if both have lasted for 1000 hours, what is the probability that at least one of them will (1) last for 1000 hours more, (2) last for not more than another 1000 hours?

6.16 The duration of snow falls in a certain place is approximately a Gamma variate with parameters $a=2$ and $\beta=10$. If snow is falling from 8 a.m. on a particular morning what is the probability that it will stop on or before 8.30 a.m., time being measured in minutes?

6.17 For the same problem in 6.16, if snow began to fall at 8 a.m. and if a person has waited for 15 minutes what is the probability that he will have to wait at least 15 minutes more before the snow fall stops?

6.18 For the problem in 6.16 what is that probability that on a particular day (1) a snow fall will last for more than 10 minutes but not exceeding 20 minutes, (2) a snow fall will last for exactly 21 minutes?

6.19 Evaluate the distribution of (1) $Y=2X+3$, (2) $T=3X-2$, if the density of X is

$$f(x) = \begin{cases} 2x, & 0 < x < 1 \\ 0, & \text{elsewhere.} \end{cases}$$

6.20 Evaluate the density of $Y=2X+4$ if the density of X is

$$f(x) = \begin{cases} x, & 0 < x < 1 \\ \dfrac{3-x}{4}, & 1 \leqslant x < 3 \\ 0, & \text{elsewhere.} \end{cases}$$

6.21 *Probability Integral Transformation.* If $f(x)$ is the density function of a s.v. X show that

$$Y = \int_{-\infty}^{X} f(x)\, dx$$

has a rectangular distribution in the interval $(0, 1)$.

6.22 If X is a standard Normal variate, that is $N(0, 1)$, show that X^2 has a Gamma distribution with $\alpha = \frac{1}{2}$ and $\beta = 2$.

6.23 If X is a Beta variate with parameters $\alpha = \dfrac{m}{2}$, $\beta = \dfrac{n}{2}$ show that

$$Y = \frac{nX}{m(1-X)}$$ has an F-distribution (See Table 6.1 for these distributions).

6.24 If X has a rectangular distribution in $(0, 1)$ and if $Y = \log X$ (the natural logarithm of X), then show that $-Y$ has an exponential distribution with the parameter $\theta = 1$ and vice versa.

6.25 *Linear Exponential Family of Distributions.* If the probability function of a s.v. X is given by

$$f(x) = \begin{cases} \dfrac{b(x)\, e^{\theta x}}{h(\theta)}, & \text{for } x \in A \subset R \ (R\text{-real line}) \\ 0, & \text{elsewhere,} \end{cases}$$

where $h(\theta) = \int b(x)\, e^{\theta x}\, dx$ if X is continuous
$\qquad\qquad = \Sigma\, b(x)\, e^{\theta x}$ if X is discrete, then

$f(x)$ is called a linear exponential family of distributions. For a linear exponential family, show that (1) $\mu = E(X) = \dfrac{h'(\theta)}{h(\theta)}$

where $h'(\theta) = \dfrac{d}{d\theta}\, h(\theta)$; (2) Obtain the following distributions as special cases of $f(x)$: (a) Normal with known variance, (b) Exponential, (c) Gamma with α known, (d) Binomial, (e) Poisson, (f) Negative Binomial, (g) Geometric, (h) Logarithmic series.

6.26 If X is a $N(\mu, \sigma^2)$ show that $aX + b$ is $N(a\mu + b, a^2\sigma^2)$.

CHAPTER SEVEN

Sampling Distributions

7.1 Independent Stochastic Variables

In Chapter 2 we defined the independence of two events A and B. That is, A and B are said to be independent if,

$$P(A \cap B) = P(A)\, P(B).$$

Consider two stochastic variables X and Y having a joint distribution and let the events A and B be,

$$A = \{x \mid a < x < b\} \text{ and } B = \{y \mid c < y < d\},$$

then,

$$A \cap B = \{(x, y) \mid a < x < b,\ c < y < d\}$$

which is shown in Figure 7.1.

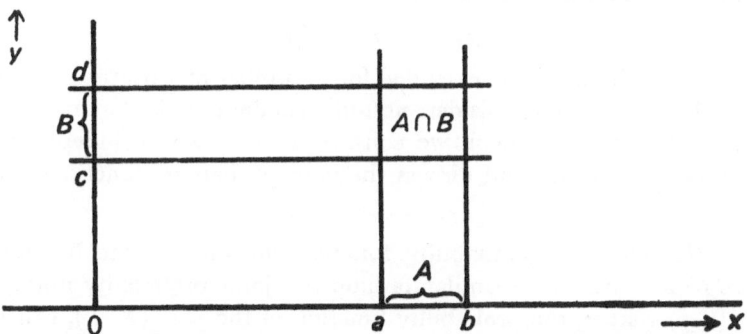

FIGURE 7.1 The intersection of two events A and B

Thus the two events A and B will be independent if the probability over the shaded region is the product of the probabilities of A and B. If X and Y are continuous variates with density functions $f(x)$ and $g(y)$ respectively then,

$$P(A) = \int_a^b f(x)\, dx$$

and
$$P(B) = \int_c^d g(y)\, dy$$

That is,

$$P(A)\, P(B) = \int_a^b f(x)\, dx \int_c^d g(y)\, dy$$

$$= \int_a^b \int_c^d f(x)\, g(y)\, dy\, dx.$$

If $h(x, y)$ denotes the density function of the pair (X, Y) (it is also called the *joint density function* of X and Y) then we can easily see that if $h(x, y) = f(x)\, g(y)$, that is, if $h(x, y)$ is factorized into the density functions of X and Y then for all such events A and B, $P(A)\, P(B) = P(A \cap B)$. So we will define the independence of two stochastic variables X and Y in the following way:

Definition. Two s.v.s X and Y are said to be statistically independent if their joint probability function factorizes into the probability functions of X and Y respectively. That is, if $h(x, y)$ denotes the joint probability function and if $f(x)$ and $g(y)$ are the individual probability functions of X and Y, (also known as the *marginal probability functions*) then X and Y are independent if,

$$h(x, y) = f(x)\, g(y).$$

This definition can be extended for a number of variates. In this book we will not consider non-independent variables in detail. In the rest of the book we consider mainly sets of independent variables. So we will discuss the joint probability functions very briefly.

We defined a probability function for one variate by using some axioms. In a similar fashion the joint probability function $h(x, y)$ that is, the probability function of the pair (X, Y), can be defined as that function of two variables X and Y such that,

(1) $h(x, y) \geqslant 0$ for all x and y

(2) $\int_x \int_y h(x, y)\, dx\, dy = 1$ if X and Y are continuous

$\sum_x \sum_y h(x, y) = 1$ if X and Y are discrete,

and if $h(x, y)$ is the joint probability function then the marginal probability functions, $f(x)$ and $g(y)$, that is, the probability functions of X and Y respectively, can be obtained as

$$f(x) = \int_y h(x, y) \, dy \text{ if } Y \text{ is continuous}$$

$$= \sum_y h(x, y) \text{ if } Y \text{ is discrete}$$

and $$g(y) = \int_x h(x, y) \, dx \text{ if } X \text{ is continuous}$$

$$= \sum_x h(x, y) \text{ if } X \text{ is discrete.}$$

The conditional probability of X given $Y=a$, denoted by $K(x \mid y=a)$ is defined as

$$K(x \mid y=a) = \frac{h(x, y)}{g(y)} \text{ at } y=a$$

$$= \frac{h(x, a)}{g(a)} \text{ if } g(a) \neq 0.$$

In a similar way the conditional probability of Y given $X=b$, denoted by $J(y \mid x=b)$, is defined as

$$J(y \mid x=b) = \frac{h(x, y)}{f(x)} \text{ at } x=b \text{ if } f(b) \neq 0.$$

In these expressions $h(x, y)$ denotes the joint probability functions and $f(x)$ and $g(y)$ are the marginal probability functions of X and Y respectively.

Since the joint probability function factorizes into the marginals when the variables are independent, it is easy to see that $K(x \mid y)=f(x)$ and $J(y \mid x)=g(y)$ when X and Y are independent. In other words, when the variables are independent, the condition imposed on one variable has no bearing on the other variable.

Example 7.1.1 If $h(x,y)=e^{-x-y}$ is a probability function check whether X and Y are statistically independent or not.

Solution. $h(x, y)=e^{-x-y}=(e^{-x})(e^{-y})=f(x) g(y)$ where $f(x)=e^{-x}$ and $g(y)=e^{-y}$ satisfy the conditions for probability functions. Hence X and Y are independent.

Comment. In our discussion, 'independence' means 'statistical independence'. In this problem $f(x)$ and $g(y)$ have the same functional form and we know that X and Y have the same exponential distributions. In other words we say that X and Y here are independently and identically distributed as exponential variates with parameter $\theta=1$. Also in this example we may notice that the conditional densities of X and Y are the marginal densities themselves. This is due to the independence of the variables here.

Example 7.1.2 If X and Y are independent Poisson variates with parameters λ_1 and λ_2 obtain the joint probability function of X and Y.

Solution. If X and Y are independent then their joint probability function $h(x, y)$ is the product of the marginal probability functions. That is,

$$h(x, y) = \begin{cases} \dfrac{\lambda_1{}^x}{x!} \dfrac{\lambda_2{}^y}{y!} e^{-(\lambda_1+\lambda_2)} \text{ for } x, y=0, 1, 2,\ldots \\ 0, \quad \text{elsewhere.} \end{cases}$$

Example 7.1.3 Check whether X and Y are independent if the joint probability function $h(x, y)$ is given as

$$h(x, y) = \begin{cases} 2, 0<x<1, \ 0<x<y<1 \\ 0, \text{ elsewhere.} \end{cases}$$

Solution. Since $h(x, y)$ is not an explicit function of x and y it is difficult to factorize $h(x, y)$ unless we determine the marginal probability functions. Since $0<x<y$, $x<y<1$ and both x and y are such that $0<x<y<1$ the marginal density of Y, that is $g(y)$ is given as,

$$g(y) = \int_x h(x, y)dx = \int_0^y 2\, dx$$

$$= \begin{cases} 2y, \ 0<y<1 \\ 0, \text{ elsewhere,} \end{cases}$$

and the marginal density of X is

$$f(x) = \int_y h(x, y)\, dy = \int_x^1 2\, dy$$

$$= \begin{cases} 2(1-x), \ 0<x<1 \\ 0, \quad \text{elsewhere.} \end{cases}$$

But $f(x)\, g(y) = 2(1-x)\, 2y \neq 2 = h(x, y)$. Hence X and Y are not independent.

Exercises

7.1 Consider an experiment of throwing a balanced coin twice. Let X denote the number of heads in the outcomes and Y denote the number of tails respectively. Show that the joint probability distribution is given as

y \ x	0	1	2	
0	0	0	$\frac{1}{4}$	$\frac{1}{4}$
1	0	$\frac{1}{2}$	0	$\frac{1}{2}$
2	$\frac{1}{4}$	0	0	$\frac{1}{4}$
	$\frac{1}{4}$	$\frac{1}{2}$	$\frac{1}{4}$	

where the entries are the probabilities. Let the marginal probability functions be $f(x)$ and $g(y)$. Write down $h(x, y)$, $f(x)$ and $g(y)$. Compare $f(x)$ and $g(y)$ with the probabilities in the margins.

7.2 In the problem 7.1 represent the joint distribution by a two dimensional bar diagram.

7.3 In problem 7.1 show that X and Y are not independent.

7.4 Evaluate the marginal probabilities from the following joint distribution.
$$h(x, y) = \begin{cases} \frac{1}{4}, & x=1, y=0 \\ \frac{1}{4}, & x=2, y=3 \\ \frac{1}{2}, & x=3, y=5 \\ 0, & \text{elsewhere.} \end{cases}$$

7.5 Evaluate the marginal distributions, from the following joint distribution.
$$h(x, y) = \begin{cases} \frac{1}{8}, & x=0, y=0 \\ \frac{1}{8}, & x=0, y=1 \\ \frac{1}{8}, & x=1, y=0 \\ \frac{5}{8}, & x=1, y=1 \\ 0, & \text{elsewhere.} \end{cases}$$

7.6 Show that X and Y are independent in the following distribution.
$$h(x, y) = \begin{cases} 2e^{-x-2y}, & x > 0, y > 0 \\ 0, & \text{elsewhere.} \end{cases}$$

7.7   Check whether X and Y are independent in the following problem.

$$h(0, 1)=\tfrac{1}{27}, \quad h(0, 2)=\tfrac{5}{27}, \quad h(0, 3)=\tfrac{6}{27}$$
$$h(1, 1)=\tfrac{2}{27}, \quad h(1, 2)=\tfrac{4}{27}, \quad h(1, 3)=\tfrac{4}{27}$$
$$h(2, 1)=\tfrac{1}{27}, \quad h(2, 2)=\tfrac{2}{27}, \quad h(2, 3)=\tfrac{2}{27}$$

and $h(x, y)=0$ elsewhere.

7.8   Represent the distribution in 7.7 in a two dimensional bar diagram.

7.9   Is the following a joint probability function? If so, check whether the variables are independent or not.

$$h(x, y) = \begin{cases} \tfrac{2}{3}\,(x+1)\,e^{-y}, & 0 < x < 1, y > 0 \\ 0, & \text{elsewhere.} \end{cases}$$

7.10 Are X and Y independent in the following problem?

$$h(x, y) = \frac{1}{2\pi}\,e^{-(x^2+y^2)/2}, -\infty < x < \infty, -\infty < y < \infty.$$

7.2 A RANDOM SAMPLE FROM A THEORETICAL POPULATION

In Chapter 1 we defined a random sample from a numerical population. Here we will define a random sample from a theoretical population, that is, a population designated by a stochastic variable, such as a Normal population, a Binomial population and so on. The ideas of ' independence ', introduced in this chapter, will help us in defining a random sample from a theoretical population.

Definition. A set of n s.v.'s X_1, X_2, ..., X_n which are independently and identically distributed is said to be a simple random sample of size n from $f(x)$ if $f(x)$ is the common probability function.

That is, if we have two s.v.'s X_1 and X_2 which are independently distributed as an exponential variate with the same probability function $f(x) = \dfrac{1}{\theta}\,e^{-x/\theta}$ then (X_1, X_2) is a simple random sample of size 2 from $f(x)$. Both X_1, X_2 have the same distribution (functional form of the probability function and the parameters are all the same). For a simple random sample of size 2 from an exponential population, the joint distribution of the *sample values* (that is, X_1 and X_2) is

$$h(x_1, x_2) = \left(\frac{1}{\theta}\, e^{-x_1/\theta}\right) \left(\frac{1}{\theta}\, e^{-x_2/\theta}\right)$$

$$= \begin{cases} \dfrac{1}{\theta^2}\, e^{-(x_1+x_2)/\theta}, & x_1 > 0,\ x_2 > 0 \\ 0, & \text{elsewhere.} \end{cases}$$

Further if $(X_1, X_2, ..., X_n)$ is a simple random sample of size n from a population designated by a probability function $f(x)$ the joint probability function of the sample values is

$$h(x_1, x_2, ..., x_n) = f(x_1)\, f(x_2) \cdots f(x_n) = \prod_{i=1}^{n} f(x_i).$$

Now we will find an interpretation for the statement that a particular number, say 5, is an observation from a theoretical population. If the population is discrete and if the corresponding s.v. takes the value 5 with non-zero probability, then evidently 5 is a value assumed by the stochastic variable. If the population is continuous with a density function $f(x)$ then the probability that X assumes the value 5 is,

$$P\{x = 5\} = \int_{5}^{5} f(x)\, dx = 0.$$

A number ' a ' is said to be an observation from a continuous population designated by a density function $f(x)$, if the probability that x falls in the interval $\left(a - \dfrac{\delta}{2},\ a + \dfrac{\delta}{2}\right)$, for $\delta \geqslant 0$, is given by,

$$P\left\{a - \frac{\delta}{2} < x < a + \frac{\delta}{2}\right\} = \int_{a-\delta/2}^{a+\delta/2} f(x)\, dx$$

where $\left(a - \dfrac{\delta}{2},\ a + \dfrac{\delta}{2}\right)$ is a neighbourhood of a. This will be the interpretation of the statement that a continuous s.v. X assumes the value a. Now we can give an interpretation of a numerical random sample (observed random sample) of size n from a theoretical population.

Definition. A set of numbers $x_1, x_2, ..., x_n$ is said to be an observed random sample of size n from a theoretical population if $x_1, ..., x_n$ are one set of values assumed by $X_1, X_2, ..., X_n$, where $X_1 ..., X_n$ is a simple random sample of size n from the same theoretical population.

Thus the numbers 2 and -4 are said to be an observed random sample of size 2 from a Normal population $\mathcal{N}(\mu, \sigma^2)$ if 2 is a value assumed by X_1 and -4 is a value assumed by X_2 where (X_1, X_2) is a simple random sample of size 2 from $\mathcal{N}(\mu, \sigma^2)$. Thus a simple random sample is a set of stochastic variables whereas observed random sample is a set of numbers.

7.21 A Statistic

Definition. Any function, of the random sample $X_1, X_2, ..., X_n$, which is again a s.v. is called a statistic. (Plural of statistic is statistics which is different from the Science Statistics or a collection of data). It is often convenient to make a distinction between statistics which are completely observable (that is, functions of sample values which do not contain any unknown parameters) and functions of sample values containing some parameters (often known as pivotal quantities as against statistics). But we will not make a distinction in our discussion.

Example 7.21.1 *The Sample Mean* $\overline{X} = \dfrac{(X_1 + X_2 + ... + X_n)}{n}$ is a statistic and is called the sample mean. That is, $\overline{X}$ is a s.v. having its own distribution.

The Sample Variance. $S^2 = \displaystyle\sum_{i=1}^{n} \dfrac{(X_i - \overline{X})^2}{n}$ is a statistic having its own distribution.

Comment. From a given simple random sample we can construct an infinite number of statistics. In fact, the original variable itself is a statistic since it can be considered to be a linear function of the sample values, that is, for example,

$$X_i = X_1 + OX_2 + OX_3 + ... + OX_n$$

If $(1, 5, 9)$ is an observed sample then we can get an observed value of $\overline{X}$ as,

$$\overline{X} = \frac{1+5+9}{3} = 5.$$

We can derive a few interesting results regarding the expected values of $\overline{X}$ and S^2.

Theorem 7.1 If $(X_1, X_2,..., X_n)$ is a simple random sample from a population with finite mean value μ then,
$$E(\overline{X})=\mu.$$

Proof. $E(\overline{X}) = E\left(\dfrac{X_1+X_2+...+X_n}{n}\right) = \dfrac{1}{n}\left\{E(X_1) + E(X_2) + ... + E(X_n)\right\}$

(Since the expected value of a sum is the sum of the expected values).

$$=\frac{1}{n}\{\mu+\mu+...+\mu\} \text{ (Since } X_1,..., X_n \text{ are identi-}$$

cally distributed)

$$= \mu.$$

Definition. If T is a statistic and if $ET=\theta$ (some parameter), then T is said to be unbiased for θ. That is, for example, the sample mean is unbiased for the population mean. In the theory of estimation a statistic is also known as an *estimator* and, thus, we can say that $\overline{X}$ is an unbiased estimator of μ.

Theorem 7.2 If two s.v.'s X and Y are independent and if $\phi(X)$ and $\psi(Y)$ are two functions of X and Y respectively then,
$$E(\phi(X)\,\psi(Y))=[E(\phi(X))]\,[E(\psi(Y))].$$

That is, expectation of a product is the product of the expectations when the variates are independent and whenever the expectations exist.

Proof. Let X and Y be continuous and let $h(x, y)$ be the joint density function. Since X and Y are assumed to be independent
$$h(x, y) = f(x)\,g(y)$$
where $f(x)$ and $g(y)$ are the marginal densities. Now adopting the same techniques used in section 4.2 to define the expected values, we have,

$$E(\phi(X)\,\psi(Y)) = \int_x \int_y \phi(x)\,\psi(y)\,h(x, y)\,dy\,dx$$

$$= \int_x \int_y \phi(x)\,\psi(y)\,f(x)\,g(y)\,dy\,dx \text{ (since}$$
$$h(x, y)=f(x)\,g(y)),$$

$$= \left(\int_x \phi(x)\,f(x)\,dx\right)\left(\int_y \psi(y)\,g(y)\,dy\right)$$

$$= [E(\phi(X))]\,[E(\psi(Y))]. \text{ (follows from the definitions in section 4.2 of Chapter 4).}$$

The proofs in the other cases are left to the reader. This result can also be extended to a number of variates, $X_1, X_2, ..., X_n$.

Definition. The Covariance: *The covariance between two variates X and Y, denoted by Cov (X, Y), is defined as,*

$$\text{Cov } (X, Y) = E[(X - E(X)) (Y - E(Y))] = E(XY) - (E(X)) (E(Y)).$$

Other usual notations are, $C(x, y)$, C_{xy}, σ_{xy}, etc.

Example 7.21.2 Evaluate the covariance between X and Y from the following joint density function of X and Y;

$h(x, y) = 2$, $0 < x < y < 1$ and $h(x, y) = 0$, elsewhere.

Solution. Cov $(X, Y) = E(XY) - (E(X)) (E(Y))$. In this case it is easier to use this simplified formula rather than evaluating the expression $E(X - E(X)) (Y - E(Y))$.

$$E(XY) = \int_x \int_y xy \, h(x, y) \, dy \, dx, \text{ when } X \text{ and } Y \text{ are continuous,}$$

where $h(x, y)$ denotes the joint density function.

$$= \int_0^1 [\int_0^y 2 \, xy \, dx] \, dy = \int_0^1 [2y \int_0^y x \, dx] \, dy$$

$$= \int_0^1 y^3 \, dy = \tfrac{1}{4}.$$

$E(X) = \int_x x f(x) dx$, (where $f(x) = \int_y h(x, y) \, dy$, is the marginal density of X),

$$= \int_0^1 x f(x) \, dx = \int_0^1 [x \int_y h(x, y) \, dy] \, dx$$

$$= \int_0^1 x \, (\int_x^1 2 \, dy) \, dx = \int_0^1 2x(1 - x) \, dx = \tfrac{1}{3}.$$

$E(Y) = \int_0^1 y \, g(y) dy$, (where $g(y) = \int_x h(x, y) \, dx$, is the marginal density of Y).

Hence the covariance between X and Y, that is,

$$\text{Cov } (X, Y) = E(XY) - (E(X)) (E(Y)) = (\tfrac{1}{4}) - (\tfrac{1}{3}) (\tfrac{2}{3}) = (\tfrac{1}{36}).$$

Comment. From the definition of the covariance, it is seen that the covariance can be negative, zero or positive. Further, when the variables are independent the covariance between them is zero. This follows from theorem 7.2. That is, when the variables are independent,

$$\text{Cov } (X, Y) = E\left[(X-E(X))\ (Y-E(Y))\right] = [E(X-E(X))]$$
$$[E(Y-E(Y))] \text{ (due to independence)}$$
$$= (0)\ (0) = 0.$$

But however, Cov $(X, Y) = 0$ does not imply that X and Y are independent in general. From the definition itself it is apparent that Cov (X, Y) measures the joint variation of X and Y in some sense. In order to measure the linearity relationship between two variables X and Y, often we use a measure called the *linear correlation coefficient* ρ between X and Y and is defined as

$$\rho = \text{Cov } (X, Y)\ /\ [\text{Var } (X)\ \text{Var } (Y)]^{1/2}$$

It can be shown that $-1 \leqslant \rho \leqslant 1$, that is, $\rho^2 \leqslant 1$ and further, when there is a linear relationship between X and Y, that is, when $Y = aX + b$ where a and b are constants, $\rho^2 = 1$ and when X and Y are independent $\rho = 0$. In this sense ρ measures the linearity relationship between X and Y. The covariance and the linear correlation coefficient are important in Correlation and Regression Analysis in Statistics. Here we will not discuss correlation and covariance in detail. For some problems on these topics see the exercises at the end of this section and the additional problems at the end of this chapter.

Theorem 7.3 Whenever the population variance σ^2 is finite the variance of the sample mean is $\dfrac{\sigma^2}{n}$.

Proof.
$$\text{Var } (\overline{X}) = E((\overline{X} - E(\overline{X}))^2) = E((\overline{X} - \mu)^2)$$

$$= E\left(\left(\frac{X_1 + \ldots + X_n}{n} - \mu\right)^2\right)$$

$$= E\left(\frac{1}{n^2}\left\{(X_1-\mu) + \ldots + (X_n-\mu)\right\}^2\right)$$

$$= E\left(\frac{1}{n^2}\left\{(X_1-\mu)^2 + \ldots + (X_n-\mu)^2\right.\right.$$
$$\left.\left. + 2\sum_{i<j} (X_i-\mu)\ (X_j-\mu)\right\}\right)$$

$$= \frac{1}{n^2}\left\{E((X_1-\mu)^2) + \ldots + E((X_n-\mu)^2)\right.$$
$$\left. + 2\sum_{i<j} E((X_i-\mu)\ (X_j-\mu))\right\}$$

But since X_i and X_j are independent for all i and j, $i \neq j$,

$$E((X_i-\mu)\,(X_j-\mu)) = [E(X_i-\mu)]\,[E(X_j-u)] = 0,$$

since $E(X-E(X))=0$ for any s.v. X.

That is, $\quad \text{Var } (\overline{X}) = \dfrac{1}{n^2} \left\{ E((X_1-\mu)^2)+\ldots+E((X_n-\mu)^2) \right\}$

$$= \dfrac{1}{n^2} \left\{ \sigma^2+\sigma^2+\ldots+\sigma^2 \right\} = \dfrac{\sigma^2}{n}$$

(Since $X_1, \ldots, X_n$ are identically distributed).

Definition. Standard Error. The positive square root of the variance of any statistic is called the standard error of the statistic.

For example, the standard error of the sample mean $\overline{X}$ is $\sqrt{\dfrac{\sigma^2}{n}} = \dfrac{\sigma}{\sqrt{n}}$ where σ^2 is the population variance. When the sample is of size one the standard error of the sample mean is σ itself.

Theorem 7.4 $\quad E(S^2) = \dfrac{n-1}{n} \sigma^2$ whenever σ^2 is finite.

Proof. Consider $\displaystyle\sum_{i=1}^{n} (X_i-\overline{X})^2$

$$E\left(\sum_{i=1}^{n} (X_i-\overline{X})^2 \right) = E\left(\sum_{i=1}^{n} (X_i-\mu+\mu-\overline{X})^2 \right)$$

$$= E\left(\sum_{i=1}^{n} (X_i-\mu)^2+n(\mu-\overline{X})^2+2(\mu-\overline{X}) \sum_{i=1}^{n} (X_i-\mu) \right)$$

$$= E\left(\sum (X_i-\mu)^2+n(\mu-\overline{X})^2-2n(\mu-\overline{X})^2 \right)$$

$$= E\left(\sum (X_i-\mu)^2-n(\mu-\overline{X})^2 \right)$$

$$= E\left(\sum (X_i-\mu)^2-n(\overline{X}-\mu)^2 \right)$$

$$= \sum E((X_i-\mu)^2) - E(n(\overline{X}-\mu)^2)$$

$$= n\,\sigma^2-\sigma^2 = (n-1)\,\sigma^2$$

(Since $E(\overline{X}) = \mu$ and Var $(\overline{X}) = \sigma^2/n$, $E(n(\overline{X}-\mu)^2) = \sigma^2$)

Hence $\quad E(S^2) = \displaystyle\sum_{i=1}^{n} E\left(\dfrac{(X_i-\overline{X})^2}{n} \right) = \dfrac{n-1}{n} \sigma^2$

Comment. It should be noticed that,

$$E\left(\sum \dfrac{(X_i-\overline{X})^2}{n-1} \right) = \sigma^2.$$

That is, $\sum \dfrac{(X_i-\bar{X})^2}{(n-1)}$ is unbiased for σ^2 whereas $\sum \dfrac{(X_i-\bar{X})^2}{n}$ is not unbiased for σ^2. Due to this reason and due to the fact that unbiasedness is a desirable property for estimators some authors define the sample variance as $\sum \dfrac{(X_i-\bar{X})^2}{(n-1)}$. But we will use the definition given before.

Example 7.21.3 If (X_1, X_2, X_3) is a simple random sample of size 3 from a population with mean value 5 and with variance 4, evaluate the expected values and the standard errors of the following statistics. (1) $T_1=(X_1+X_2-X_3)$; (2) $T_2=(2X_1+X_2-3X_3)$.

Solution. (1) $T_1 = (X_1+X_2-X_3)$

$$E(T_1) = E(X_1)+E(X_2)-E(X_3)=\mu+\mu-\mu=\mu=5.$$

$$\begin{aligned}
\mathrm{Var}(T_1) &= E[(T_1-E(T_1))^2]=E([(X_1-\mu)+(X_2-\mu)\\
&\quad -(X_3-\mu)]^2)\\
&= E((X_1-\mu)^2)+E((X_2-\mu)^2)+E((X_3-\mu)^2)\\
&\quad + 2E((X_1-\mu)(X_2-\mu))-2E((X_1-\mu)(X_3-\mu))\\
&\quad -2E((X_2-\mu)(X_3-\mu))\\
&= E((X_1-\mu)^2)+E((X_2-\mu)^2)+E((X_3-\mu)^2)
\end{aligned}$$

(The cross product terms vanish because the variates are independent).

$$=\sigma^2+\sigma^2+\sigma^2=3\sigma^2=3(4)=12.$$

Hence the standard error $=\sqrt{12}$.

(2) $\quad E(T_2) = E(2X_1+X_2-3X_3)=2E(X_1)+E(X_2)-3E(X_3)$
$$= 2\mu+\mu-3\mu=0.$$

$$\begin{aligned}
\mathrm{Var}(T_2) &= E((T_2-ET_2)^2)=E((T_2-0)^2)\\
&= E([2(X_1-\mu)+(X_2-\mu)-3(X_2-\mu)]^2)\\
&= 4E((X_1-\mu)^2)+E((X_2-\mu)^2)+9E((X_2-\mu)^2)+0\\
&= 14\sigma^2=14(4)=56.
\end{aligned}$$

The standard error (S.E.) of T_2 is $\sqrt{56}$.

7.22 *The Sampling Distributions*

Definition. The distribution of a statistic is called a sampling distribution.

For example, the distribution of the sample mean is a sampling distribution and the distribution of the sample variance is a sampling distribution. We will be interested in some sampling distributions when the parent population is the Normal population because these distributions appear in a good number of problems of statistical decision making. These will be discussed in section 7.3.

7.23 *The Central Limit Theorem*

This is a very important theorem in the theory of sampling distributions. This theorem makes the Normal distribution a very important distribution. It can be stated in different ways depending upon the conditions. One such statement is given here.

Theorem 7.5 Let $X_1, ..., X_n$ be a simple random sample from a population, continuous or discrete, with finite variance σ^2 and mean value μ. Then the standardized sample mean, that is,

$$\Upsilon = \frac{\overline{X} - \mu}{\sigma / \sqrt{n}}$$

is approximately distributed as a standard Normal distribution $\mathcal{N}(0, 1)$, when the sample size n is large.

Proof. We have

$$\Upsilon = \sum_{i=1}^{n} \frac{(X_i - \mu)}{(\sigma \sqrt{n})}$$

The moment generating function of $\dfrac{(X_i - \mu)}{\sigma}$ is given by $e^{t^2/2}$ and therefore the moment generating function $M(t)$ of $\dfrac{(X_i - \mu)}{(\sigma \sqrt{n})}$ is given by

$$M(t) = e^{t^2/(2n)} = 1 + \frac{t^2}{2n} + \text{terms of higher order in } \frac{t}{\sqrt{n}}.$$

$$= 1 + \frac{t^2}{2n} + o\left(\frac{t^2}{n}\right)$$

But

$$M_\Upsilon(t) = [M(t)]^n.$$

Hence

$$M_\Upsilon(t) = \left[1 + \frac{t^2}{2n} + o\left(\frac{t^2}{n}\right)\right]^n.$$

Since $\lim\limits_{n\to\infty} o\left(\dfrac{t^2}{n}\right)=0$ and $\lim\limits_{n\to\infty}\left(1+\dfrac{1}{n}\right)^n=e$,

we have

$$M_T(t) = \lim_{n\to\infty}\left[1+\frac{t^2}{(2n)}+o\left(\frac{t^2}{n}\right)\right]^n$$

$$= \lim_{n\to\infty}\left(1+\frac{t^2}{(2n)}\right)^n = e^{t^2/2}.$$

This proves the theorem.

A good approximation is available for n as small as 30 if the parent distribution is symmetric. This theorem in effect says, for example, the standardized sample mean is approximately a standard normal when the parent distribution is the Exponential distribution or a Binomial distribution or a Gamma distribution and so on.

Exercises

7.11 If $(1, 5, -3)$ is an observed sample from a population, obtain a value of the sample mean and the sample variance.

7.12 If (X_1, X_2, X_3) is a simple random sample from a population show that the following statistics are unbiased for the population mean.
   (1) $T_1=(X_1+X_2+X_3)/3$; (2) $T_2(3X_1+5X_2-7X_3)$;
   (3) $T_3=(X_1+5X_2+X_3)/7$; (4) $T_4(4X_1-X_3)/3$.

7.13 If $(X_1,..., X_n)$ is a simple random sample from a population with mean value μ and variance σ^2 and if $a_1, a_2,..., a_n$ are some constants show that

   (1) $E(a_1X_1+a_2X_2+...+a_nX_n) = \left(\sum\limits_{i=1}^{n} a_i\right)\mu$

   (2) $\text{Var}\left[\sum\limits_{i=1}^{n} a_iX_i\right] = \left(\sum\limits_{i=1}^{n} a_i^2\right)\sigma^2$

7.14 Evaluate the covariance between x and y if the joint probability function $h(x, y)$ is, $h(0, 0)=\frac{1}{4}$, $h(0, 1)=\frac{1}{4}$, $h(2, 0)=\frac{1}{4}$, $h(2, 1)=\frac{1}{4}$ and $h(x, y)=0$ elsewhere.

7.15 For any s.v.s $X_1,..., X_n$ show that

$$\text{Var}\left[\sum_{i=1}^{n} a_iX_i\right]= \sum_{i=1}^{n} a_i^2\,\text{Var}\,(X_i)+ \sum_{i\neq j} a_ia_j\,\text{Cov}\,(X_i, X_j).$$

7.16 Show that Cov $(X, Y) = E(XY) - (E(X))(E(Y))$.

7.17 If X and Y are two independent variates and if $M_1(t)$ and $M_2(t)$ denote their M.G.F.'s show that the M.G.F. of the sum $X + Y$ is the product of $M_1(t)$ and $M_2(t)$.

7.18 If $X_1, X_2, ..., X_n$ are independent then show that the M.G.F. of the linear combination $a_1 X_1 + ... + a_n X_n$ is $\prod_{i=1}^{n} M_i(a_i t)$ where $M_i(t)$ is the M.G.F. of X_i.

7.19 If $(X_1, ..., X_n)$ is a simple random sample from a population with M.G.F. $M(t)$ show that the M.G.F. of $(X_1 + ... + X_n)$ is $[M(t)]^n$.

7.20 (a) Using the uniqueness property of M.G.F., show that the distribution of the sample mean $\overline{X}$, when the sample is from a $N(\mu, \sigma^2)$, is a $N\left(\mu, \dfrac{\sigma^2}{n}\right)$.

(b) By using the M.G.F.'s show that the distribution of the sample mean is Gamma when the population is exponential or Gamma.

7.3 SAMPLING DISTRIBUTIONS ASSOCIATED WITH THE NORMAL POPULATION

Since the Normal population is very important in Statistics, the sampling distributions associated with the Normal population are of great importance. The most important sampling distributions are (1) the Student-t distribution, (2) the Chi-square distribution and (3) the F-distribution.

7.31 *The Distribution of the Sample Mean*

If $X_1, X_2, ..., X_n$ is a simple random sample of size n from a Normal population $N(\mu, \sigma^2)$ it can be shown that the sample mean $\overline{X}$ is again distributed as a Normal variate with mean value μ and with variance σ^2/n, that is $N(\mu, \sigma^2/n)$. (see problem 7.20). That is, if

$$X : N(\mu, \sigma^2) \text{ with } f(x) = \frac{1}{\sigma\sqrt{2\pi}} e^{-\frac{(x-\mu)^2}{2\sigma^2}}$$

then, $$\overline{X} : N\left(\mu, \frac{\sigma^2}{n}\right) \text{ with } f(\bar{x}) = \frac{\sqrt{n}}{\sigma\sqrt{2\pi}} e^{\frac{-n(\bar{x}-\mu)^2}{2\sigma^2}}$$

That is, as n increases $\dfrac{\sigma^2}{n}$ decreases and hence the Normal distributions will become more and more peaked and finally it coincides
with the ordinate at $x=\mu$ (degenerate). Figure 7.2 gives the distributions of the sample mean for various values of n, when the
population is Normal.

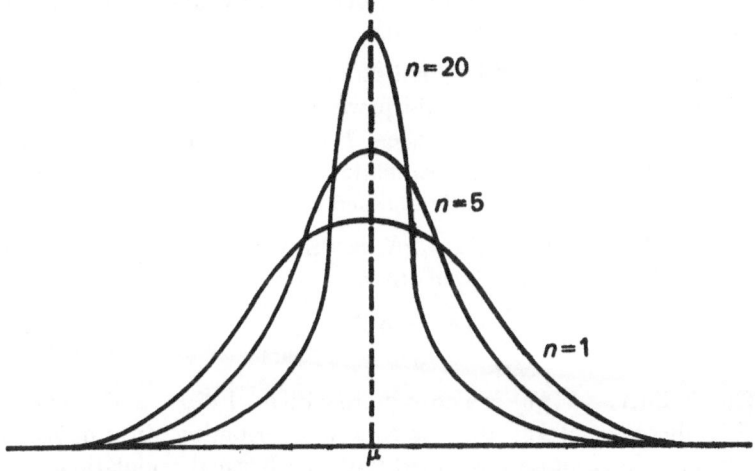

FIGURE 7.2 Distribution of $\bar{X}$, when X is $\mathcal{N}(\mu, \sigma^2)$, for different n

7.32 The Chi-Square Distribution

If $Y_1, Y_2, ..., Y_k$ are independently and identically distributed as
a standard Normal distribution $\mathcal{N}(0, 1)$ then the sum of squares
$Z = Y_1^2 + Y_2^2 + ... + Y_k^2$ is said to have a chi-square distribution
with k degrees of freedom, where k is the number of independent
standard Normal variates involved in Z. The term ' degrees of
freedom ' occurs in statistical literature in different contexts.
Here it means the number of free (independent) standard normal
variables contained in Z. In tests of hypotheses the degrees of
freedom of a hypothesis H_0 will mean the number of independent
restrictions imposed by H_0. The density function of Z is given in
Table 6.1 and which can be derived by using the result in problem
7.18. The density function is,

$$f(z) = \begin{cases} \dfrac{z^{k/2-1}\ e^{-z/2}}{2^{k/2}\ \Gamma\left(\dfrac{k}{2}\right)}, & z>0, \text{ and } k \text{ is a positive integer,} \\[4mm] 0, \text{ elsewhere,} \end{cases}$$

which is a Gamma distribution with parameter $a=k/2$ and $\beta=2$. We will state the following theorems without proofs. (These can be proved with the help of the results in problems 7.17 and 7.18).

Theorem 7.32.1 If X and Y are two independent chi-square variates with m and n degrees of freedom respectively then $X+Y$ is a chi-square variate with $m+n$ degrees of freedom. This can be extended to a number of variates.

Theorem 7.32.2 If X is a chi-square variate with m degrees of freedom and if $X+Y$ is a chi-square variate with $m+n$ degrees of freedom (d.f.) where $n\geqslant1$, then Y is a chi-square variate with n d.f. if X and Y are independent. A chi-square variate with k degrees of freedom is usually denoted as χ_k^2.

Theorem 7.32.3 If $X_1, X_2,..., X_n$ is a simple random sample from a Normal population $N(\mu, \sigma^2)$ then

$$\sum_{i=1}^{n} \frac{(X_i - \bar{X})^2}{\sigma^2} : \chi^2_{n-1}$$

That is $\Sigma (X_i-\bar{X})^2/\sigma^2$ is a chi-square with $n-1$ degrees of freedom. This theorem makes the chi-square distribution an important sampling distribution associated with the Normal population.

7.33 *The Student-t Distribution*

If X and Y are two independent variates where X is a standard Normal $N(0, 1)$ and Y is a chi-square with k degrees of freedom then

$$t = \frac{X}{\sqrt{Y/k}}$$

is distributed as a student-t with k degrees of freedom. Here the denominator of t is the square root of a chi-square variate divided by its degrees of freedom. The density function is given in Table 6.1 which is,

$$f(t) = \frac{\Gamma\left(\frac{k+1}{2}\right)}{\sqrt{k}\ \sqrt{\pi}\ \Gamma\left(\frac{k}{2}\right)} \left(1 + \frac{t^2}{k}\right)^{-\frac{(k+1)}{2}},$$

$-\infty < t < \infty$, and k is a positive integer.

Theorem 7.33.1 If $X_1, X_2, ..., X_n$ is a simple random sample of size n from a Normal population $\mathcal{N}(\mu, \sigma^2)$ then

$$t = \frac{(\overline{X} - \mu)}{S'/\sqrt{n}} = \frac{(\overline{X} - \mu) / (\sigma/\sqrt{n})}{\sqrt{S'^2 / \sigma^2}} = \frac{\Upsilon}{\sqrt{\chi^2_{n-1} / (n-1)}} ;$$

$$\Upsilon : \mathcal{N}(0, 1),$$

is distributed as a student-t with $n-1$ degrees of freedom, where,

$$S'^2 = \sum_{i=1}^{n} \frac{(X_i - \overline{X})^2}{n - 1}.$$

This theorem makes Student-t distribution an important sampling distribution. A Student-t with k degrees of freedom is usually denoted as t_k. W.S. Gosset who found out this distribution used the pen name 'Student' and hence the distribution is known as a Student-t distribution. If the sample is from a Normal population it can be shown that the sample mean and the sample variance are independently distributed. Further, this property is enjoyed only by the Normal population and hence it is a characteristic property of the Normal distribution. There are other such characteristic properties for different distributions which we will not discuss here. The degrees of freedom of the Student-t variate is the degrees of freedom associated with the chi-square which is appearing in the denominator of t.

7.34 *The F-Distribution*

Let χ_m^2 and χ_n^2 be two independent chi-square variates with m and n degrees of freedom respectively. Then the ratio,

$$F = \frac{\chi_m^2/m}{\chi_n^2/n}$$

that is, the ratio of the chi-square variates divided by their degrees of freedom, is called an F-statistic with m and n degrees of freedom, denoted by $F_{m,n}$ and its distribution is called an F-distribution. The density is given in Table 6.1 which is,

$$f(x) = \begin{cases} \dfrac{\Gamma \dfrac{(m+n)}{2}}{\Gamma \left(\dfrac{m}{2}\right) \Gamma \left(\dfrac{n}{2}\right)} \left(\dfrac{m}{n}\right)^{m/2} \dfrac{x^{m/2-1}}{\left(1 + \dfrac{m}{n}x\right)^{\left(\frac{m+n}{2}\right)}}, & x > 0, m, n \text{ are positive integers.} \\ 0, & \text{elsewhere.} \end{cases}$$

Theorem 7.34.1 If $X_1, X_2, ..., X_m$ and $Y_1, Y_2, ..., Y_n$ are two independent random samples of sizes m and n respectively, from two independent normal populations $N(\mu_1, \sigma^2)$ and $N(\mu_2, \sigma^2)$ then,

$$F_{m-1,\, n-1} = \frac{\displaystyle\sum_{i=1}^{m} (X_i - \overline{X})^2/(m-1)}{\displaystyle\sum_{i=1}^{n} (Y_i - \overline{Y})^2/(n-1)}$$

is distributed as an F with $m-1$ and $n-1$ degrees of freedom. This theorem makes an F-distribution an important sampling distribution. It may be noticed that, since

$$\sum_{i=1}^{m} \frac{(X_i - \overline{X})^2}{\sigma^2} : \chi^2_{m-1}$$

and

$$\sum_{i=1}^{n} \frac{(Y_i - \overline{Y})^2}{\sigma^2} : \chi^2_{n-1}$$

according to theorem 7.32.3, the above expression

$$\frac{\displaystyle\sum_{i=1}^{m} (X_i - \overline{X})^2/(m-1)}{\displaystyle\sum_{i=1}^{n} (Y_i - \overline{Y})^2/(n-1)} = \frac{\chi_{m-1}^2/(m-1)}{\chi_{n-1}^2/(n-1)} = F_{m-1,\, n-1}$$

where the two chi-squares are independent. Graphical representations of the Student-t distribution for a t_v, a chi-square distribution for χ_k^2, and an F-distribution for an $F_{m,\,n}$, are given in Figures 7.3, 7.4 and 7.5.

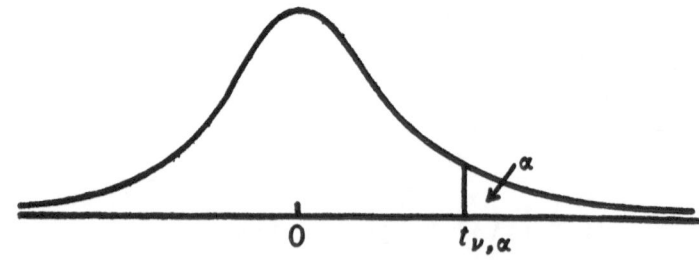

FIGURE 7.3   The distribution of a standard-t with v d. f.

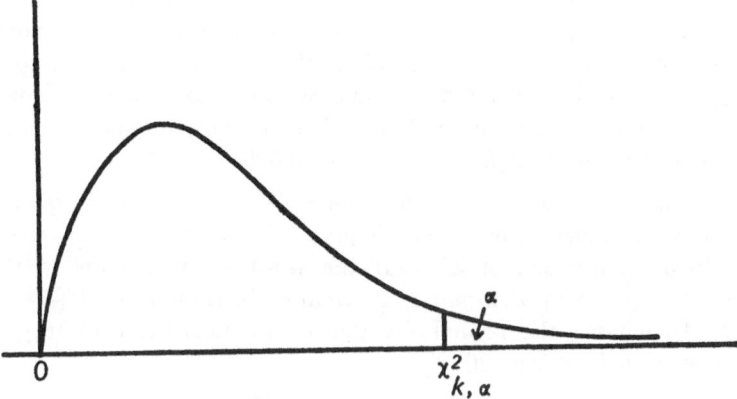

FIGURE 7.4 The distribution of a χ_k^2

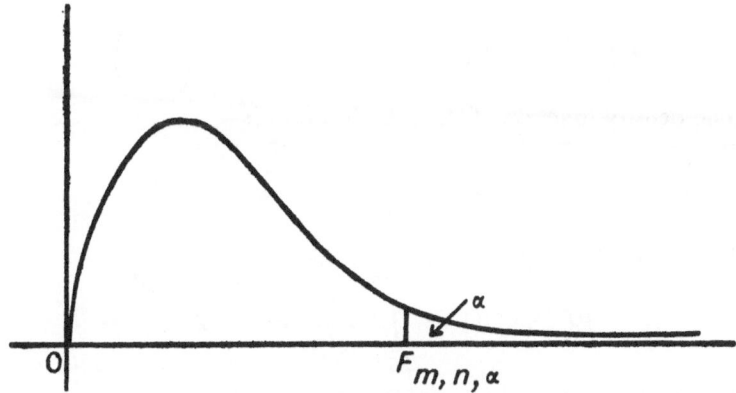

FIGURE 7.5 The distribution of an $F_{m, n}$

Here $t_{v, a}$, $\chi_{k, a}^2$, $F_{m, n, a}$ denote the points corresponding to the tail area a. That is,

$$\int_{t_{v, a}}^{\infty} f(t_v) \, dt_v = a; \quad \int_{\chi_{k, a}^2}^{\infty} g(\chi_k^2) \, d\chi_k^2 = a; \quad \int_{F_{m, n, a}}^{\infty} h(F_{m, n}) \, dF_{m, n} = a,$$

where $f(t_v)$, $g(\chi_k^2)$ and $h(F_{m, n})$ denote the density functions of a Student-t with v degrees of freedom, a chi-square with k degrees of freedom and an F with m and n degrees of freedom respectively. Tables are available for these tail areas which are known as a Student-t table, a chi-square table and an F-table respectively.

Extracts of these tables are given at the end of this book. A Student-t table gives $t_{v, a}$ for various values of v and a; a chi-square table gives $\chi_{k, a}^2$ for various values of k and a; an F-table gives $F_{m, n, a}$ for various values of m, n and a. Since there are three quantities m, n and a usually an F-table is given only for some selected values of a, for example $a=0.025$, $a=0.005$.

Example 7.3.1 A dress-maker made the following observations. The waist measurements of 16 girls of a particular age group yielded an average of 20″. If the waist measurements in this age group are approximately Normally distributed as $N(\mu=22, \sigma^2=4)$, what is the probability that for the next batch of 16 girls the average is at least 20″ ?

Solution. The parent distribution is a $N(\mu=22, \sigma^2=4)$ and the sample size is 16 and hence the distribution of the sample mean is a

$$N\left(\mu, \frac{\sigma^2}{n}\right) = N\left(\mu = 22, \frac{\sigma^2}{n} = \frac{4}{16}\right) = N\left(22, \frac{1}{4}\right).$$

The density function of the sample mean is

$$f(\bar{x}) = \frac{2}{\sqrt{2\pi}} \, e^{\dfrac{-4(\bar{x}-22)^2}{2}}$$

The required probability is,

$$P\{\bar{x} \geqslant 20\} = \int\limits_{20}^{\infty} \frac{2}{\sqrt{2\pi}} \, e^{\dfrac{-4(\bar{x}-22)^2}{2}} \, d\bar{x}$$

Now standardizing the variable, that is, putting

$$y = \frac{\bar{x} - \mu}{\sigma / \sqrt{n}} = 2(\bar{x} - 22)$$

we have,

$$dy = 2 \, d\bar{x}$$

and when $\bar{x}=20$, $y=-4$ and when $\bar{x}=\infty$, $y=\infty$, and the density function of $\bar{x}$ reduces to the density function of a $N(0, 1)$. That is,

$$P\{\bar{x} \geqslant 20\} = P\{y \geqslant -4\} = \int\limits_{-4}^{\infty} \frac{1}{\sqrt{2\pi}} \, e^{-\dfrac{y^2}{2}} \, dy$$

$$= 0.9999 \text{ approx. (From Normal tables).}$$

Example 7.3.2 The yield of a particular variety of corn is assumed to be a $\mathcal{N}(\mu, \sigma^2=25)$. A random sample of 25 experimental plots yielded an average of 22 quintals. Will you accept the hypothesis that $\mu=20$, assuming that we are ready to accept the hypothesis if the probability of getting a sample mean larger than the observed one is at least 0.05.

Solution.
$$Z = \frac{\overline{X}-\mu}{\sigma/\sqrt{n}} : \mathcal{N}(0, 1).$$

If the hypothesis is true, that is, $\mu=20$, then an observed value of z is

$$= \frac{22-20}{5/\sqrt{25}} = 2.$$

The probability of getting a sample mean larger than 22 is the same as the probability of getting a z larger than 2

$$= P\{z \geqslant 2\}$$

where Z is a standard Normal variable.

But,

$$P\{z \geqslant 2\} = 0.0227 < 0.05. \text{ (From a Normal table)}$$

Hence we cannot accept the hypothesis. Tests of Statistical Hypotheses will be discussed in detail in Chapter 9.

Example 7.3.3 A random sample of 17 salmon who could jump over a dam are caught and weighed. A sample variance of 10 units is observed. If the previous knowledge substantiates the assumption that the weight distribution is Normal $\mathcal{N}(\mu, \sigma^2)$, will you accept the hypothesis that $\sigma^2=7$ assuming that we will accept the hypothesis if the probability of getting a chi-square larger than the observed one is at least 0.2 ?

Solution. We are given the sample size n and the sample variance, and the hypothesis gives the population variance. But we know that if we have a random sample from a $\mathcal{N}(\mu, \sigma^2)$ then

$$\sum_{i=1}^{n} \frac{(X_i-\overline{X})^2}{\sigma^2} : \chi_{n-1}^2 \text{ (a chi-square with } n-1 \text{ d.f.)}$$

An observed value of,

$$S^2 = \frac{\Sigma(X_i-\overline{X})^2}{n}$$

is given to be 10 and hence an observed value of $\Sigma(X_i-\overline{X})^2$

$$= nS^2 = 17 \times 10 = 170.$$

Under the hypothesis, that is, $\sigma^2 = 7$, an observed value of a χ^2 with $n-1 = 16$ d.f. is,

$$= \frac{\Sigma(x_i - \bar{x})^2}{\sigma^2} = \frac{170}{7} = 24.3.$$

The probability of getting a χ_{16}^2 larger than 24.3

$$= P\{\chi_{16}^2 \geqslant 24.3\} < 0.2 \quad \text{(Observed from a chi-square}$$

table corresponding to 16 degrees of freedom).

Hence, according to our criterion, we cannot accept the hypothesis.

Example 7.3.4 Bags are filled with potatoes with an expected weight of 50 lbs. The automatic device used to fill the bags can only count the potatoes. A random sample of 4 bags give the following results:

$\Sigma x_i = 212$ lbs, $\Sigma x_i^2 = 11254$ lbs, where x_i denotes the weight of the ith bag, $i = 1, 2, 3, 4$. A customer picks up 4 bags at random. What is the probability that they will weigh on the average at least 55 lbs?

Solution. Here nothing is given about the distribution of weights. Assuming that the chances of getting a negative deviation from the expected weight is the same as the chances of getting a positive deviation, we may assume, without much loss of generality, that the weight distribution is $N(\mu, \sigma^2)$ with $\mu = 50$. Our inference based on the following procedure will be correct only if this assumption is justifiable. We are given Σx_i and Σx_i^2. From these we can construct the sample mean and the sample variance. A distribution connecting the population mean, the sample mean and the sample variance is the Student-t distribution. That is,

$$t_{n-1} = \frac{(\bar{x} - \mu)}{S'/\sqrt{n}}$$

where $S'^2 = \Sigma(x_i - \bar{x})^2 / (n-1)$

But $\Sigma(x_i - \bar{x})^2 = \Sigma x_i^2 - n\bar{x}^2$ and $\bar{x} = (\Sigma x_i)/n$

Therefore, $\bar{x} = \dfrac{212}{4} = 53.$

$$\Sigma(x_i - \bar{x})^2 = 11254 - 4(53)^2 = 18$$

Hence, $\dfrac{\Sigma(x_i - \bar{x})^2}{n-1} = \dfrac{18}{3} = 6.$

When $\mu=50$, an observed value of a Student-t with $n-1$ d.f. can now be found out. Hence the required probability is

$$P\{\bar{x} \geqslant 55\} = P\left\{\frac{\bar{x}-\mu}{S/\sqrt{n}} \geqslant \frac{55-50}{\sqrt{6/4}}\right\}$$

$$= P\{t_3 \geqslant 10/\sqrt{6}\} \approx 0 \text{ (from a Student-t table}$$

corresponding to 3 degrees of freedom). Hence the required probability is negligibly small.

Comment. If the population variance σ^2 was known we would have used a standard Normal Statistic instead of a Student-t Statistic. In a practical situation an appropriate statistic is selected by examining all the available information. If the normality assumption is not justifiable, for large sample cases, we can use the Central Limit Theorem to make some probability statements in such situations. In order to find justifications for the assumption of normality one has to look into the various characterization theorems for the normal distribution and examine the experimental situations.

Example 7.3.5 Two random samples of size 13 each of students are taken and their heights are measured. Assuming that the height measurement is distributed as a $N(\mu, \sigma^2)$, what is the probability that the ratio of the sample variances (first to the second) is at least 2.69?

Solution. Here nothing is known about μ and σ^2 and we have to make a decision on the ratio of the sample variances. We know that,

$$\frac{\Sigma(X_i-\bar{X})^2/(n_1-1)}{\Sigma(Y_i-\bar{Y})^2/(n_2-1)} : F_{n_1-1,\ n_2-1}$$

where $X_1, ..., X_{n_1}$ and $Y_1, ..., Y_{n_2}$ are two independent random samples from two populations $N(\mu_1, \sigma^2)$ and $N(\mu_2, \sigma^2)$. The required probability is,

$$P\left\{\frac{s_1^2}{s_2^2} \geqslant 2\cdot69\right\} = ?,$$

where $s_1^2 = \Sigma(x_i-\bar{x})^2/13$ and $s_2^2 = \Sigma(y_i-\bar{y})^2/13$

Hence, $\dfrac{s_1^2}{s_2^2} = \dfrac{\Sigma(x_i-\bar{x})^2}{\Sigma(y_i-\bar{y})^2} = \dfrac{\Sigma(x_i-\bar{x})^2/(n_1-1)}{\Sigma(y_i-\bar{y})^2/(n_2-1)} = F_{12,12}$

because n_1 and n_2 get cancelled since they are equal in this problem.

That is, the required probability

$$= P\{F_{12,\,12} \geqslant 2.69\} = 0.05 \text{ (From an } F\text{-table).}$$

Comment. From all these examples it is evident that for a particular experimental situation the appropriate statistic is selected by examining the experimental conditions, in the sense, the available information about the underlying distribution. Since a Student-t distribution approximates a standard Normal distribution for large d.f. one can look into a $\mathcal{N}(0, 1)$ table for a Student-t value for d.f. $\geqslant 30$. Similarly when the d.f. is large a chi-square variate approximates to a Normal variate. If X is a chi-square variate with k degrees of freedom then $\sqrt{2x} - \sqrt{2k}$, when k is large, approximates to a standard Normal variate. A good approximation is available for k as small as 30. In the following table we will list some important sampling distributions.

TABLE 7.1

SOME SAMPLING DISTRIBUTIONS WHEN THE SAMPLES ARE FROM A

$\mathcal{N}(\mu, \sigma^2)$

Statistic	*Distribution*
1. X_i for any i	$\mathcal{N}(\mu, \sigma^2)$
2. $X_1 + ... + X_n$	$\mathcal{N}(n\mu, n\sigma^2)$
3. $\dfrac{(X_1 + ... + X_n)}{n}$	$\mathcal{N}(\mu, \sigma^2/n)$
4. $\dfrac{(X_i - \mu)}{\sigma}$ for any i	$\mathcal{N}(0, 1)$
5. $\dfrac{(\overline{X} - \mu)}{\left(\dfrac{\sigma}{\sqrt{n}}\right)}$	$\mathcal{N}(0, 1)$
6. $\dfrac{(\overline{X} - \mu)}{\left(\dfrac{S'}{\sqrt{n}}\right)}$ where $S'^2 = \Sigma(X_i - \overline{X})^2/(n-1)$	Student-t with $n-1$ degrees of freedom (d.f.)
7. $(X_i - \mu)^2/\sigma^2$ for any i	Gamma with $\alpha = \frac{1}{2}, \beta = 2; (\chi^2_1)$
8. $\sum\limits_{i=1}^{n} (X_i - \mu)^2/\sigma^2$	Gamma with $\alpha = n/2, \beta = 2; (\chi^2_n)$

TABLE 7.1 (*Continued*)

Statistic	Distribution
9. $\Sigma(X_i-\overline{X})^2/\sigma^2$	χ^2_{n-1} (chi-square with $n-1$ d.f.)
10. $\chi^2_{k_1} + \chi^2_{k_2} +...+ \chi^2_{k_n}$, where the chi-squares are independent.	$\chi^2_{k_1+k_2+...+k_n}$
11. $\dfrac{\dfrac{\chi^2_m}{m}}{\dfrac{\chi^2_n}{n}}$ where χ^2_m and χ^2_n are independent.	$F_{m, n}$ (F with m and n d.f.)
12. $\dfrac{S_1'^2}{S_2'^2}$ where $S_1'^2 = \Sigma(X_i-\overline{X})^2/(n_1-1)$ and $S_2'^2 = \Sigma(Y_i-\overline{Y})^2/(n_2-1)$ where $X_1,..., X_{n_1}$ and $Y_1,..., Y_{n_2}$ are two random samples from two independent $N(\mu_1, \sigma^2)$ and $N(\mu_2, \sigma^2)$ or from the same $N(\mu, \sigma^2)$.	F_{n_1-1, n_2-1}

Exercises

7.21 A random sample of size 20 is taken from a $N(\mu, \sigma^2)$ with $\mu=30$ and $\sigma^2=16$. What is the probability that the sample mean will lie between 25 and 35?

7.22 From a Normal population with $\sigma^2=25$ a random sample of size 20 is taken. What is the probability that the sample mean will not differ from the population mean by more than 2 in absolute value?

7.23 If $\bar{x}$ denotes the sample mean what is the probability that $3(\bar{x}-\mu)\geqslant 4$ if a random sample of size 15 is taken from a $N(\mu, \sigma^2)$ with $\sigma^2=4$?

7.24 If a random sample of size 10 is taken from a $N(\mu, \sigma^2=25)$ what is the probability that the sample variance is larger than 36.7 or smaller than 8.3?

7.25 What is the probability that the ratio of the sample variance to the population variance will not exceed 0.51 if a random sample of size 20 is taken from a $N(\mu, \sigma^2=100)$.

7.26 A random sample of size 17 from a $N(\mu, \sigma^2)$ yielded a sample variance of 25. What is the probability that the sample mean will not differ from the population mean by 2.18 in absolute value?

7.27 If the height measurements of people in a certain age group is assumed to be a $N(\mu, \sigma^2)$ and if a random sample of 26 persons in this set have an average height of 65″ with a standard deviation of 2″, is it reasonable to take a decision that the expected height is 64″?

7.28 Two independent random samples of sizes 20 and 25 from $N(\mu_1, \sigma^2)$ and $N(\mu_2, \sigma^2)$ have means 12.45 and 10.00 and variances 16 and 18 respectively. Is it reasonable to take a decision that $\mu_1=\mu_2$ based on these observations?

7.29 If two random samples of sizes 15 and 25 are taken from a $N(\mu, \sigma^2)$ what is the probability that the ratio of the sample variances (first to the second) does not exceed 2.05?

7.30 A machine part with a specified diameter of 10 units is produced by a machine. A random sample of 5 parts have the diameter 10.01, 9.99, 9.98, 10.002, 10.001. If such samples of size 5 are taken, what is the probability that 2 out of every 3 samples will have the average diameter between 9.99 and 10.01?

7.4 CONTROL CHARTS

When large numbers of a certain item are produced in a short interval of time it is practically impossible to check each and every item for quality specifications. There will be slight variations in the quality of goods produced even if we use the best type of machinery available. Unless some process of controlling the quality is introduced the goods produced may not be useful for the consumer. For example, if bullets with a fixed diameter are produced they may not be good if the diameter goes below a certain limit or above a certain limit. In this case there are two limits and an item is tolerably good if its quality lies between these limits. In a problem of checking the fraction defectives in lots, a lot is acceptable to the customer if the fraction defective is less than a certain proportion. Here the lower bound is evidently zero. Also from the discussions so far, we know that for a Normal variate X,

$$P\left\{\mu-\frac{\sigma}{\sqrt{n}}<\bar{x}<\mu+\frac{\sigma}{\sqrt{n}}\right\}=0.68 \text{ approx.}$$

$$P\left\{\mu-\frac{2\sigma}{\sqrt{n}}<\bar{x}<\mu+\frac{2\sigma}{\sqrt{n}}\right\}=0.95 \text{ approx.}$$

$$P\left\{\mu-\frac{3\sigma}{\sqrt{n}}<\bar{x}<\mu+\frac{3\sigma}{\sqrt{n}}\right\}=0.99 \text{ approx.}$$

If nothing is known about the variate then some inequalities can be obtained by using Chebyshev's theorem. Then we can say that the above probabilities will be at least 1, 0.75, 0.9 respectively. By using these results, quality control engineers construct control charts and by using these control charts they control the quality of the outgoing items.

7.41 *Control Chart for Means*

This is a control chart based on the sample means. Figure 7.6 gives a 3σ level control chart for means. The probability of getting a sample mean between $\mu-3\sigma/\sqrt{n}$ and $\mu+3\sigma/\sqrt{n}$ is at least 0.9 (By Chebyshev's inequality) or 0.997 if X is a Normal variate.

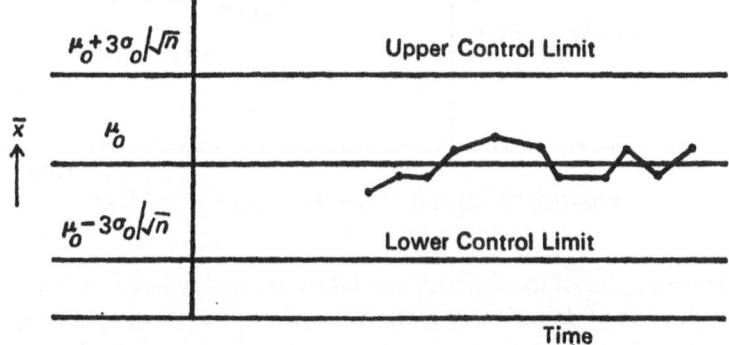

FIGURE 7.6 Control Chart for means, 3σ level

The control inspector will take random samples of a fixed size at regular intervals of time and plot the sample means. If a point falls above the upper control limit or below the lower control limit then the process is said to be *out of control* otherwise it is called *under control*. μ_0 and σ_0 are either preassigned values or values estimated from past experience, depending upon the experimental situation. μ denotes the mean value and σ^2 denotes the variance and thus $\sigma/\sqrt{n}$ denotes the standard error of the sample mean. This chart is said to be at 3σ level because the upper and lower control limts are at 3 times the standard error away from the line μ. Similarly we can use 2σ level or 1σ level depending upon the need in a particular problem.

7.42 *Control Charts for Proportions*

In a Binomial probability situation with parameters N and p we know that the expected value and the variance of the sample proportion X/N are p and $p(1-p)/N$ respectively. Figure 7.7 gives a 2σ level control chart for proportions.

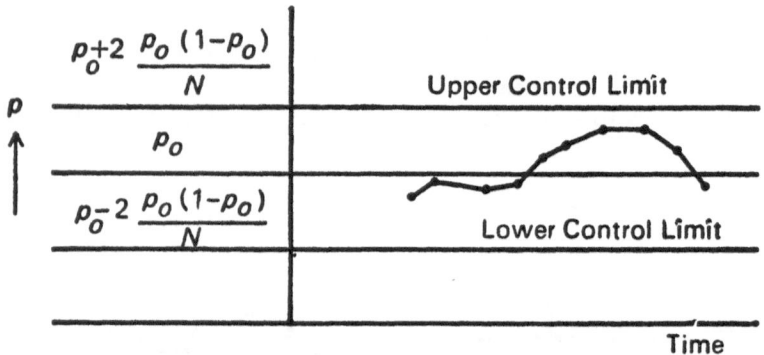

FIGURE 7.7 Control Charts for proportions, 2σ level

Here samples of fixed size N are taken at regular intervals and the proportion of defectives is plotted. Again p_0 and N are preassigned values or values estimated from previous experience. Depending upon the nature of the problem, either both upper and lower control limits are used or only one control limit is used in a particular situation.

Exercises

7.31 Taking $\mu_0 = 5$ and $\sigma_0 = 0.15$ construct a 2σ and a 3σ control charts for the means. Plot the following data and check whether the process is 'out of control' at any time. The observations are the sample means of samples of size 4 taken at half an hour interval. 4.5, 4.9, 4.87, 5, 5.11, 5.13, 5, 5.1, 5.9. 4.85, 4.84, 4.9.

7.32 Taking the values of p_0 as 0.10, construct a 2σ control chart for the proportion of defectives and plot the following data. These observations are the numbers of defectives in random samples of 20 each 1, 2, 1, 3, 5, 8, 6, 2, 4, 3, 5, 7, 10, 8, 7, 6, 1, 2, 1, 5, 1, 2.

7.5 MULTIVARIATE DISTRIBUTIONS

In this section, we shall discuss change of variables, the multinomial distributions and conditional distributions.

7.51 *Change of Variables*

The result of Section 6.5 can easily be extended to functions of k-dimensional random variables and is stated below without proof.

Theorem 7.51.1 Let $(x_1, \ldots, x_k)$ be a k-dimensional continuous s.v. with p.d.f. $f(x_1, \ldots, x_k)$. Let $y_i = \psi_i(x_1, \ldots, x_k)$, $i = 1, \ldots, k$ possess continuous first derivatives such that the Jacobian $\mathcal{J} \equiv \left| \dfrac{\partial(x_1, \ldots, x_k)}{\partial(y_1, \ldots, y_k)} \right|$ $\neq 0$ in some open domain A of the space of the x's and that $x_i = \psi_i^{-1}(y_1, \ldots, y_k)$, $i = 1, \ldots, k$ is unique. Let the range of A in the space of y's be denoted by B. Then $(y_1, \ldots, y_k)$ is a continuous s.v. having p.d.f. at a point in B given by

$$g(y_1, \ldots, y_k) = f(x_1, \ldots, x_k) \mathcal{J}.$$

Also

$$\int_A f \, dx_1 \ldots dx_k = \int_B f \cdot \mathcal{J} \, dy_1 \ldots dy_k$$

In the above results, the Jacobian $\mathcal{J}$ is to be evaluated at $x_i = \psi_i^{-1}(y_1, \ldots, y_k)$.

7.52 *The Multinomial Distribution*

This distribution is a generalization of the binomial distribution. Consider the case of N repeated independent trials where each trial can have one of the k mutually exclusive outcomes $E_1, E_2, ..., E_k$ whose respective probabilities are $p_1, p_2, ..., p_k$ with $p_i \geqslant 0$ and $p_1 + p_2 + ... + p_k = 1$. Using arguments similar to those given in Section 5.2 for binomial distribution, the probability $f(x_1, x_2, ..., x_k)$ that in N trials E_1 occurs x_1 times, E_2 occurs x_2 times, etc., is given by

$$f(x_1, x_2, ..., x_k) = \frac{N!}{x_1! \, x_2! \, ... x_k!} \, p_1{}^{x_1} \cdot p_2{}^{x_2} ... p_k{}^{x_k} \qquad (7.52.1)$$

where the x_i are arbitrary non-negative integers subject to the condition

$$x_1 + x_2 + ... + x_k = N$$

The result in (7.52.1) is called the multinomial distribution since it is the general term in the expansion of $(p_1 + p_2 + ... + p_k)^N$. For $k=2$, (7.52.1) reduces to the binomial distribution.

Example 7.52.1 A fair die is tossed 8 times. Find the probability of obtaining the faces 3 and 4 twice and each of the others once.

Solution. The required probability is

$$\frac{8!}{2! \, 2! \, (1!)^4} \cdot (\tfrac{1}{6})^2 \cdot (\tfrac{1}{6})^2 \cdot (\tfrac{1}{6})^4 = .006$$

7.53 *Conditional Distributions*

If $f(x, y)$ is the joint probability function of X and Y and if $f(x)$ and $g(y)$ are the marginal probability functions, then the *conditional probability functions*

$h(x\,|\,y=a)$ — conditional probability function of X given $Y=a$

$k(y\,|\,x=b)$ — conditional probability function of Y given $X=b$

are defined as

$$h(x\,|\,y=a) = \left[\frac{f(x, y)}{g(y)}\right]_{y=a} \quad \text{if } g(a) \neq 0$$

and

$$k(y\,|\,x=b) = \left[\frac{f(x, y)}{f(x)}\right]_{x=b} \quad \text{if } f(b) \neq 0$$

If X and Y are independent, then

$$h(x\,|\,y{=}a) = f(x) \text{ and } k(y\,|\,x{=}b) = g(y)$$

This concept can be easily extended to a general case as follows:

Let $f(x_1,\dots,x_n)$ be the probability function of $X_1,\dots,X_n$ and let $f_{1,\dots,m}(x_1,\dots,x_m)$ be the marginal probability functions of $X_1,\dots,X_m$. Then the conditional probability function of $(x_{m+1},\dots,x_n\,|\,x_1,\dots,x_m)$ is defined as $f(x_{m+1},\dots,x_n\,|\,x_1,\dots,x_m) = \dfrac{f(x_1,\dots,x_n)}{f_{1,\dots,m}(x_1,\dots,x_m)}$ provided the denominator does not vanish.

If $X_1,\dots,X_m$ and $X_{m+1},\dots,X_n$ are independent, then $f(x_{m+1},\dots,x_n\,|\,x_1,\dots,x_m) = f_{m+1,\dots,n}(x_{m+1},\dots,x_n)$.

Exercises

7.33 If $f(x,y) = e^{-x-y},\ x,y \geqslant 0$
$\qquad\qquad = 0$, otherwise,

find the joint p.d.f. of the s.v.s $U{=}X{+}Y,\ V{=}Y/X$.

7.34 If $X,\ Y,\ Z$ have the joint p.d.f. given by $f(x,y,z){=}(2\pi)^{-3/2}$ $\exp\,[-(x^2{+}y^2{+}z^2)/2],\ -\infty{<}x,y,z<\infty$; find the joint p.d.f. of $U{=}X{+}Y{+}Z,\ V{=}X{+}Y{-}2Z$ and $W{=}X{-}Y$.

7.35 The probability that a certain type of light bulb will burn out in less than 400 hours is 0.5, the probability that it will burn out in less than 700 but more than 400 hours is 0.3, and the probability that it will last more than 700 hours is 0.2. Find the probability that among 10 such light bulbs, 4 will burn out in less than 400 hours, 4 will burn out in less than 700 but more than 400 hours, while 2 will last more than 700 hours.

7.36 In rolling 12 dice, find the probability of getting each face twice.

7.37 A company produces 50% red, 30% blue and 20% green toys. In a sample of 5 toys, find the probability that 2 are red, 2 are green and 1 is blue.

7.38 Show that for a bivariate normal distribution the two conditional distributions are also normal.

Additional Exercises

7.1 Prove theorem 7.32.1.

7.2 Prove theorem 7.32.2.

7.3 Prove theorem 7.32.3.

7.4 Prove theorem 7.33.1.

7.5 Prove theorem 7.34.1.

7.6 If (X_1, X_2) is a random sample of size 2 from a $\mathcal{N}(0, 1)$, show that the sample mean and the sample variance are independently distributed.

7.7 If $\int\limits_{F_{m,n,a}}^{\infty} f(x)\,dx = a$ where $f(x)$ is the density function of an F statistic with m and n degrees of freedom show that

$$F_{m, n, a} = \frac{1}{F_{n, m, 1-a}}.$$

7.8 If X and Y are two independent s.v.s show that aX and bY are also independent where a and b are some non-zero constants.

7.9 If $X_1, \ldots, X_k$ are independent Normal variates with unit variance and if $\sum\limits_{i=1}^{n} a_i b_i = 0$, show that $a_1 X_1 + \ldots + a_k X_k$ and $b_1 X_1 + \ldots + b_k X_k$ are independent. *Hint.* For Normal variates X and Y, Cov $(X, Y) = 0$ implies independence. In general Cov $(X, Y) = 0$ does not imply independence of X and Y.

7.10 Let ρ denote the linear correlation coefficient between X and Y. That is,

$$\rho = \frac{\text{Cov } (X, Y)}{\sqrt{\text{Var } (X) \text{ Var } (Y)}}$$

where X and Y are non-degenerate. Show that if $Y = aX + b$ where a and b are constants then $|\rho| = 1$ (Hence ρ is called a *linear correlation coefficient*. There are also other types of correlation coefficients such as partial correlation, multiple correlation, curve-linear correlation, serial correlation, biserial correlation, etc.)

7.11 Evaluate the correlation between X and Y if the joint density is given as,
$$f(x, y) = \begin{cases} 2, & 0 < x < 1,\ 0 < x < y < 1 \\ 0, & \text{otherwise.} \end{cases}$$

7.12 Evaluate ρ if the joint density of X and Y is given as
$$f(x, y) = \begin{cases} e^{-x-y}, & x > 0,\ y > 0 \\ 0, & \text{elsewhere.} \end{cases}$$

7.13 Show that $\rho = 0$ when X and Y are independent. (The converse need not be true).

7.14 If the sample covariance is defined as $\displaystyle\sum_{i=1}^{n} \frac{(x_i - \bar{x})\ (y_i - \bar{y})}{n} = c_{xy}$

and if the sample correlation is defined as

$$\gamma = \frac{c_{xy}}{\sigma_x\ \sigma_y} \text{ where } \sigma_x{}^2 = \frac{\Sigma(x_i - \bar{x})^2}{n} \text{ and } \sigma_y{}^2 = \frac{\Sigma(y_i - \bar{y})^2}{n}$$

evaluate the sample correlation coefficient for the following set of observations which are height and weight measurements.

Height	10	11	12	11	12	13	15
Weight	20	22	27	23	28	31	38

7.15 Evaluate the sample correlation between x and y from the following sample which is a single sample from a bivariate population.

x	1	3	5	6	7
y	2	7	6	7	8

7.16 *The Rank Correlation.* Consider the ranks obtained by n students in 2 different subjects. (These will be numbers 1 to n in some order). Show that (1) the mean ranks are $\dfrac{n+1}{2}$ each and (2) the rank correlation

$$\gamma = 1 - \frac{6 \displaystyle\sum_{i=1}^{n} d_i{}^2}{n(n^2 - 1)}$$

where d_i is the difference of the ranks of the ith student.

7.17 Evaluate the following conditional distributions

(1) $h(x \mid y=2)$, (2) $k(y \mid x=1)$ if $f(x, y)=e^{-x-y}$, $x>0, y>0$.

7.18 Evaluate the conditional distributions $h(x \mid y = \frac{1}{2})$ and $k(y \mid x = \frac{1}{4})$ if,

$$f(x, y) = \begin{cases} 2, & 0 < x < 1, \ 0 < x < y < 1 \\ 0, & \text{otherwise.} \end{cases}$$

7.19 Evaluate the conditional probability that $P\{x \geqslant 1 \mid y = 2\}$ if the joint probability function is given as
$f(0, 0)=\frac{1}{8}, f(1, 0)=\frac{1}{8}, f(2, 0)=\frac{1}{8}, f(0, 2)=\frac{1}{8}, f(1, 2)=\frac{1}{8},$
$f(2, 2)=\frac{3}{8}$ and $f(x, y)=0$ elsewhere.

7.20 Given the joint density of X, Y and Z as

$$f(x, y, z) = \begin{cases} \frac{1}{2} x \, e^{-y-z} & \text{for } 0 < x < 2, \ y > 0, \ z > 0 \\ 0 & \text{elsewhere,} \end{cases}$$

obtain the expectation of Y given $X=x$ and $Z=z$ for some x and z.

7.21 Show that the sample correlation coefficient r is such that $-1 \leqslant r \leqslant 1$.

7.22 Show that the correlation coefficient between X and Y is such that $-1 \leqslant \rho \leqslant 1$. *Hint.* Consider $\text{Var}\left(\dfrac{X}{\sigma_1} + \dfrac{Y}{\sigma_2}\right)$ and $\text{Var}\left(\dfrac{X}{\sigma_1} - \dfrac{Y}{\sigma_2}\right)$.

7.23 *Conditional expectation.* The expectation of $\phi(X)$ in the conditional distribution of X is called the conditional expectation of $\phi(X)$. If c is a constant and if X and Y are variables, show that

(1) $E(c \mid X) = c$,   (2) $E(cY \mid X) = cE(Y \mid X)$.

7.24 Show that (1) $E((X+Y) \mid Z) = E(X \mid Z) + E(Y \mid Z)$

(2) $E(X) = E(E(X \mid Y))$ where Y is any other variate and $E(X \mid Y)$ is treated as a s.v.

(3) $\text{Var}(Y) = E(\text{Var}(Y \mid X)) + \text{Var}[E(Y \mid X)]$.

7.25 Evaluate the conditional expectation of x given $y=2$ if the joint density of X and Y is

$$f(x, y) = \begin{cases} 3x \, e^{-3y/2}, & 0 < x < 2, \ y > 0 \\ 0, & \text{elsewhere.} \end{cases}$$

7.26 Calculate the rank correlation if the following table gives the ranks of 5 students in 2 different subjects.

Ranks

Subject 1	2	1	3	5	4
Subject 2	3	4	2	1	5

7.27 Prove the central limit theorem for the cases where the M.G.F. exist.

7.28 Five test firings of a certain type of rocket, with expected range 3000 km, gives an average range of 3000 km with a standard deviation of 10 km. What is the probability that the average of the next 5 tests will be at least 3020? Assume a Normal distribution for the range of the rocket.

7.29 In problem 7.28 is it reasonable to assume that the expected range of the rocket is 3010 km?

7.30 The diameters of a random sample of 5 pipes, produced by a machine, are 0.21, 0.25, 0.27, 0.19, 0.18. Assuming that the distribution of the diameter is approximately a $\mathcal{N}(\mu, \sigma^2)$ what is the probability that the sample variance in another sample of 5 is at least 0.0005?

7.31 In problem 7.30 is it reasonable to assume that the population variance $\sigma^2 = 0.0004$?

7.32 If X and Y are independent Gamma variables show that $U = X + Y$ and $V = X/(X + Y)$ are independently distributed and further U is a Gamma and V is a Beta variates.

7.33 If random samples $X_1, ..., X_{n_1}$ and $Y_1, ..., Y_{n_2}$ are taken from two independent populations with mean values μ_1 and μ_2 and with variances σ_1^2 and σ_2^2 respectively, show that, $E(\overline{X} - \overline{Y}) = \mu_1 - \mu_2$ and $\text{Var}(\overline{X} - \overline{Y}) = \sigma_1^2/n_1 + \sigma_2^2/n_2$.

7.34 If two independent random samples of sizes 10 and 12 are taken from a $\mathcal{N}(\mu_1, \sigma_1^2)$ and a $\mathcal{N}(\mu_2, \sigma_2^2)$ respectively, where $\sigma_1^2 = 4$ and $\sigma_2^2 = 9$, what is the probability that $\overline{X} - \overline{Y}$ will not differ from $\mu_1 - \mu_2$ by 3 in absolute value?

7.35 If two random samples of size 21 each are taken from two independent $\mathcal{N}(\mu_1, \sigma^2)$ and $\mathcal{N}(\mu_2, \sigma^2)$, what is the probability that the ratio of the sample variances is at least 2.12?

7.36   If the joint density of X and Y, is

$$h(x, y) = \begin{cases} \dfrac{1}{\Gamma(\alpha)\,\Gamma(\beta)}\, x^{\alpha-1}\,(y-x)^{\beta-1}\,e^{-y}, & 0 < x < y < \infty \\ 0, & \text{elsewhere}, \end{cases}$$

show that X and Y are Gamma variates.

7.37   If $X_1, \ldots, X_n$ are independently and identically distributed as $f(x)=2$, $\frac{1}{2} < x < 1$ and $f(x)=0$, elsewhere, show that $-\log(X_1 \ldots X_n)$ has a Gamma distribution.

7.38   Show that $E([Y - \phi(x)]^2)$ is a minimum over all functions $\phi(x)$ of x when $\phi(x)=E(Y\,|\,X)$. *Hint.* Use the property that $E((Y-a)^2)$ is a minimum when $a=E(Y)$.

7.39   Let $X_1, \ldots, X_k$ have the joint p.d.f., $f(x_1, \ldots, x_k)$. The moment generating function $M_{X_1, \ldots, X_k}(t_1, \ldots, t_k)$ of $X_1, \ldots, X_k$ whenever it exists is defined as $M_{X_1, \ldots, X_k}(t_1, \ldots, t_k) = E(\exp.$ $(t_1 X_1 + \ldots + t_k X_k))$ where E denotes ' mathematical expectation '. Evaluate the moment generating function in the following cases:

(a) $f(x_1, x_2)=e^{-x_1-x_2}$, $x_1, x_2 > 0$ and $f(x_1, x_2)=0$ elsewhere.

(b) $f(x_1, x_2)=2$ for $0 < x_1 < x_2 < 1$ and zero elsewhere.

7.40   If $X_1, \ldots, X_k$ are independent stochastic variables with the moment generating functions $M_{X_j}(t)$, $j = 1, \ldots, k$ and if $Y = a_1 X_1 + \ldots + a_k X_k$ then show that $M_Y(t) = \prod_{j=1}^{k} M_{X_j}(a_j\, t)$ where $a_1, \ldots, a_k$ are constants.

7.41   By using the uniqueness of the moment generating functions obtain the density function of $Y = 2X_1 - X_2 + X_3$ if X_1, X_2, X_3 are independent normal variables with mean values 3, 8, -1 and variances 4, 9, 1 respectively.

Part III

Statistical Inference

CHAPTER EIGHT

Statistical Estimation

There are practical situations where one would like to estimate some parameters of an underlying statistical distribution. For example, a toothpaste manufacturer would like to make a statement such as his new toothpaste would reduce cavities by 21 to 49%. Here the manufacturer is interested in two numbers such as 21% and 49% such that the true proportion of reduction in cavities is somewhere between 21% and 49%. A weather bureau may forecast the temperature variation in a forthcoming month as between $a°F$ and $b°F$ and it may claim that the forecast will be correct in 95% of the cases. A design engineer would like to get an estimate of the average weight of the type of passengers who are likely to fly, when designing an aircraft. In all these problems one would like to get an estimate of an unknown quantity. In statistical estimation problems, one is interested in getting an estimate of either an unknown parameter or an unknown probability statement. Two types of estimates for a parameter are usually sought for. They are the point estimates and the interval estimates.

Let $f(x, \theta)$ denote a probability function where θ denotes a parameter. If a parameter θ is estimated by a single quantity or if a single number is given as an estimate of a parameter then the estimate is called a *point estimate* such as the estimate of the average income of people in a certain profession. If two numbers θ_0 and θ_1 are given such that the unknown parameter is said to be somewhere in the interval (θ_0, θ_1) then such an estimate is called an *interval estimate*. So the problem of estimation can be classified into *Point Estimation* and *Interval Estimation*.

8.1 The Methods of Point Estimation

There are several methods used for getting point estimates for the parameters of a given probability distribution. These are the method of moments, the method of maximum likelihood, the method of minimum chi-square, the method of minimum risk, the method of minimum dispersion, invariance method, Bayes procedure, the method of least squares and so on. We will consider the method of moments and the method of maximum likelihood. The method of Least Squares will be discussed in Chapter 11.

8.11 *The Method of Moments*

This is one of the classical methods and the motivation comes from the fact that the sample moments are in some sense estimates for the population moments. Thus, according to this principle the sample moments are equated to the corresponding population moments and by using these equations the parameters in a given population are estimated. That is, the parameters are estimated by using the relations,

$$m'_r = \mu' , \quad r=1, 2,\ldots$$

where m'_r and μ'_r are the sample and population moments about the origin respectively.

Example 8.11.1 By using the method of moments estimate the parameters in (1) Poisson distribution, (2) a $\mathcal{N}(\mu, \sigma^2)$, if $x_1, x_2,\ldots, x_n$ is an observed sample.

Solution. (1) If X is a Poisson variate with parameter λ then $E(X)=\lambda=\mu'_1$. The sample moment $m'_1=(x_1+x_2+\ldots+x_n)/n=\bar{x}$. Hence if, $\qquad\qquad m'_1=\mu'_1$

then, $\mu'_1=\lambda=m'_1=\bar{x}$.

Hence an estimate of λ is $\bar{x}$. For example, if $x_1=1$, $x_2=5$, and $x_3=4$ then λ is estimated by $(1+5+3)/3=3$.

(2) For a Normal variate X, $E(X)=\mu$ and $E(X^2)=\mu'_2$. Since there are only 2 parameters μ and σ^2 perhaps 2 equations may be sufficient to estimate μ and σ^2. That is, if,

$$m'_1=\mu'_1,$$
$$m'_2=\mu'_2,$$

then $\qquad\qquad \mu'_1=\mu=m'_1=\bar{x}$

and $\sigma^2=\mu'_2 - (\mu'_1)^2=m'_2 - (m'_1)^2= \dfrac{\Sigma x_i^2}{n} -\bar{x}^2= \dfrac{\Sigma(x_i-\bar{x})^2}{n}=s^2.$

Hence μ is estimated by $\bar{x}$ and σ^2 is estimated by the sample variance.

Comment. The estimate of a parameter θ is usually denoted by $\hat{\theta}$. Hence in the above problem

$$\hat{\lambda}=\bar{x}, \; \hat{\mu}=\bar{x}, \; \hat{\sigma}^2=\Sigma(x_i-\bar{x})^2/n=s^2.$$

Corresponding to these estimates we have the statistics, $\bar{X}$, and $S^2=\Sigma(X_i-\bar{X})^2/n$, where $(X_1, X_2,..., X_n)$ is a random sample from the corresponding population. Naturally $\bar{x}$ and s^2 can be considered to be values assumed by the statistics $\bar{X}$ and S^2. The statistics which are used to estimate a parameter are called *estimators*. That is, $\bar{X}$ and S^2 are estimators in the above problem. Hence an estimator is a s.v. whereas an estimate is a number. Estimators have their own sampling distributions.

8.12 *The Method of Maximum Likelihood*

If $X_1, X_2,..., X_n$ is a random sample from a population with probability function $f(x,\theta)$ then the joint probability function L of $X_1,..., X_n$ is,

$$L=f(x_1, \theta)\,f(x_2, \theta)...f(x_n, \theta) = \prod_{i=1}^{n} f(x_i, \theta),$$

since $X_1,..., X_n$ are independently and identically distributed. The motivation of the method of maximum likelihood comes from the desire to pick up that value for θ such that the joint probability L is a maximum for a given set of observed values for $X_1,..., X_n$. L is often known as the likelihood function. Hence the principle of maximum likelihood suggests us to maximize L and choose that value of θ which maximizes L. That is, we may choose $\hat{\theta}$ such that,

$$\left[\frac{\partial L}{\partial \theta}\right]_{\hat{\theta}} = 0$$

$$\text{and} \quad \left[\frac{\partial^2 L}{\partial \theta^2}\right]_{\hat{\theta}} < 0$$

where $\frac{\partial}{\partial \theta} L$ denotes the partial derivative of L with respect to θ.

Example 8.12.1 Obtain the maximum likelihood estimate of the parameters in a $N(\mu, \sigma^2)$.

Solution. Let $x_1, x_2, ..., x_n$ be an observed sample. The likelihood function L is,

$$L = \prod_{i=1}^{n} f(x_1, \theta) = \prod_{i=1}^{n} \frac{1}{\sigma\sqrt{2\pi}} \exp\left[\frac{-(x_i - \mu)^2}{2\sigma^2}\right]$$

$$= \frac{1}{(\sigma\sqrt{2\pi})^n} \exp\left[-\sum_{i=1}^{n} \frac{(\lambda_i - \mu)^2}{2\sigma^2}\right].$$

Since $\hat{\theta}$ which maximizes L also maximizes $\log L$, we will maximize $\log L$ for convenience.

$$\log L = -\frac{n}{2} \log \sigma^2 - n \log \sqrt{2\pi} - \sum_{i=1}^{n} \frac{(x_i - \mu)^2}{2\sigma^2}$$

Now we will differentiate $\log L$ partially with respect to μ and σ^2. That is,

$$\sum_{i=1}^{n} \frac{(x_i - \mu)}{\sigma^2} = 0, \ (1) \ \text{and} \ -\frac{n}{2\sigma^2} + \sum_{i=1}^{n} \frac{(x_i - \mu)^2}{2(\sigma^2)^2} = 0, \ (2)$$

Equation (1) yields,

$$\mu = \frac{\Sigma x_i}{n} = \bar{x}$$

and substituting this value in equation (2) we get

$$\hat{\sigma}^2 = \frac{\Sigma(x_i - \bar{x})^2}{n}.$$

Thus the maximum likelihood estimates of μ and σ^2 are the sample mean and the sample variance respectively.

Comment. The technique of differentiation is a convenient tool but it is not needed in all the problems. Equations (1) and (2) give only the turning points but we can take the second derivatives and show that at these turning points $\log L$ is a maximum, that is, L is a maximum. Whenever it is convenient, instead of maximizing L we may take a one-to-one function of L and maximize it. In the above problem if we had differentiated $\log L$ with respect to μ and σ instead of μ and σ^2 we would have arrived at the same result. It is because, if $\hat{\theta}$ is the maximum likelihood estimate (MLE) of θ then $\phi(\hat{\theta})$ is the MLE of $\phi(\theta)$ where $\phi(\theta)$ is a non-degenerate function of θ. This is due to the result that

$$\theta = \frac{\partial}{\partial\theta} L = \left[\frac{\partial L}{\partial\phi(\theta)}\right] \left[\frac{d\phi(\theta)}{d\theta}\right]$$

and if $\frac{d}{d\theta} \phi(\theta) \neq 0$ then $\frac{\partial L}{\partial\theta}$ and $\frac{\partial L}{\partial\phi(\theta)}$, vanish together. If there are a number of maxima in a particular problem the one corresponding to the largest ordinate is taken as the MLE. The principle of maximum likelihood as well as the likelihood function is controversial. Here we have taken L as the joint probability function for given sample values.

Exercises

8.1 If 2, 1, 4, 5 is an observed random sample obtain the point estimates of the parameters of (1) μ and σ^2 in a $N(\mu, \sigma^2)$; (2) α and β in a Gamma distribution, by the method of moments.

8.2 Obtain MLE of the parameters in a $N(\mu, \sigma^2)$ for the same numerical sample in 8.1.

8.3 Compare the point estimates of θ in the distribution

$$f(x) = \begin{cases} \dfrac{1}{\theta}, & 0 < x < \theta \\ 0, & \text{elsewhere} \end{cases}$$

obtained by the method of moments and by the method of maximum likelihood.

8.4 Inspection of a random sample of 20 radios from a shipment, shows that 2 are defectives. Obtain an estimate of the proportion of defectives in the shipment.

8.5 The annual expenditures in excess of Rs. 2000 of the people in a community is approximately negative exponentially distributed. A random sample of 3 shows the expenditures Rs. 10,000, Rs. 8,000 and Rs. 3,000. Obtain an estimate of the average expenditure of people in this community by the ethod of maximum likelihood.

8.6 A private phone received 2 and 3 wrong calls in two randomly selected days. Assuming a Poisson distribution obtain a point estimate of the expected number of wrong calls in 4 days.

8.7 After the appointment of a new sales-girl, the sales in a shop increased. Assuming the increase to be exponentially distributed and if the provincial sales tax is 5% obtain a point estimate of the expected increase in the sales tax return if on four randomly selected days the increases are Rs. 100, Rs. 400, Rs. 300 and Rs. 600 respectively.

8.8　Obtain the MLE of α and β in

$$f(x) = \begin{cases} \dfrac{1}{\beta - \alpha}, & \alpha < x < \beta \\ 0 & \text{elsewhere.} \end{cases}$$

Assume that a random sample is given.

8.2 Some Properties of Estimators

If the parameter θ of a rectangular population

$$f(x) = \begin{cases} \dfrac{1}{\theta}, & 0 < x < \theta \\ 0, & \text{elsewhere,} \end{cases}$$

is estimated by the two methods, the method of moments and the method of maximum likelihood, then we get the estimates $\hat{\theta} = 2\bar{x}$ by the method of moments and $\hat{\theta} =$ the largest of the observations, by the method of maximum likelihood. So it is natural to get different estimates for the same parameter, if different methods are used to estimate the parameter. Now the question arises; which method should be selected in a particular problem? This can be answered to some extent by considering some desirable properties of the estimators. But there is no unique set of desirable properties which are good for all experiments. Hence in this section we will point out some properties and the choice of a particular method is left to the experimenter. He should choose that method which will give him an estimator having the properties he is seeking for in an estimator. Hence none of the properties listed below can be said to be the most desirable property. These properties will be discussed only very briefly here.

Unbiasedness. If T is an estimator for a parameter θ and if $E(T) = \theta$ then T is said to be unbiased for θ.

Consistency. If T_n is an estimator for θ based on a sample of size n and if $T_n \to \theta$ in probability when $n \to \infty$ then T_n is said to be consistent for θ. That is,

$$P\{|T_n - \theta| \to 0\} \to 1, \quad \text{as } n \to \infty.$$

Relative Efficiency. If T_1 and T_2 are two estimators for the same parameter θ, based on a given sample of size n then

$$e = \frac{E(|T_1-\theta|^2)}{E(|T_2-\theta|^2)}$$

is said to be the relative efficiency of T_2 with respect to T_1.

That is, if $e<1$, T_1 is said to be *more efficient* than T_2. If $e<1$ only when $n \to \infty$ then T_1 is said to be *asymptotically more efficient* than T_2. There are other measures of relative efficiency which will not be discussed here.

Minimum variance unbiasedness. If T is an estimator of θ and if $E(T)=\theta$ and Var(T) is a minimum compared to any other unbiased estimator for θ then T is said to be a minimum variance unbiased estimator for θ.

Sufficiency. Let $X_1,\dots,X_n$ be the sample values and let T be an estimator for θ. If the conditional distribution of $X_1,\dots,X_n$, given $T=t$ (for some given t), is independent of θ then T is said to be sufficient for θ.

Again the term ' sufficient ' is a little misleading. If T is sufficient for θ then we say that T contains all the relevant information about θ, available from the given sample. It can be shown that when there exists a single sufficient statistic T for θ and if the range of the variable does not depend on θ, then the likelihood function L factorizes into two functions L_1 and L_2, that is,

$$L=L_1 \, (T, \, \theta) \, L_2 \, (x_1,\dots, x_n)$$

where L_1 is a function of T and θ alone and L_2 is independent of θ. If the conditional distribution of $X_1,\dots,X_n$, given T, does not depend upon θ then naturally no inference about θ can be made from this conditional distribution. In this sense we say that T contains all the relevant information about θ in the sample.

Example 8.2.1 Show that the MLE of μ in a $N(\mu, 1)$ is (1) unbiased, (2) consistent, (3) sufficient, for μ.

Solution. It is seen from example 8.11.1 that the MLE of μ in a $N(\mu, \sigma^2)$ is the sample mean $\overline{X}$.

(1) $E\overline{X}=\mu$ whatever be the population and hence $\overline{X}$ is unbiased for μ.

(2) Var $(\overline{X})=\dfrac{\sigma^2}{n}=\dfrac{1}{n}$ in this case because $\sigma^2=1$. By Chebyshev's inequality we have,

$$P\{|T-ET| \geqslant k\} \leqslant \frac{\text{Var }(T)}{k^2}$$

Hence,

$$P\{|\bar{x}-\mu| \geqslant k\} \leqslant \frac{\mathrm{Var}\ (\overline{X})}{k^2} = \frac{1}{k^2 n} \to 0 \text{ as } n \to \infty$$

Therefore,

$$P\{|\bar{x}-\mu| \leqslant k\} \to 1 \text{ as } n \to \infty, \text{ true for all } k,$$

however small it may be. That is,

$$P\{|\bar{x}-\mu| \to 0\} \to 1 \text{ as } n \to \infty.$$

(3) The likelihood function,

$$L = \prod_{i=1}^{n} f(x_i) = \left(\frac{1}{\sqrt{2\pi}}\right)^n \exp\left[-\sum_{i=1}^{n} \frac{(x_i-\mu)^2}{2}\right]$$

But $\Sigma(x_i-\mu)^2 = \Sigma(x_i-\bar{x}+\bar{x}-\mu)^2 = \Sigma(x_i-\bar{x})^2 + n(\bar{x}-\mu)^2$

That is,

$$L = \left\{\left(\frac{1}{\sqrt{2\pi}}\right)^n \exp\left[-n\frac{(\bar{x}-\mu)^2}{2}\right]\right\}\left\{\exp\left[-\sum_{i=1}^{n}(x_i-\bar{x})^2\right]\right\} = L_1 L_2$$

where

$$L_1 = \left(\frac{1}{\sqrt{2\pi}}\right)^n \exp\left[-n\frac{(\bar{x}-\mu)^2}{2}\right]$$

is a function of $\bar{x}$ and μ alone and L_2 is independent (does not contain) μ. That is, L is factorizable into two factors L_1 and L_2 satisfying the conditions for a sufficient statistic. Further, the range of x, that is, $-\infty < x < \infty$, is independent of μ and hence $\bar{x}$ is sufficient for μ.

Comment. The above discussion is not enough to understand the importance of all the properties discussed above. For further discussion, see books on Advanced Statistics.

Exercises

8.9   If $E(T) = \theta$ and Var $(T) \to 0$ as $n \to \infty$ show that T is consistent for θ.

8.10   By using the factorization principle, show that the sample mean is sufficient for the population mean in (1) Poisson distribution; (2) exponential distribution.

8.11   If X_1, X_2, X_3 is a random sample of size 3 from a population with mean value μ and variance σ^2 show that T_1 and T_2 are both unbiased for μ but T_1 is more efficient than T_2, where, $T_1=(X_1+X_2-X_3)$ and $T_2=2X_1+3X_3-4X_2$.

8.12 If T is a consistent estimator for θ show that $nT/(n+1)$ is also consistent for θ.

8.13 In a Binomial probability situation, show that the sample proportion is the MLE which is unbiased and consistent for the population proportion p.

8.14 If T is sufficient for θ show that any one-to-one function $\phi(T)$ of T is also sufficient for θ. Hence show that $2\bar{X}+3$ is sufficient for $2\mu+3$ in a $N(\mu, \sigma^2)$.

8.15 If L is the likelihood function and if θ is the parameter under consideration then

$$I=E\left(\frac{\partial}{\partial\theta}\log L\right)^2$$

is known as the Fisher's measure of information about θ, in the sample. Show that

$$I=E\left(\left(\frac{\partial}{\partial\theta}\log L\right)^2\right)=-E\left(\frac{\partial^2}{\partial\theta^2}\log L\right)=nE\left(\left[\frac{\partial}{\partial\theta}\log f(x,\theta)\right]^2\right)$$

$$=-nE\left(\frac{\partial^2}{\partial\theta^2}\log f(x,\theta)\right)$$

where $f(x,\theta)$ is the density function. (The above expected values are all assumed to exist).

8.3 Interval Estimation

If X is a $N(\mu, \sigma^2)$ then we know that the sample mean $\bar{X}$ is such that,

$$P\left\{\mu-1.96\,\frac{\sigma}{\sqrt{n}}\leqslant\bar{x}\leqslant\mu+1.96\,\frac{\sigma}{\sqrt{n}}\right\}=0.95,$$

where the number 1.96 is obtained from a Normal probability table. The same probability statement can be written in a different way because the following sets of inequalities.

$$\mu-1.96\,\frac{\sigma}{\sqrt{n}}\leqslant\bar{x}\leqslant\mu+1.96\,\frac{\sigma}{\sqrt{n}}$$

and $$\bar{x}-1.96\,\frac{\sigma}{\sqrt{n}}\leqslant\mu\leqslant\bar{x}+1.96\,\frac{\sigma}{\sqrt{n}}$$

are equivalent.   Therefore we may write

$$P\left\{\bar{x}-1.96\,\frac{\sigma}{\sqrt{n}}\leqslant\mu\leqslant x+1.96\,\frac{\sigma}{\sqrt{n}}\right\}=0.95.$$

If σ is known then when we have an observed sample we know the

values of $\bar{x}-1.96\,\dfrac{\sigma}{\sqrt{n}}$ and $\bar{x}+1.96\,\dfrac{\sigma}{\sqrt{n}}$.   That is,   we will be

able to make a probability statement that the unknown quantity

μ is such that the known interval $\left(\bar{x}-1.96\,\dfrac{\sigma}{\sqrt{n}},\,\bar{x}+1.96\,\dfrac{\sigma}{\sqrt{n}}\right)$ covers

μ with a probability of 0.95.   In an interval estimation problem,
we will be seeking for two known numbers θ_0 and θ_1 such that

$$P\{\theta_0\leqslant\theta\leqslant\theta_1\}=1-\alpha$$

in the sense that the known interval (θ_0, θ_1) covers the unknown
parameter θ with a probability of $1-\alpha$.   Here (θ_0, θ_1) is called a
$100(1-\alpha)\%$ *confidence interval* for the parameter θ and $1-\alpha$ is called
the confidence coefficient.   This has the meaning that if samples
of the same size n are taken and if the interval (θ_0, θ_1) is constructed
for every sample then in the long-run $100(1-\alpha)\%$ of the intervals
will cover the unknown parameter θ and hence with a confidence
of $100(1-\alpha)\%$ we can say that θ lies on the interval (θ_0, θ_1).   In
other words, if we make a statement that θ is on the interval
(θ_0, θ_1) then in the long-run we will be correct in $100(1-\alpha)\%$ of
the cases.   The probability statement is on the random interval
(θ_0, θ_1) rather than on the parameter θ. From the following examples,
we will see that we will be successful in constructing a confidence
interval for θ if we can get a statistic involving the sample values
and the parameter θ but whose distribution is independent of the
parameter.   Such a statistic is called a *pivotal quantity*.   In a practical problem, for convenience, we will look for a pivotal quantity.
Logically, an estimate of an interval covering the unknown
parameter θ has more meaning than a point estimate, because
$P\{(\hat{\theta}=\theta\}=0$, if the population is continuous, where $\hat{\theta}$ denotes
the estimator of θ.

Example 8.3.1 A random sample of 16 experimental animals
showed an average increase of 20 lb when a new diet is administered.   Assuming that the increase in weight is approximately
a $\mathcal{N}(\mu, \sigma^2)$ with $\sigma^2=4$ construct a 95% confidence interval for the
expected increase in weight due to this new diet.

Solution. We wish to construct a 95% confidence interval for μ in a $N(\mu, \sigma^2=4)$. We are given the sample size and the sample mean. We can get a pivotal quantity which is a function of $\overline{X}$, n and μ, namely,

$$Z = \frac{\overline{X} - \mu}{\sigma/\sqrt{n}} = \frac{\overline{X} - \mu}{2/\sqrt{n}}.$$

We know that the distribution of Z is a $N(0, 1)$ which does not contain μ. But,

$$P\left\{ -1.96 \leqslant \frac{x - \mu}{\sigma/\sqrt{n}} \leqslant 1.96 \right\} = 0.95.$$

(This is obtained from a normal probability table). That is,

$$P\left\{ -1.96 \frac{\sigma}{\sqrt{n}} \leqslant \bar{x} - \mu \leqslant 1.96 \frac{\sigma}{\sqrt{n}} \right\} = 0.95$$

or $P\left\{ x - 1.96 \frac{\sigma}{\sqrt{n}} \leqslant \mu \leqslant \bar{x} + 1.96 \frac{\sigma}{\sqrt{n}} \right\} = 0.95$

But $\bar{x}=20$, $n=16$, $\sigma=2$ and therefore,
$$P\{19.02 \leqslant \mu \leqslant 20.98\} = 0.95.$$
Hence the required confidence interval is (19.02, 20.98).

Comment. Here the confidence coefficient is 0.95. The probability statement,
$$P\{19.02 \leqslant \mu \leqslant 20.98\} = 0.95$$
does not mean that μ is a s.v. and the probability that it lies on the interval (19.02, 20.98) is 0.95. But it means that the probability that the interval (19.02, 20.98) covers μ is 0.95. In other words (19.02, 20.98) is an observed value of a random interval. In general if
$$P\{\theta_0 \leqslant \theta \leqslant \theta_1\} = 1 - \alpha$$
θ_0 and θ_1 are known as the *lower and upper confidence limits* respectively. It may be noticed that, for example,

$$P\left\{ -\infty \leqslant \frac{x - \mu}{\sigma/\sqrt{n}} \leqslant 1.65 \right\} = 0.95.$$

That is, $P\{-\infty \leqslant \mu \leqslant 20.825\} = 0.95.$

That is, we get another 95% confidence interval for μ, from the same sample, as $(-\infty, 20.825)$. In other words a $100(1-\alpha)\%$ confidence interval is not unique. It may be desirable to look for that interval which is shorter than any other interval with the same confidence coefficient. There are other desirable properties such as 'shortest on the average', 'most selective' and so on, which we will not discuss here. If a confidence interval is constructed by omitting equal tail areas then we get what is known as the *central interval*. In a symmetric distribution, it can be shown that the central intervals are the shortest. Figure 8.1 gives the central interval whereas in Figure 8.2 the interval is not central.

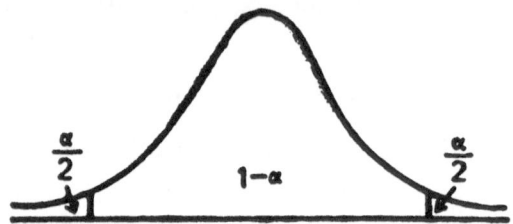

FIGURE 8.1 $100(1-\alpha)\%$ Confidence interval

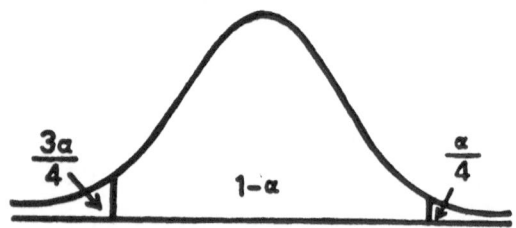

FIGURE 8.2 $100(1-\alpha)\%$ Confidence interval

Example 8.3.2 A dressmaker obtained an average of 36″ with a standard deviation of 2″ in a random sample of 9 women's bust measurements. Assuming the bust measurements to be approximately Normally distributed consturct a 99% confidence interval for the expected value of the measurement.

Solution. Here the population is a $N(\mu, \sigma^2)$. We know the sample mean and the sample variance and we want a confidence interval for μ where σ^2 is unknown. Hence the pivotal quantity is

$$t_{n-1} = \frac{\overline{X} - \mu}{S'/\sqrt{n}} \quad \text{where} \quad S'^2 = \frac{\Sigma(X_i - \overline{X})^2}{(n-1)}$$

and the distribution of t_{n-1} is a Student-t distribution with $n-1$ degrees of freedom and further, the density function does not contain μ as well as the *nuisance parameter* σ^2. Here we are interested in making a statement on μ and not on σ^2. But σ^2 is another parameter in a $N(\mu, \sigma^2)$, in which we are not interested. Such parameters are often called ' nuisance parameters ' in statistical literature. An observed value of,

$$\bar{x} = 36 \text{ and } \Sigma(x_i - \bar{x})^2/n = 4 \text{ and } n = 9.$$

Therefore,

$$s'^2 = \Sigma(x_i - \bar{x})^2/(n-1) = 9(4)/8 = 4.5.$$

Omitting tail areas $(1 - 0.99)/2 = 0.005$ at both tails we have,

$$P\left\{ -3.355 \leqslant \frac{\bar{x} - \mu}{s'/\sqrt{n}} \leqslant 3.355 \right\} = 0.99,$$

where the number 3.355 is obtained from a Student-t table corresponding to $n-1=8$ degrees of freedom. That is,

$$P\left\{ \bar{x} - 3.355 \frac{s'}{\sqrt{n}} \leqslant \mu \leqslant \bar{x} + 3.355 \frac{s'}{\sqrt{n}} \right\} = 0.99.$$

That is,

$$P\{33.655 \leqslant \mu \leqslant 38.365\} = 0.99.$$

Hence a 99% confidence interval for μ is (33.655, 38.365).

Comment. The above interval (33.655, 38.365) is the central interval and since the Student-t distribution is symmetric this interval can be expected to be the shortest interval among the 99% confidence intervals that can be constructed from the same sample by using the same principles.

Example 8.3.3 The thickness of a random sample of 20 metal pieces gives a standard deviation of 0.2 cm. Assuming an approximate Normal distribution for the thickness of such metal pieces, evaluate a 95% confidence interval for the true variance.

Solution. Here the population is a $N(\mu, \sigma^2)$ and we want a confidence interval for σ^2. We are given the sample standard deviation. A pivotal quantity connecting the sample variance and the population variance is the chi-square statistic. That is,

$$\chi_{n-1}^2 = \frac{\Sigma(X_i - \bar{X})^2}{\sigma^2}$$

which has a chi-square distribution with $n-1$ d.f. and the density is independent of σ^2. Omitting equal tail areas of 0.025 each we can get the numbers, 8.91 and 32.85 from a chi-square table corresponding to $n-1=19$ d.f., such that,

$$P\left\{ 8.91 \leqslant \frac{\Sigma(x_i - \bar{x})^2}{\sigma^2} \leqslant 32.85 \right\} = 0.95.$$

That is,

$$P\left\{\frac{1}{32.85} \leqslant \frac{\sigma^2}{\Sigma(x_i-\bar{x})^2} \leqslant \frac{1}{8.91}\right\} = 0.95.$$

Therefore,

$$P\left\{\frac{\Sigma(x_i-\bar{x})^2}{32.85} \leqslant \sigma^2 \leqslant \frac{\Sigma(x_i-\bar{x})^2}{8.91}\right\} = 0.95.$$

But we are given that, $\dfrac{\Sigma(x_i-\bar{x})^2}{n} = 0.04$ which implies that,

$$\Sigma(x_i-\bar{x})^2 = n(0.04) = 20(0.04) = 0.8.$$

Hence $P\left\{\dfrac{0.8}{32.85} \leqslant \sigma^2 \leqslant \dfrac{0.8}{8.91}\right\} = 0.95.$

That is, a 99% confidence interval for σ^2 is $\left(\dfrac{0.8}{32.85}, \dfrac{0.8}{8.91}\right)$.

Comment. Since a chi-square distribution is not symmetric, the above interval which is the central interval need not be the shortest one. For convenience, we obtained the central interval.

Example 8.3.4 In a random sample of 100 articles 10 are found to be defective. Obtain a 95% confidence interval for the true proportion of defectives in the population of such articles.

Solution. The population is Binomial with parameters N and p. We are given N and the sample proportion $(X/N) = \hat{p}$. But there is no pivotal quantity connecting p and $\hat{p}$. But we can show that,

$$\frac{\hat{p}-p}{\sqrt{\dfrac{p(1-p)}{N}}} \approx N(0, 1), \text{ when } N \text{ is large.}$$

Hence by using a Normal table we can get the number 1.96 such that

$$P\left\{-1.96 \leqslant \frac{\hat{p}-p}{\sqrt{\dfrac{p(1-p)}{N}}} \leqslant 1.96\right\} = 0.95.$$

By solving the quadratic equation $\left\{\dfrac{\hat{p}-p}{\sqrt{\dfrac{p(1-p)}{N}}}\right\}^2 = (1.96)^2$

we can get two numbers which are functions of $\hat{p}$ and these numbers will be the confidence limits. We can prove further that if N is very large then,

$$\frac{\hat{p}-p}{\sqrt{\dfrac{\hat{p}(1-\hat{p})}{N}}} \approx N(0, 1).$$

We will use this approximation to get an approximate confidence interval for p. That is,

$$P\left\{-1.96 \leqslant \frac{\hat{p}-p}{\sqrt{\dfrac{\hat{p}(1-\hat{p})}{N}}} \leqslant 1.96\right\} = 0.95,$$

or

$$P\left\{\hat{p}-1.96\sqrt{\frac{\overline{\hat{p}(1-\hat{p})}}{N}} \leqslant p \leqslant \hat{p}+1.96\sqrt{\frac{\overline{\hat{p}(1-\hat{p})}}{N}}\right\}=0.95.$$

But $\hat{p} = \dfrac{10}{100} = 0.1$ and hence a 95% confidence interval is,

$$\left(\hat{p}-1.96\sqrt{\frac{\overline{\hat{p}(1-\hat{p})}}{N}}, \hat{p}+1.96\sqrt{\frac{\overline{\hat{p}(1-\hat{p})}}{N}}\right)=(0.0412, 0.1588).$$

Comment. If N is large or when N is very large we can use the Normal approximations to construct the confidence interval but for small values of N the following technique can be used. If we need a 95% confidence interval (in general $100(1-a)\%$) and if the sample proportion is $\dfrac{k}{N}$ then we are looking for two values p_0 and p_1 such that the true proportion p is such that

$$P\{p_0 \leqslant p \leqslant p_1\}=0.95 \quad \text{(in general, } 1-a)$$

Consider the equations,

$$\sum_{x=0}^{k} \binom{N}{x} p^x(1-p)^{N-x}=0.025 \quad \text{(in general } a/2)$$

and

$$\sum_{x=k}^{N} \binom{N}{x} p^x(1-p)^{N-x}=0.025.$$

Here k is known but p is unknown. Look into the Binomial probability table for various values of p so that the above equations are satisfied. We will naturally get 2 values of p, one from each equation which will be p_1 and p_0 respectively. Since it is a discrete case we may not be able to get an exact probability statement as $a/2 = 0.025$. In that case we may modify the equations as,

$$\sum_{x=0}^{k} \binom{N}{x} p^x (1-p)^{N-x} \leqslant 0.025 \qquad \text{(a)}$$

$$\sum_{x=k}^{N} \binom{N}{x} p^x (1-p)^{N-x} \leqslant 0.025 \qquad \text{(b)}$$

It is obvious that the same value of p will not satisfy the equations (a) and (b). More accurate values can be obtained by using a more elaborate Binomial table. Usually the Binomial probabilities are tabulated for only up to $p=0.5$. By using the same table the Binomial probability for $p>0.5$ can be evaluated, because

$$\sum_{x=0}^{k} \binom{N}{x} p^x q^{N-x} = \sum_{y=N-k}^{N} \binom{N}{y} q^y (1-q)^{N-y} \text{ where } q=1-p.$$

For convenience, in the following table we give the pivotal quantities and the corresponding confidence intervals (only the central ones are given) for a number of problems. The following notations $z_{a/2}$, $t_{v,\,a/2}$, $X_{k,\,a/2}^2$, $X_{k,\,1-a/2}^2$, (which are illustrated in figures 8.3, 8.4 and 8.5),

$$S^2 = \sum_{i=1}^{n} (X_i - \overline{X})^2/n, \quad S'^2 = \sum_{i=1}^{n} (X_i - \overline{X})^2/(n-1),$$

$$S''^2 = \frac{(n_1 S_1^2 + n_2 S_2^2)}{n_1 + n_2 - 2}$$

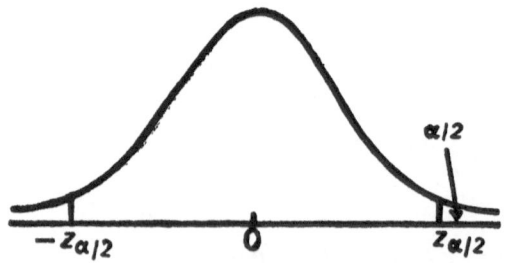

FIGURE 8.3 $\mathcal{N}(0, 1)$

where, $S_1{}^2 = \sum_{i=1}^{n_1} (X_i - \overline{X})^2/n_1$ and $S_2{}^2 = \sum_{i=1}^{n_2} (Y_i - \overline{Y})^2/n_2$, are used.

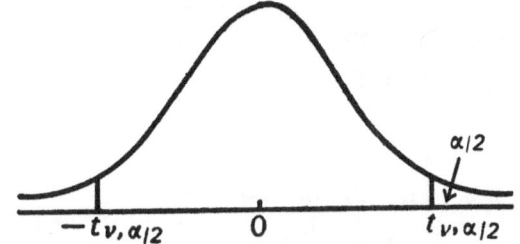

FIGURE 8.4 Student-t

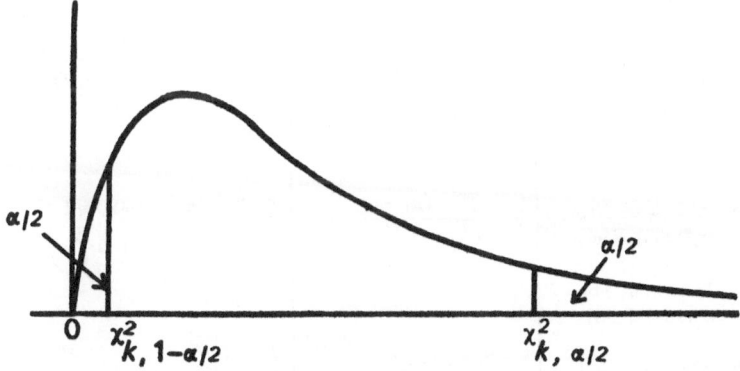

FIGURE 8.5 Chi-square

TABLE 8.1

PIVOTAL QUANTITIES AND CONFIDENCE INTERVALS

For the parameter	Pivotal quantity	$100(1-a)\%$ confidence interval
1. μ in $\mathcal{N}(\mu, \sigma^2)$ (σ^2 known)	$(\overline{X} - \mu)/(\sigma/\sqrt{n}) : \mathcal{N}(0, 1)$	$\bar{x} \pm z_{a/2} \dfrac{\sigma}{\sqrt{n}}$
2. μ in $\mathcal{N}(\mu, \sigma^2)$ (σ^2 unknown but the sample size n is large)	$(\overline{X} - \mu)/(S/\sqrt{n}) \approx \mathcal{N}(0, 1)$	$\bar{x} \pm z_{a/2} \dfrac{s}{\sqrt{n}}$
3. μ in $\mathcal{N}(\mu, \sigma^2)$ (σ^2 unknown, n small)	$(\overline{X} - \mu)/(S'/\sqrt{n}) : t_{n-1}$	$\bar{x} \pm t_{n-1,\,a/2} \dfrac{s'}{\sqrt{n}}$

TABLE 8.1 (*continued*)

For the parameter	*Pivotal quantity*	$100(1-\alpha)\%$ *confidence interval*
4. $\mu_1-\mu_2$ in independent $\mathcal{N}(\mu_1,\sigma_1^2)$ and $\mathcal{N}(\mu_2,\sigma_2^2)$ (σ_1^2 and σ_2^2 known)	$\dfrac{(\bar{X}-\bar{Y})-(\mu_1-\mu_2)}{\left(\dfrac{\sigma_1^2}{n_1}+\dfrac{\sigma_2^2}{n_2}\right)^{\frac{1}{2}}}$ $:\mathcal{N}(0,1)$	$\bar{x}-\bar{y}\pm z_{\alpha/2}\left(\dfrac{\sigma_1^2}{n_1}+\dfrac{\sigma_2^2}{n_2}\right)^{\frac{1}{2}}$
5. $\mu_1-\mu_2$ in $\mathcal{N}(\mu_1,\sigma_1^2)$ and $\mathcal{N}(\mu_2,\sigma_2^2)$ (σ_1^2 and σ_2^2 unknown, large sample sizes)	$\dfrac{(\bar{X}-\bar{Y})-(\mu_1-\mu_2)}{\left(\dfrac{S_1^2}{n_1}+\dfrac{S_2^2}{n_2}\right)^{\frac{1}{2}}}$ $\approx \mathcal{N}(0,1)$	$\bar{x}-\bar{y}\pm z_{\alpha/2}\left(\dfrac{s_1^2}{n_1}+\dfrac{s_2^2}{n_2}\right)^{\frac{1}{2}}$
6. $\mu_1-\mu_2$ in $\mathcal{N}(\mu_1,\sigma_1^2)$ and $\mathcal{N}(\mu_2,\sigma_2^2)$ $\sigma_1=\sigma_2=\sigma$, σ-unknown, $n_1+n_2-2>30$	$\dfrac{(\bar{X}-\bar{Y})-(\mu_1-\mu_2)}{S''\left(\dfrac{1}{n_1}+\dfrac{1}{n_2}\right)^{\frac{1}{2}}}$ $\approx \mathcal{N}(0,1)$	$\bar{x}-\bar{y}\pm z_{\alpha/2}\ s''\left(\dfrac{1}{n_1}+\dfrac{1}{n_2}\right)^{\frac{1}{2}}$
7. $\mu_1-\mu_2$ in $\mathcal{N}(\mu_1,\sigma_1^2)$ and $\mathcal{N}(\mu_2,\sigma_2^2)$ $\sigma_1=\sigma_2=\sigma$ unknown, n_1+n_2-2 small ($\leqslant 30$)	$\dfrac{(\bar{X}-\bar{Y})-(\mu_1-\mu_2)}{S''\left(\dfrac{1}{n_1}+\dfrac{1}{n_2}\right)^{\frac{1}{2}}}$ $:t_{n_1+n_2-2}$	$\bar{x}-\bar{y}\pm t_{n_1+n_2-2,\,\alpha/2}$ $\times\ s''\left(\dfrac{1}{n_1}+\dfrac{1}{n_2}\right)^{\frac{1}{2}}$
8. σ^2 in $\mathcal{N}(\mu,\sigma^2)$	$n\,B^2/\sigma^2:\chi_{n-1}^2$	$\left(\dfrac{n\,s^2}{\chi_{n-1,\,\alpha/2}^2},\ \dfrac{n\,s^2}{\chi_{n-1,\,1-\alpha/2}^2}\right)$
9. σ in $\mathcal{N}(\mu,\sigma^2)$	''	(Take the square root of the numbers)
10. p in a Binomial situation (large N, $Np>5$, $Np(1-p)>5$)	$\dfrac{\hat{p}-p}{\sqrt{\hat{p}(1-\hat{p})/N}}\approx \mathcal{N}(0,1)$	$\hat{p}\pm z_{\alpha/2}\ \sqrt{\hat{p}(1-\hat{p})/N}$
11. p_1-p_2 in two independent Binomial populations (large number of trials each)	$\dfrac{(\hat{p}_1-\hat{p}_2)-(p_1-p_2)}{\left[\dfrac{\hat{p}_1(1-\hat{p}_1)}{N_1}+\dfrac{\hat{p}_2(1-\hat{p}_2)}{N_2}\right]^{\frac{1}{2}}}$ $\approx \mathcal{N}(0,1)$	$\hat{p}_1-\hat{p}_2\pm z_{\alpha/2}$ $\times\left[\dfrac{\hat{p}_1(1-\hat{p}_1)}{N_1}+\dfrac{\hat{p}_2(1-\hat{p}_2)}{N_2}\right]^{\frac{1}{2}}$

In the Binomial case when N is small, see the comments in Example 8.3.4 and also see problem 8.21 in the additional set of problems at the end of this chapter.

Exercises

8.16 For a $N(\mu, \sigma^2)$ with $\sigma^2=4$, construct a 95% confidence interval for $2\mu+3$ if a random sample size of 25 gives a sample mean of 20.

8.17 The average annual income of a random sample of 25 citizens in a country is $10,000 with a standard deviation of $200. Assuming a Normal distribution, obtain a 90% interval estimate for the average income of the citizens in that country.

8.18 A random sample of 35 days shows an average daily sale of $50 with a standard deviation of $10, in a particular hot dog shop. Assuming a Normal distribution, construct a 95% interval estimate for the expected sale per day.

8.19 Ten bullets from an enemy gun show an average diameter of 5 units with a standard deviation of 0.02 units. Obtain a 99% confidence interval for the diameter of the enemy gun barrel, taking the diameter of the gun barrel as $\mu+0.01$ when μ is the expected diameter of the bullet. (Assume a Normal distribution). Is it possible to get an interval estimate if (1) only one bullet is available, (2) only one bullet is available but the true variance is known?

8.20 A farmer can sell his chicken at the rate of $0.25 per pound. He picked up 40 chickens at random and weighed. An average of 4 lb with a standard deviation of 0.5 lb is observed. Construct a 99% confidence interval for his expected income if he has 10,000 chickens and if the assumption of an underlying Normal distribution is justifiable.

8.21 Independent random samples of 10 girls and 12 boys in a certain age-group have the average weights 104 lb and 103 lbs with standard deviation 1.1 lbs and 1.2 lb respectively. Construct a 95% confidence interval for the expected difference in the weights, if the assumption of Normal populations with the same variance is appropriate.

8.22 A random sample of 40 test runs shows the average output of 30 units by machine A and 35 units by machine B with standard deviations 5 as 7 respectively. Under the assumption of Normality, construct a 99% confidence interval for the expected difference in the outputs.

8.23   Independent random samples of 50 and 70 experimental animals under 2 different diets show the average weights of 100 lb and 80 lb with standard deviations of 5 lb and 2 lb respectively. Assuming Normality, construct a 95% interval estimate for the expected difference in the weights. *Hint.* Use a Normal approximation.

8.24   A random sample of 100 sea-gull eggs collected from an island shows that 10 of them will not hatch. If there are 10,000 eggs on that island, obtain a 99% interval estimate for the expected number of chicks.

8.25   A random sample of 40 microscopes from a shipment shows that 5 of them do not meet the quality specifications. Construct a 99% interval estimate for the expected number of defectives in a shipment of 10,000 such microscopes.

8.26   Two different drugs are administered on 2 different sets of 100 patients suffering from the same disease. Their survival rates are found to be 90 and 95 respectively. Construct a 99% confidence interval for the expected difference in the survival rates.

8.27   From a shipment of oranges 15 of them are taken at random and 2 of them are found to be spoiled. Construct a 95% interval estimate for the true proportion of spoiled oranges in the shipment.

8.28   A random sample of 10 suburban housewives shows an average weight of 135 lb with a standard deviation of 5 lb. Assuming Normality obtain a 99% interval estimate for the true variance σ^2.

8.29   A random sample of 12 married teenage girls shows an average I.Q. of 90 with a standard deviation of 2. Assuming a Normal distribution for the I.Q.'s construct a 99% confidence interval for 2σ where σ^2 is the true variance.

8.30   The thickness of a random sample of 20 metal plates shows a standard deviation of 0.2 mm. Assuming the thickness to be approximately a $N(\mu, \sigma^2)$ obtain a 95% interval estimate for σ.

Additional Exercises

8.1   Obtain the MLE of the parameters in a Gamma distribution .

8.2 Obtain the estimates of the parameters of the distributions in Table 5.1 by (1) method of moments, (2) the method of maximum likelihood.

8.3 Obtain the estimates of the parameters of the distributions in Table 6.1, by (1) the method of moments, (2) the method of maximum likelihood.

8.4 Obtain an unbiased estimator of p^2+2 where p is the parameter of a Binomial distribution, with N known. *Hint.*: Evaluate the second factorial moment.

8.5 If $E(T)=\theta$ and $\dfrac{\partial}{\partial\theta}\log L=k\,(T-\theta)$ then it can be shown that T is a minimum variance unbiased estimator (MVUE) for θ and in this case $k=1/\mathrm{Var}\,(T)$. Show that the sample mean is the MVUE for λ in a Poisson distribution with parameter λ and evaluate the variances of the estimator.

8.6 Show that, if $\overline{X}$ is the sample mean, (1) $5\overline{X}+7$ is a sufficient statistic for μ in a $\mathcal{N}(\mu,\,1)$. (2) $\overline{X}$ and $\dfrac{n+a}{n+b}\,\overline{X}$, $(a,\,b>0$ and finite) are consistent for μ, in a $\mathcal{N}(\mu,\,1)$.

8.7 (a) Show that the sample mean is an unbiased and consistent estimator of $\theta+\tfrac{1}{2}$ for the following distribution.
$$f(x) = \begin{cases} 1, & \theta<x<\theta+1 \\ 0, & \text{elsewhere.} \end{cases}$$
(b) An unbiased estimator can be meaningless sometimes: Show that $(-4)^X$ is unbiased for $e^{-5\lambda}$ where λ is a Poisson parameter and X follows a Poisson distribution. (Note that $e^{-5\lambda}$ is never negative).

8.8 By using Schwartz's inequality or otherwise, show that
$$[\mathrm{Cov}(X,\,Y)]^2 \leqslant \mathrm{Var}\,(X)\,\mathrm{Var}\,(Y).$$

8.9 *Cramer-Rao Inequality.* Let $ET=\theta+b(\theta)$ and let $b'(\theta)=\dfrac{d}{d\theta}b(\theta)$ then it can be shown, under some general conditions, that,
$$\mathrm{Var}\,(T)\geqslant[1+b'(\theta)]^2/I$$
where $I=\left(\dfrac{\partial}{\partial\theta}\log L\right)^2$. Prove this result under the conditions that all the expected values appearing in the inequality exist and the range of X is independent of θ. [*Note.* This I is known as Fisher's measure of 'Information' about θ.]

8.10 Give an example of an estimator for the parameter θ of an exponential distribution, which is (1) not unbiased; (2) not consistent; (3) not sufficient, for θ.

8.11 By using the approximation that $\dfrac{\hat{p}-p}{\sqrt{p(1-p)/N}} \approx \mathcal{N}(0, 1)$ write down a $100(1-\alpha)\%$ confidence interval for p, by solving the quadratic equation,

$$\left[\frac{\hat{p}-p}{\sqrt{p(1-p)/N}}\right]^2 = (\pm z_{\alpha/2})^2.$$

8.12 If $X_1, X_2, ..., X_n$ is a random sample from a $\mathcal{N}(\mu, \sigma^2)$ it can be shown that,

$$S' = [\Sigma(X_i - \bar{X})^2/(n-1)]^{1/2}$$

is approximately normally distributed with mean σ and with variance $\sigma^2/2n$. Construct a $100(1-\alpha)\%$ confidence interval for σ by using this approximation.

8.13 In a random sample of 10 business executives 2 of them have stomach ulcer. Construct a 90% interval estimate for the true proportion of business executives who have stomach ulcer.

8.14 Construct the interval estimate in problem 8.13 if the following observations are made (1) 5 in a random sample of 100 have stomach ulcer; (2) 5 in a random sample of 30 have stomach ulcer.

8.15 Construct a 90% interval estimate for the parameter λ of a Poisson distribution based on a random sample $x_1,...,x_n$ when (1) n is large, (2) n is small.

8.16 Twenty measurements of specific gravity of a certain metal shows an average specific gravity of 2.5 with a standard deviation of 0.01. Construct a 99% interval estimate for the true standard deviation.

8.17 The following are the lengths of 8 nails produced by a machine: 1.00, 1.10, 1.12, 1.15, 1.14, 1.10, 1.11, 1.08. Assuming an underlying Normal distribution construct a 95% interval estimate for (1) the expected length, (2) the true standard deviation.

8.18 A lawn-mower is test run five times on one litre of fuel. The following durations are recorded: 101, 102, 101, 98, 99. Construct a 90% interval estimate for (1) the expected duration, (2) the true standard deviation of duration (Assume normality).

8.19 I.Q's of 40 boys and 45 girls give the averages 110 and 108 with standard deviations 1 and 2 respectively. Construct a 90% interval estimate for the expected difference in the I.Q's of boys and girls (Assume normality).

8.20 Two random samples of 10 and 12 people in two different professions have average annual incomes of $10,000 and $15,000 with standard deviations $500 and $200 respectively. Construct a 95% interval estimate for the expected difference in the incomes, under the assumption of normality (with population variances equal).

8.21 *A general procedure.* Let $\hat{\theta}$ be an estimator with the density $g(\hat{\theta}, \theta)$. Then we can determine two quantities h_1 and h_2 such that,

$$P[h_1(\theta) < \hat{\theta} < h_2(\theta)] = \int_{h_1(\theta)}^{h_2(\theta)} g(\hat{\theta}, \theta)\,d\hat{\theta} = 1 - a,$$

where h_1 and h_2 depend on θ. That is, for given values of θ, h_1 and h_2 are completely determined. If $\hat{\theta}$ is plotted against θ then $h_1(\theta)$ and $h_2(\theta)$ will be two curves on this graph. Corresponding to an observed sample, $\hat{\theta}$ is available, say, $\hat{\theta}'$. Then the line $\hat{\theta} = \hat{\theta}'$ cuts the curves $h_2(\theta)$ and $h_1(\theta)$ at two points and the corresponding points θ_0 and θ_1 on the x-axis are such that,

$$P\{\theta_0 < \theta < \theta_1\} = 1 - a$$

and thus (θ_0, θ_1) is a $100(1 - a)$ % confidence interval for θ, because

$$\int_{h_2(\theta)}^{\infty} g(\hat{\theta}, \theta)\,d\hat{\theta} = a/2 = \int_{\hat{\theta}'}^{\infty} g(\hat{\theta}, \theta)\,d\hat{\theta}$$

and

$$\int_{-\infty}^{h_1(\theta)} g(\hat{\theta}, \theta)\,d\hat{\theta} = a/2 = \int_{-\infty}^{\hat{\theta}'} g(\hat{\theta}, \theta_1)\,d\hat{\theta}.$$

In the discrete case replace the integrals by sums and ($= a/2$) by ($\leqslant a/2$). By using this technique construct confidence intervals for (a) Binomial parameter p, (b) Poisson parameter λ, (c) exponential parameter θ, in small samples.

Tests of Statistical Hypotheses

In Chapter 8 we considered the problem of estimating the parameters in a statistical distribution. Here we will consider another inference problem, namely, the problem of testing statistical hypotheses. This is a remarkable aspect of Statistics but at the same time this has led to many misinterpretations and misuses of statistical methods. Some people claim that anything and everything can be proved by the methods of Statistics. But it should be pointed out that we do not prove anything by Statistical Inference, but we will only make some probability statements regarding some unknown statistical quantities. We are familiar with different types of hypotheses all of which are not statistical hypotheses.

9.1 STATISTICAL AND NON-STATISTICAL HYPOTHESES

Here we will examine a few statements which the laymen call hypotheses and see which ones can be considered to be statistical hypotheses.

1. A physicist claims that every particle attracts every other particle.
2. A biologist hypotheses that there is life on Mars.
3. A religious preacher asserts that the only way to go to heaven is through his religion.
4. A farmer claims that fertilizer A is better than fertilizer B, as far as the yield of corn is concerned.
5. A toothpaste manufacturer claims that his toothpaste reduces cavities by 21 to 49%.
6. A bird watcher claims that birds on the average lay more eggs in Quebec than in Ontario.

The first two hypotheses are conjuctures just like many other hypotheses in physical sciences. A conjecture can be proved to be faulty if a counter-example is given. If two particles are found which do not attract each other then the conjecture is disproved. Since the third one involves a number of undefined quantities we do not even take it as a conjecture. The hypotheses 4, 5 and 6 say something about the behaviour of stochastic variables. If a hypothesis is concerned with the behaviour of a s.v. then it is called a *statistical hypothesis*. All other hypotheses are called non-statistical. Sometimes it is possible to convert a non-statistical hypothesis into a statistical one by inducing our ignorance, regarding the various factors affecting the hypothesis, on the experimental set-up. There are some problems in Astronomy which fall into this category which we will not discuss here. If X and Y denote the effects of the two fertilizers (may be measured in terms of the yield of corn) then the 4th hypothesis says that $EX > EY$. This hypothesis says something about the parameters in statistical distributions.

Parametric Hypotheses. A parametric statistical hypothesis is a restriction on an estimable (for which unbiased estimators exist) parametric function.

Here ' restriction ' means that the parameters are constrained by the hypothesis or the hypothesis will enable us to specify or eliminate a number of parameters. For example, if we have a hypothesis that $\mu = 10$, where μ is the parameter in $\mathcal{N}(\mu, \sigma^2)$ then the hypothesis is a parametric hypothesis. If we have the hypothesis that the weight of a person is independent of (has nothing to do with) his intelligence, or a set of numbers are normally distributed and so on, then these types of hypotheses are not basically concerned with restrictions on estimable parametric functions. Such hypotheses are called *non-parametric hypotheses*. In this chapter we will discuss parametric hypotheses and in the next chapter we will discuss some non-parametric and *distribution-free* (with no basic assumption of an underlying distribution) tests.

9.11 *Simple and Composite Hypotheses*

If a statistical hypothesis specifies a distribution completely (functional form as well as the parameters) then it is called a *simple hypothesis* and otherwise it is called *composite*. If the hypothesis is on a parameter θ, say $\theta = 2$ and if this hypothesis is tested, naturally, we will be testing this against an alternative hypothesis that either

$\theta \neq 2$, or $\theta > 2$ or $\theta < 2$. Evidently an alternative hypothesis is implied in every hypothesis that is tested. A hypothesis that is tested by using a random sample from some population is called a *null hypothesis*. The following is a list of some simple and composite hypotheses where H_0 denotes the null hypothesis (hypothesis under consideration) and H_1 denotes the alternative (hypothesis against which H_0 is tested). Let us assume that a probability function $f(x, \theta)$ is known except for the parameter θ and we are testing a hypothesis on θ. θ_0 and θ_1 denote specified values or given numbers.

H_0: $\theta = \theta_0$ (simple), H_1: $\theta = \theta_1$ (simple)
H_0: $\theta = \theta_0$ (simple), H_1: $\theta < \theta_0$ (composite, one-sided alternative)
H_0: $\theta = \theta_0$ (simple), H_1: $\theta \neq \theta_0$ (composite, two-sided alternative)
H_0: $\theta \leqslant \theta_0$ (composite), H_1: $\theta > \theta_0$ (composite).

For example if we test a hypothesis that H_0: $\mu = 5$ in a $\mathcal{N}(\mu, 1)$ then the hypothesis H_0 is simple because already the functional form is known (Normal) and H_0 now specifies the only unknown parameter and hence H_0 is a simple hypothesis whereas if the population is $\mathcal{N}(\mu, \sigma^2)$ where μ and σ^2 are unknown, and if H_0: $\mu = 5$ then H_0 is not simple because even though the functional form is known to be normal and μ is specified by H_0, since σ^2 is unknown, $\mathcal{N}(\mu, \sigma^2)$ is not completely specified. Hence this is a composite hypothesis. In a $\mathcal{N}(\mu, \sigma^2)$ the hypothesis H_0: $\mu = -6$ and $\sigma^2 = 4$ is a simple hypothesis.

9.2 Type I and Type II Errors

When a hypothesis is tested and a decision is taken either to accept it or to reject it then one of two types of errors is possible. That is the error of rejecting a correct hypothesis or accepting a wrong hypothesis. These are called type I and type II errors respectively.

Hypothesis / Decision	H_0 is true	H_0 is not true
Accept H_0	Correct decision	Type II error
Reject H_0	Type I error	Correct decision

The probabilities of type I and type II errors are denoted by α and β respectively. Consider a Normal distribution $\mathcal{N}(\mu, 1)$. Consider the hypothesis H_0: $\mu = 8$ and H_1: $\mu = 12$. Suppose that the hypothesis is accepted based on the following criterion: A single observation from $\mathcal{N}(\mu, 1)$ is taken and if that observation is less than or equal to 9 the hypothesis H_0 is accepted otherwise H_0 is rejected and H_1 is accepted.

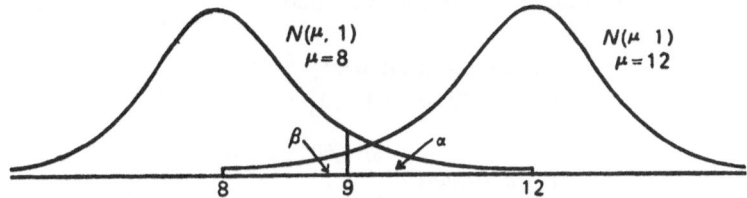

FIGURE 9.1   Probabilities α and β of type I and type II errors

In this example,

α = Probability of type I error
= Probability of rejecting H_0 when it is true
= $P\{x \geqslant 9$ given H_0 is true$\}$
= $P\{x \geqslant 9 \,|\, \mathcal{N}(\mu = 8, \sigma^2 = 1)\}$
$$= \int_9^\infty \frac{1}{\sqrt{2\pi}} e^{-(x-8)^2/2} \, dx = \int_1^\infty \frac{1}{\sqrt{2\pi}} e^{-t^2/2} \, dt = 0.1586$$

β = Probability of type II error
= Probability of accepting H_0 when it is not true
= Probability of accepting H_0 when H_1 is true (since we assumed that rejection of H_0 is equivalent to acceptance of H_1 and vice versa).
= $P\{x < 9 \,|\, \mathcal{N}(\mu = 12, \sigma^2 = 1)\}$
$$= \int_{-\infty}^9 \frac{1}{\sqrt{2\pi}} e^{-(x-12)^2/2} \, dx = \int_{-\infty}^{-3} \frac{1}{\sqrt{2\pi}} e^{-t^2/2} \, dt = 0.0014.$$

Example 9.2.1   The daily consumption of milk in a particular township is assumed to be approximately exponentially distributed. Suppose that a hypothesis H_0: expected consumption is 10,000 gallons, is tested against the hypothesis that it is 20,000. Suppose that the criterion is as follows: A day is selected at random. If the consumption on that day is 16,000 gallons or more H_0 is rejected in favour of H_1. Evaluate α and β.

Solution. The criterion is 'reject H_0 if $x > 16,000$, otherwise accept H_0', where x denotes the consumption on any particular day. Since X is distributed as,

$$f(x) = \begin{cases} \dfrac{1}{\theta} e^{-x/\theta}, & x > 0 \\ 0, & \text{elsewhere,} \end{cases}$$

where $\theta = E(X)$, the hypotheses are,

$H_0: \theta = 10,000$ and $H_1: \theta = 20,000$

Hence,

$$\begin{aligned} a &= P\{\text{Reject } H_0 \,|\, H_0 \text{ is true}\} \\ &= P\{x \geqslant 16,000 \,|\, \theta = 10,000\} \\ &= \int_{16,000}^{\infty} \frac{1}{10,000} e^{-x/10,000} \, dx = e^{-1.6} \\ \beta &= P\{\text{accept } H_0 \,|\, H_1 \text{ is true}\} \\ &= P\{x \leqslant 16,000 \,|\, \theta = 20,000\} \\ &= \int_0^{16,000} \frac{1}{20,000} e^{-x/20,000} \, dx = 1 - e^{-0.8} \end{aligned}$$

Example 9.2.2 Let p denotes the probability of getting a head when a given coin is tossed once. Suppose that the hypothesis $H_0: p = \frac{1}{2}$ is rejected in favour of $H_1: p = 0.6$ if 10 trials result in 7 or more heads. Calculate the probabilities of type I and type II errors.

Solution.
$$\begin{aligned} a &= P\{\text{reject } H_0 \,|\, H_0 \text{ is true}\} \\ &= P\{x \geqslant 7 \,|\, p = \tfrac{1}{2}\} \\ &= \sum_{x=7}^{10} \binom{10}{x} (\tfrac{1}{2})^x (\tfrac{1}{2})^{10-x} = 0.1719 \\ \beta &= P\{\text{accept } H_0 \,|\, H_1 \text{ is true}\} \\ &= P\{x \leqslant 6 \,|\, p = 0.6\} \\ &= \sum_{x=0}^{6} \binom{10}{x} (0.6)^x (0.4)^{10-x} = 0.6177 \end{aligned}$$

Comment. In the above discussions we assumed that acceptance of H_0 is equivalent to rejection of H_1 and vice versa. Throughout our discussion we will assume this unless otherwise stated. These are known as two-decision problems. There are other types of decision-making. The probabilities of type I and type II errors are also known as the sizes of the respective errors. That is,

Type I error   size a.   Type II error   size β.

9.21 *The Critical Region*

In example 9.2.1 the null hypothesis is rejected when the observed value or the sample point falls in the interval (16,000, ∞). In this problem, a decision is made by taking only a single observation. The set of all possible values the sample point can assume (the sample space) is the interval (0, ∞).

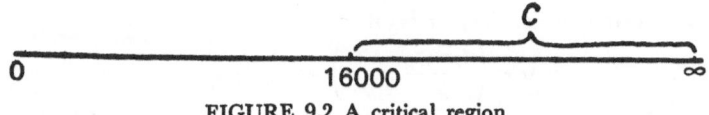

FIGURE 9.2 A critical region

If the sample point falls in C in Figure 9.2 then the hypothesis H_0 is rejected. Thus when a decision is taken, the sample space is partitioned into regions C and $\overline{C}$, such that the hypothesis is rejected, if the sample point falls in C and this C is called the *critical region*.

Critical region. It is the region in the sample space where the null hypothesis H_0 is rejected.

But the probability of rejecting a null hypothesis when it is true

$$=a=P\{x\epsilon C\,|\,H_0\}$$

where x denotes the sample point and C denotes the critical region. a is called the size of the *critical region*. In general if a decision is made by using a sample of size n, we may represent the sample space S by a Venn diagram as in Figure 9.3 where C denotes the critical region.

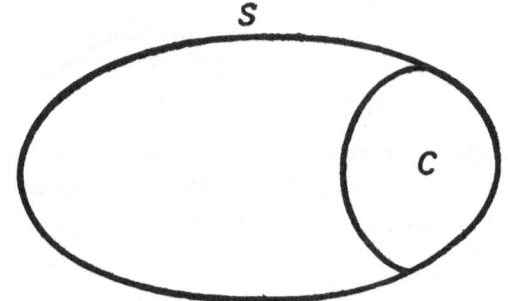

FIGURE 9.3   A sample space S and a critical region C

Size of $C=a=P\{x\epsilon C\,|\,H_0\}$ where x denotes a sample point.

Example 9.21.1 For a $\mathcal{N}(\mu, 1)$ the hypothesis $H_0:\mu=10$ is rejected based on a sample of size 2 and under the criterion ' Reject H_0 if the standardized sample mean, when H_0 is true, is greater than 2.' Find the critical region.

Solution. The standardized sample mean in this problem is

$$z = \frac{\bar{x}-\mu}{\sigma/\sqrt{n}} = \frac{\sqrt{n}\,(\bar{x}-\mu)}{\sigma} = \sqrt{2}(\bar{x}-\mu)$$

The criterion is, ' reject H_0 if

$$\sqrt{2}\left(\frac{x_1+x_2}{2} - \mu\right) \geqslant 2,$$

when H_0 is true.' That is, when $\mu=10$, we obtain the critical region as that set of all possible values of (x_1, x_2) such that,

$$\sqrt{2}\left(\frac{x_1+x_2}{2} - 10\right) \geqslant 2$$

That is, $x_1+x_2\geqslant20+2\sqrt{2}$.

This is illustrated in Figure 9.4.

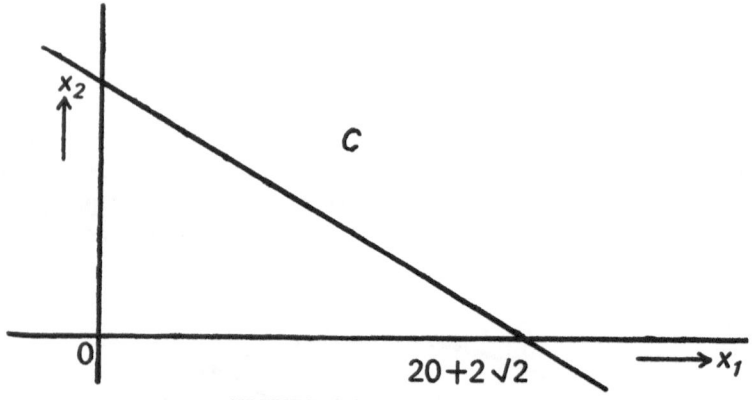

FIGURE 9.4 A critical region

The sample space S here is the entire 2 dimensional space because $-\infty<x_1<\infty$ and $-\infty<x_2<\infty$ and thus the critical region C is the region above the line $x_1+x_2=20+2\sqrt{2}$.

Exercises

9.1   A coin is tested for unbiasedness. The hypothesis that it is unbiased is rejected if 7 or more trials out of 10 trials result in heads.   In this case an alternative hypothesis that the probability of getting a head is 0.8 is accepted.   Obtain (1) the size of type II error, (2) the critical region, (3) the size of the critical region.

9.2   A box contains 10 marbles out of which θ are white and the rest are red.   We want to test the hypothesis $H_0:\theta=5$ against $H_1:\theta=4$. Suppose that H_0 is rejected if two marbles taken at random with replacement, are both red.   Calculate (1) C, (2) α, (3) β.

9.3   The daily consumption of electricity in a township is assumed to be exponentially distributed.   The hypothesis that the average consumption per day is 5,000 k.w. is rejected in favour of the hypothesis that it is 10,000 k.w., if a day selected at random has a consumption of 15,000 k.w. or more.   Obtain the critical region and the probabilities α and β of the type I and type II errors.

9.4   In a $\mathcal{N}(\mu, 1)$ the hypothesis that $\mu=25$ is rejected if the standardized sample mean of a sample of size 16, under the null hypothesis, is greater than 3.   Obtain the critical region and its size.

9.5   In a $\mathcal{N}(\mu, \sigma^2=25)$ the hypothesis $H_0:\mu=50$ is tested against $H_1:\mu=60$ by using a sample of size n and the criterion to reject H_0 if the standardized sample mean is greater than a specified quantity. How large should n be if the probabilities of type I and type II errors are $\alpha=0.025$ and $\beta=0.01$ respectively.

9.3 The Power of a Test

In our discussion we used β for the size of the type II error and therefore $1-\beta=$probability of rejecting the null hypothesis when it is not true.   This is the probability for a correct decision and this probability $1-\beta$ is often known as the *power of a test.*

Power$=1-\beta$ = Probability of rejecting H_0 when H_1 is true
$$= P\{x \in C \mid H_1\}$$

where x denotes the sample point and C denotes the critical region. For various alternative hypothesis H_1 we get a number of values for $1-\beta$ and if $1-\beta$ is plotted against the various alternative values of the parameter θ (on which the hypothesis is tested) then we get a *power-curve.*

Example 9.3.1 The hypothesis that $\theta=2$ is rejected in favour of an alternative that $\theta>2$ where θ is the parameter in an exponential distribution, if an observation from the population is greater than 6. Draw the power curve.

Solution. The size of the critical region

$$\alpha=P\{x\geqslant 6\,|\,\theta=2\}=\int_{6}^{\infty}\frac{1}{2}\,e^{-x/2}\,dx=e^{-3}$$

The power of the test,

$$1-\beta=P\{x\geqslant 6\,|\,\theta=\theta_1\}=\int_{6}^{\infty}\frac{1}{\theta_1}\,e^{-x/\theta_1}\,dx=e^{-6/\theta_1}$$

For various values of $\theta_1>\theta_0=2$ (since $H_1:\theta=\theta_1>2$) if we plot $1-\beta$ we get the power curve which is given in Figure 9.5.

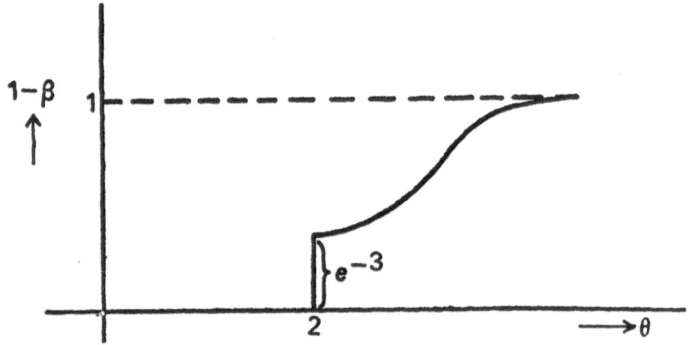

FIGURE 9.5 A power curve

Example 9.3.2 For a $\mathcal{N}(\mu,\ \sigma^2=1)$ the hypothesis $H_0:\mu=10$ is tested against the alternative that $\mu\neq 10$ by using the criterion: ' reject H_0 if the standardized sample mean of a sample of size 9, in absolute value, is greater than or equal to 1.96 under H_0. Draw the power curve.

Solution. $\alpha=P\left\{\left|\dfrac{\bar{x}-\mu}{\sigma/\sqrt{n}}\right|\geqslant 1.96 \text{ given } \mathcal{N}(\mu=10,\ \sigma^2=1)\right\}=0.05$

(from Normal probability tables).

$$=P\left\{\bar{x}\leqslant 10-\frac{1.96}{3} \text{ or } \bar{x}\geqslant 10+\frac{1.96}{3}\,\bigg|\,\mathcal{N}(\mu=10,\ \sigma^2=1)\right\}$$

(since $\mu=10$, $\sigma^2=1$ and $n=9$).

That is, we reject the null hypothesis if $\bar{x} \leqslant 10 - 1.96/3 = 9.347$ or if $\bar{x} \geqslant 10.653$. Hence the probability of rejecting H_0 when the alternative is true, is $= P\{\bar{x} \leqslant 9.347 \text{ or } \geqslant 10.653 \,|\, N(\mu, 1), \mu \neq 10\}$. That is,

$$1 - \beta = \int_{-\infty}^{9.347} \frac{3}{\sqrt{2\pi}} \exp\left[-9(\bar{x} - \mu)^2/2\right] d\bar{x} + \int_{10.653}^{\infty} \frac{3}{\sqrt{2\pi}}$$

$$\exp\left[-9(\bar{x} - \mu)^2/2\right] d\bar{x}$$

$$= \int_{-\infty}^{3(9.347 - \mu)} \frac{1}{\sqrt{2\pi}} e^{-t^2/2} \, dt + \int_{3(10.653 - \mu)}^{-\infty} \frac{1}{\sqrt{2\pi}} e^{-t^2/2} \, dt$$

(by standardization). This can be obtained from Normal probability tables for various values of μ. For example when $\mu = 9$ we have, $3(9.347 - \mu) = 1.041$ and $3(10.653 - \mu) = 4.959$ and

$$1 - \beta = \int_{-\infty}^{1.041} \frac{1}{\sqrt{2\pi}} e^{-t^2/2} \, dt + \int_{4.959}^{\infty} \frac{1}{\sqrt{2\pi}} e^{-t^2/2} \, dt = 0.8508.$$

When $\mu = 11$, $3(9.347 - \mu) = -4.959$ and $3(10.653 - \mu) = -1.041$ and

$$1 - \beta = \int_{-\infty}^{-4.959} \frac{1}{\sqrt{2\pi}} e^{-t^2/2} \, dt + \int_{-1.041}^{\infty} \frac{1}{\sqrt{2\pi}} e^{-t^2/2} \, dt$$

$$= 0.8508.$$

For various values of μ if $1 - \beta$ is plotted we get a curve as shown in Figure 9.6.

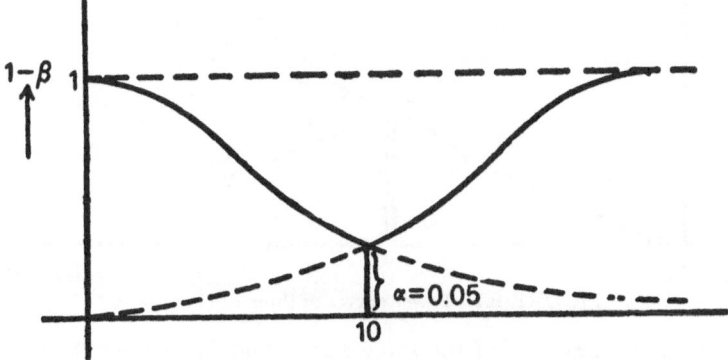

FIGURE 9.6 A power curve

If β is plotted against the various alternative values of the parameter, instead of $1-\beta$, we get an OC-curve (discussed in section 5.5) which for this problem is given by the dotted curve in Figure 9.6.

9.31 *The Best Test*

So far we were concerned with the properties of a test if a test criterion is given. Now we will consider the problem of selecting a test criterion based on power and other considerations. From Figure 9.1 it can be noticed that if α decreases then β increases and vice versa. In a particular problem of decision making we would like to take a decision by minimizing α as well as β, the probabilities of the two types of errors.

In most of the problems, this does not seem to be possible because of the phenomenon evidenced by Figure 9.1. So in practice, what is done is to take a decision by minimizing β (maximizing $1-\beta$, the power) for a fixed α. We could have taken a decision by minimizing α for a fixed β but we will not consider such a decision procedure here. In Figure 9.7 we give the power curves corresponding to three different tests A, B and C with the same size α. Since in a simple problem of testing, selection of a test criterion is equivalent to fixing a critical region and vice versa, α may be called the *size of the test*.

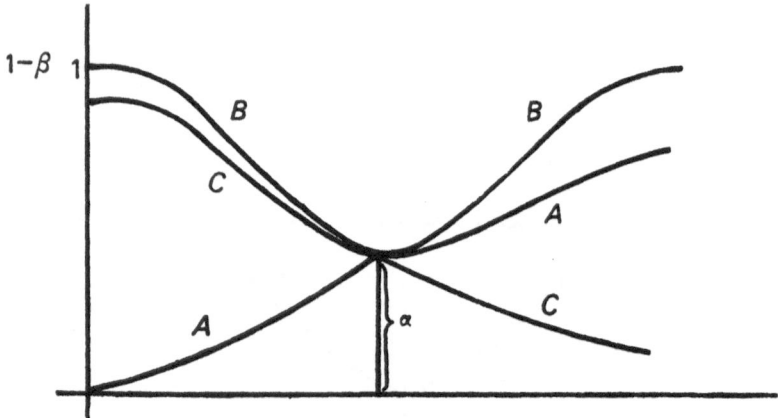

FIGURE 9.7 Power curves of three different tests

Here all the tests are of the same size α and for testing the same hypothesis. For $\theta < \theta_0$, C and A are less powerful compared to B and hence we say that C and A are non-admissible compared to B for $\theta < \theta_0$. For $\theta < \theta_0$, A is non-admissible compared to C whereas for $\theta > \theta_0$, C is non-admissible compared to A. For all θ, C and A are non-admissible compared to B, or B is *uniformly more powerful* compared to A and C. If there exists a test which is uniformly more powerful than any other test of the same size a then we call it the *uniformly most powerful test*. So in a particular problem of testing a hypothesis our aim will be to find out the uniformly most powerful test (UMPT) if it exists. The following lemma gives a technique of obtaining the most powerful test for a simple null hypothesis against a simple alternative.

Neyman-Pearson Lemma. Consider a problem of testing a simple hypothesis H_0: $\theta = \theta_0$ against a simple alternative H_1: $\theta = \theta_1$. Let $x_1, \ldots, x_n$ be an observed sample from a population with probability function $f(x, \theta)$. Let the likelihood function when H_0 is true be denoted by L_0 and that when H_1 is true be denoted by L_1. That is,

$$L_0 = \prod_{i=1}^{n} f(x_i, \theta_0)$$

and $L_1 = \prod_{i=1}^{n} f(x_i, \theta_1)$.

For example, if the population is,

$$f(x, \theta) = \frac{1}{\theta} e^{-x/\theta}$$

and if $\theta_0 = 2$ and $\theta_1 = 5$, then

$$L_0 = \frac{1}{2^n} e^{-\sum_{i=1}^{n} x_i/2} \qquad \text{and } L_1 = \frac{1}{5^n} e^{-\sum_{i=1}^{n} x_i/5}$$

Then the Neyman-Pearson lemma says that if there exists a critical region C of size a and a constant k such that,

$$\frac{L_0}{L_1} < k \text{ inside } C$$

and $\qquad \dfrac{L_0}{L_1} > k$ outside C

then C is the most powerful critical region, in the sense that the test which fixes C is the most powerful, for testing $H_0 : \theta = \theta_0$ against $H_1 : \theta = \theta_1$. This lemma gives a technique of getting the most powerful test, if it exists, for testing a simple hypothesis against a simple alternative. The set of points for which $\dfrac{L_0}{L_1}=k$ is a set of probability measure zero in the continuous case and hence we may write in such a case that,

$$\frac{L_0}{L_1} \leqslant k \text{ inside } C$$

and $\quad \dfrac{L_0}{L_1} \geqslant k \text{ outside } C.$

Proof. Let D be any other critical region of the same size a for testing the same hypothesis. Let the sample space S and the critical regions C and D be as shown in Figure 9.8.

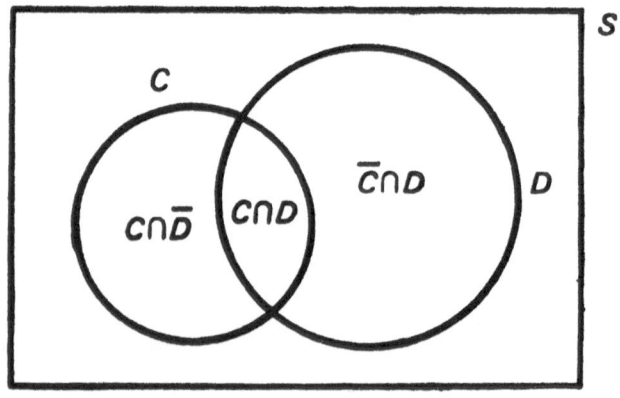

FIGURE 9.8 Proof of Neyman-Pearson lemma

$$a= P\ \{x \in C\,|\,H_0\}=P\{x \in D\,|\,H_0\}$$
$$=\int_C L_0\ dX=\int_D L_0\ dX$$

where $dX=dx_1...dx_n$, the integral stands for the multiple integral and L_0 is the likelihood function under H_0 (The case when the distribution is discrete is left to the reader).

But

$$\int_C L_0\ dX= \int_{C\cap\bar{D}} L_0\ dX+ \int_{C\cap D} L_0\ dX$$

and

$$\int_D L_0\ dX= \int_{\bar{C}\cap D} L_0\ dX+ \int_{C\cap D} L_0\ dX$$

Therefore

$$\int_{C \cap \bar{D}} L_0 \, dX = \int_{\bar{C} \cap D} L_0 \, dX \tag{1}$$

The power of $C = \int_C L_1 \, dX = \int_{C \cap \bar{D}} L_1 \, dX + \int_{C \cap D} L_1 \, dX$

$$\geqslant \int_{C \cap \bar{D}} (L_0/k) \, dX + \int_{C \cap D} L_1 \, dX \text{ (since } L_0/L_1 \leqslant k$$

inside C)

$$= \int_{\bar{C} \cap D} (L_0/k) \, dX + \int_{C \cap D} L_1 \, dX \text{ (from (1))}$$

$$\geqslant \int_{\bar{C} \cap D} L_1 \, dX + \int_{C \cap D} L_1 \, dX \text{ (since } L_0/L_1 \geqslant k$$

outside C)

$$= \int_D L_1 \, dX = \text{the power of } D.$$

This completes the proof.

Example 9.31.1 Consider a random sample of size n from a $N(\mu, \sigma^2)$ where σ^2 is known. Obtain the most powerful critical region of size a, if it exists, for testing $H_0 : \mu = \mu_0$ against $H_1 : \mu = \mu_1$ where μ_0 and μ_1 are given numbers.

Solution. The likelihood functions, under H_0 and H_1 are,

$$L_0 = \frac{1}{(\sigma \sqrt{2\pi})^n} \exp\left[- \sum_{i=1}^{n} (x_i - \mu_0)^2 / 2\sigma^2 \right]$$

and $\quad L_1 = \dfrac{1}{(\sigma \sqrt{2\pi})^n} \exp\left[- \displaystyle\sum_{i=1}^{n} (x_i - \mu_1)^2 / 2\sigma^2 \right]$

Therefore, $\dfrac{L_0}{L_1} = \exp\left[\dfrac{n}{2} (\mu_1{}^2 - \mu_0{}^2) + (\mu_0 - \mu_1) \displaystyle\sum_{i=1}^{n} x_i \right] / \sigma^2$

(obtained by simplification).

According to Neyman-Pearson lemma if C is the most powerful critical region and if there exists a constant k then,

$$\frac{L_0}{L_1} \leqslant k \text{ inside } C.$$

But, $\quad \dfrac{L_0}{L_1} \leqslant k \implies \exp\left[\dfrac{n}{2} (\mu_1{}^2 - \mu_0{}^2) + (\mu_0 - \mu_1) \displaystyle\sum_{i=1}^{n} x_i \right] / \sigma^2 \leqslant k$ for

some constant k. Taking logarithms and simplyfying the inequality, we get,

$$(\mu_0 - \mu_1) \bar{x} \leqslant k' \text{ where } k' \text{ is a constant.}$$

Case I. Let $\mu_1 > \mu_0$. Then $\mu_0 - \mu_1 < 0$ and division of the inequality $(\mu_0 - \mu_1)\bar{x} \leqslant k'$ by $\mu_0 - \mu_1$ gives

$$\bar{x} \geqslant K' \text{ where } K' \text{ is some constant.}$$

That is, the *best* (most powerful) critical region is such that, $\bar{x} \geqslant K'$ with probability a. But we can find a K' satisfying this condition. From a standard Normal table we know that

$$P\{z \geqslant z_a \mid H_0\} = a$$

where

$$Z = \frac{\bar{X} - \mu}{\sigma/\sqrt{n}}$$

and z_a is given in Figure 9.9.

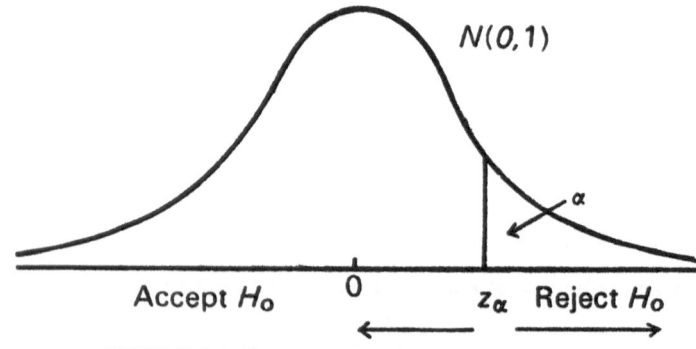

FIGURE 9.9 The critical region for a one-sided test

That is,

$$a = P\left\{z \geqslant z_a \mid H_0\right\} = P\left\{\bar{x} \geqslant \mu_0 + z_a \frac{\sigma}{\sqrt{n}}\right\}$$

Therefore, $k' = \mu_0 + z_a \dfrac{\sigma}{\sqrt{n}}$ which is completely known and hence the most powerful test is

' Reject H_0 if $\bar{x} \geqslant \mu_0 + z_a \dfrac{\sigma}{\sqrt{n}}$, otherwise accept H_0'

Case II. Let $\mu_1 < \mu_0$ then $\mu_0 - \mu_1 > 0$ and the inequality $(\mu_0 - \mu_1)\,\bar{x} \leqslant k'$ reduces to $\bar{x} \leqslant K'$ inside C. Again from a standard Normal table we have

$$P\left\{z \leqslant -z_a \mid H_0\right\} = a = P\left\{\bar{x} \leqslant \mu_0 - z_a \frac{\sigma}{\sqrt{n}}\right\}$$

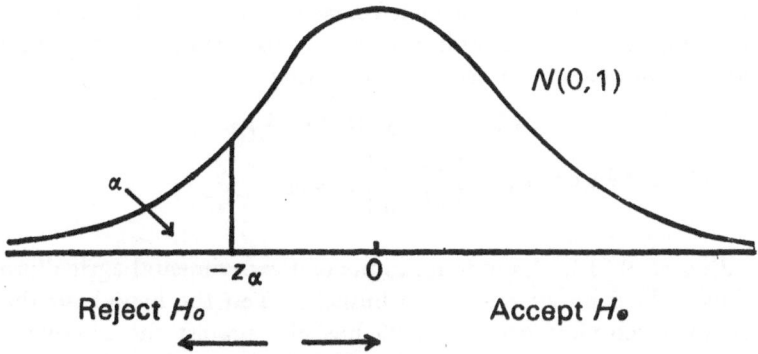

FIGURE 9.10 The critical region for a one-sided test

Therefore the criterion in this case is,

' Reject H_0 if $\bar{x} \leqslant \mu_0 - z_\alpha \dfrac{\sigma}{\sqrt{n}}$, otherwise accept H_0'.

The criteria obtained in cases *I* and *II* are true for all values of the alternative hypotheses. Hence we may write the criteria as follows.

Hypothesis   *Criterion*

$H_0 : \mu = \mu_0,\ H_1 : \mu > \mu_0$;   Reject H_0 if $\bar{x} \geqslant \mu_0 + z_\alpha \dfrac{\sigma}{\sqrt{n}}$

$H_0 : \mu = \mu_0,\ H_1 : \mu < \mu_0$;   Reject H_0 if $\bar{x} \leqslant \mu_0 - z_\alpha \dfrac{\sigma}{\sqrt{n}}$

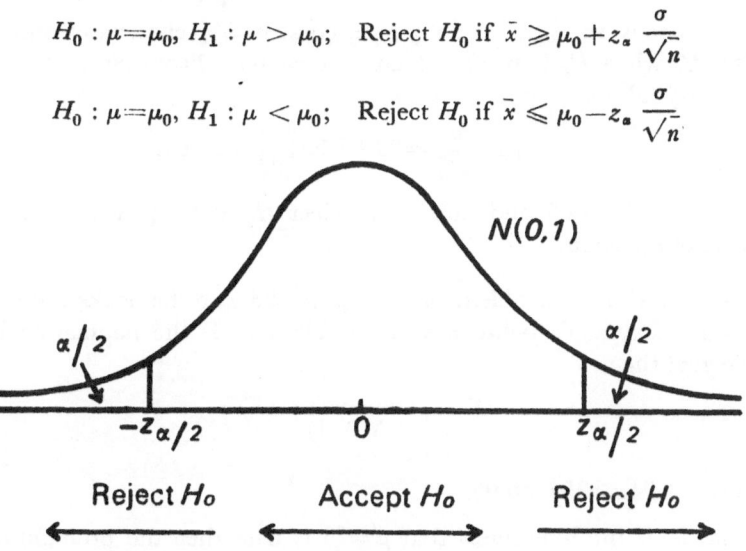

FIGURE 9.11 Critical region for a two-sided alternative

Thus we see that the Neyman-Pearson lemma can be used even for composite alternatives in this way. Now combining the above criteria we may give a criterion for testing

$$H_0 : \mu = \mu_0, \ H_1 : \mu \neq \mu_0$$

as follows ' Reject H_0 if $\left| \dfrac{\bar{x} - \mu_0}{\sigma / \sqrt{n}} \right| \geqslant z_{\alpha/2}$

Example 9.31.2 A random sample of 4 experimental agricultural plots yielded an average of 25 bushels. Test the hypothesis that the expected yield per plot is 23 bushels, against the alternative that it is more, assuming that the yield can be considered to have a distribution $N(\mu, \sigma^2 = 4)$.

Solution. The population is $N(\mu, \sigma^2 = 4)$. We have to test,

$$H_0 : \mu = 23, \ H_1 : \mu > 23$$

Given that, $\bar{x} = 25$ and $n = 4$.

$$\frac{\bar{X} - \mu}{\sigma / \sqrt{n}} : \ N(0, 1)$$

and the criterion is to reject H_0 if $\bar{x} \geqslant \mu_0 + z_\alpha \dfrac{\sigma}{\sqrt{n}}$

Let us consider a test at 95% level (that is, 5% chances of rejecting H_0 when H_0 is true). That is, $\alpha = 0.05$. From Normal probability tables, we get $z_\alpha = 1.65$.

$$\mu_0 + z_\alpha \frac{\sigma}{\sqrt{n}} = 23 + 1.65 \left(\frac{2}{2} \right) = 24.65.$$

But $\bar{x} = 25 > 24.65$ and hence we reject H_0 at 95% level or 5% level of rejection.

Comment. This problem of testing $\mu = 23$ can be looked upon from a Layman's point of view as follows. If the population is Normal then,

$$\frac{\bar{X} - \mu_0}{\sigma / \sqrt{n}} : N(0, 1)$$

and $P\{\bar{x} \geqslant 25\} < 0.05.$

That is, if the hypothesis that $\mu = 23$ is true then the probability of observing an $\bar{x}$ larger than or equal to 25 is less than 0.05. So we can say that if the hypothesis is true an improbable event has happened and hence we reject the hypothesis. In this problem the alternative is $\mu > 23$ which is a composite hypothesis. In general when H_0 or H_1 or both are composite then a uniformly most powerful test is usually available by using a principle known as the likelihood ratio criterion; see problem 9.12 below. Also, usually a one-sided alternative leads to a one-sided test (using only one-tail area) and a two sided alternative leads to a two-tail test. But this is not true in general. For the hypotheses on μ of a $\mathcal{N}(\mu, \sigma^2)$ where σ^2 is known, we see that the one-sided alternatives lead to one-tail tests and two-sided alternative leads to a two-tail test when we are using the standardized sample mean as a test statistic. Some authors use the term ' a test at 5% level ' meaning that there is 5% chances of rejecting H_0 when it is true. In this book we will call such a test ' at 95% level ' so that the reader can compare confidence statements and tests.

Exercises

9.6   By using the Neyman-Pearson lemma obtain the best test, if it exists, for testing $H_0 : \sigma = \sigma_0$, $H_1 : \sigma > \sigma_0$ in a $\mathcal{N}(\mu, \sigma^2)$, assume that a sample of size n is given and the size of the critical region is α.

9.7   By using the Neyman-Pearson lemma obtain the best test for testing $H_0 : \theta = \theta_0$, $H_1 : \theta > \theta_0$ in an exponential population. (Size of the critical region is given to be α and the sample size is n).

9.8   By using Neyman-Pearson lemma obtain the best test for testing $H_0 : p = p_0$, $H_1 : p > p_0$ in a Binomial population.

9.9   By using Neyman-Pearson lemma obtain the best test for testing $\theta = 5$ against $\theta = 6$ in the population $f(x, \theta) = \dfrac{1}{\theta}$ for $0 < x < \theta$.

9.10 In a packet of 10 articles θ are defectives. The hypothesis that $\theta = 5$ is rejected if two articles taken at random without replacement are either both good or both defectives, otherwise the hypothesis is accepted. Obtain β if H_1 is $\theta = 0$, 1, 2, 3, 4, 6, 7, 8, 9 and 10. Plot the power curve and the OC-curve.

9.11   In a $\mathcal{N}(\mu, \sigma^2)$ for the hypothesis $H_0 : \sigma^2 = 4$, $H_1 : \sigma^2 \neq 4$, draw the power curve and the OC-curve if the size of the critical region is 0.05, assuming that the test is based on a random sample of size $n = 9$.

9.12   *The likelihood ratio criterion or the λ-criterion.* In problems involving composite hypotheses, usually a uniformly most powerful test is available by using the following technique

$$\lambda = \frac{\max L_0}{\max L} < k \text{ inside } C$$

$$> k \text{ outside } C,$$

where L_0 is the likelihood function L when H_0 is true and max L_0 and max L denote L_0 and L after substituting the maximum likelihood estimates for all the unknown parameters in L_0 and L respectively. By using the λ-criterion obtain the best tests for the following problem in a $\mathcal{N}(\mu, \sigma^2)$ where σ^2 is known.

(1) $\begin{cases} H_0 : \mu = \mu_0 \\ H_1 : \mu > \mu_0 \end{cases}$;   (2) $\begin{cases} H_1 : \mu = \mu_0 \\ H_1 : \mu < \mu_0 \end{cases}$;   (3) $\begin{cases} H_0 : \mu = \mu_0 \\ H_1 : \mu \neq \mu_0 \end{cases}$

9.13   When the sample size n is large $-2 \log \lambda$ where λ is the likelihood ratio criterion, is approximately distributed as a chi-square with r degrees of freedom, where r is the number of independent parameters specified by the null hypothesis H_0. Construct a test criteria by using this approximation for the problem 9.12.

9.4 Some Practical Tests

In sections 9.1 to 9.3 we discussed some aspects of the theory of the tests of statistical hypotheses. Here we will discuss some tests mainly associated with a Normal population, such as the tests for the mean values, tests for variances, tests for differences of mean values, tests for ratio of variances and so on. In these simple tests we can easily show that, if we use the Neyman-Pearson lemma in the problems of simple H_0 versus simple H_1 and the λ-criterion in composite hypotheses, the final test will be based on the pivotal quantities given in Table 8.1 in Chapter 8. Since a one-sided alternative usually leads to a one-tail test and a two-sided alternative usually leads to a two-tail test we will discuss mainly the two-tail tests here. For the one-sided alternatives the corresponding one-tail test may be constructed in a similar fashion. A list of the usual classical tests is given in Table 9.1. The following results are used in order to construct the tests.

When $X_1, X_2, ..., X_n$ is a random sample from a $\mathcal{N}(\mu, \sigma^2)$ then

(1) $\dfrac{\overline{X}-\mu}{\sigma/\sqrt{n}} : \mathcal{N}(0, 1)$; (2) $\dfrac{\overline{X}-\mu}{S/\sqrt{n}} \approx \mathcal{N}(0, 1)$

$$\text{when } n \text{ is large } (>30);$$

(3) $\dfrac{\overline{X}-\mu}{S'/\sqrt{n}} : t_{n-1}$; (4) $\dfrac{\overline{X}-\mu}{S'/\sqrt{n}} \approx \mathcal{N}(0, 1)$

$$\text{when } n \text{ is large } (>30),$$

(5) $\dfrac{\Sigma(X_i-\overline{X})^2}{\sigma^2} : \chi_{n-1}^2$, where $S^2 = \dfrac{\Sigma(X_i-\overline{X})^2}{n}$

$$\text{and } S'^2 = \dfrac{\Sigma(X_i-\overline{X})^2}{n-1}$$

When $X_1, ..., X_{n_1}$ and $Y_1, ..., Y_{n_2}$ are two random samples from two independent populations $\mathcal{N}(\mu_1, \sigma_1^2)$ and $\mathcal{N}(\mu_2, \sigma_2^2)$ and if,

$$S_1^2 = \frac{\Sigma(X_i-\overline{X})^2}{n_1}, \quad S_2^2 = \frac{\Sigma(Y_i-\overline{Y})^2}{n_2}, \quad S'_1^2 = \frac{\Sigma(X_i-\overline{X})^2}{n_1-1}$$

$$S'_2^2 = \frac{\Sigma(Y_i-\overline{Y})^2}{n_2-1}, \quad S^2 = \frac{n_1 S_1^2 + n_2 S_2^2}{n_1+n_2-2}, \text{ then we have}$$

(1) $\dfrac{(\overline{X}-\overline{Y})-(\mu_1-\mu_2)}{\left(\dfrac{\sigma_1^2}{n_1} + \dfrac{\sigma_2^2}{n_2}\right)^{\frac{1}{2}}} : \mathcal{N}(0, 1)$;

(2) $\dfrac{(\overline{X}-\overline{Y})-(\mu_1-\mu_2)}{\left(\dfrac{S_1^2}{n_1} + \dfrac{S_2^2}{n_2}\right)^{\frac{1}{2}}} \approx \mathcal{N}(0, 1)$

when n_1 and n_2 are large $(n_1>30, n_2>30)$.

(3) $\dfrac{(\overline{X}-\overline{Y})-(\mu_1-\mu_2)}{S\left(\dfrac{1}{n_1} + \dfrac{1}{n_2}\right)^{\frac{1}{2}}} : t_{n_1+n_2-2}$ when $\sigma_1=\sigma_2=\sigma$;

(4) $\dfrac{(\overline{X}-\overline{Y})-(\mu_1-\mu_2)}{S\left(\dfrac{1}{n_1} + \dfrac{1}{n_2}\right)^{\frac{1}{2}}} : \mathcal{N}(0, 1)$ when $\sigma_1=\sigma_2=\sigma$,

$$\text{and } n_1+n_2-2>30;$$

TABLE 9.1

TEST CRITERIA FOR THE DIFFERENT CLASSICAL HYPOTHESES

	Population	Hypothesis	Statistic	Criterion Reject H_0 if	Illustration
1. (a)	$N(\mu, \sigma^2)$ σ^2-known	$H_0 : \mu = \mu_0$ $H_1 : \mu > \mu_0$	$\dfrac{\bar{X} - \mu}{\sigma/\sqrt{n}} : N(0, 1)$	$\dfrac{\bar{x} - \mu_0}{\sigma/\sqrt{n}} > z_\alpha$	Illustration A
1. (b)	"	$H_0 : \mu = \mu_0$ $H_1 : \mu < \mu_0$	"	$\dfrac{\bar{x} - \mu_0}{\sigma/\sqrt{n}} < -z_\alpha$	Illustration B
1. (c)	"	$H_0 : \mu = \mu_0$ $H_1 : \mu \neq \mu_0$	"	$\left| \dfrac{\bar{x} - \mu_0}{\sigma/\sqrt{n}} \right| > z_{\alpha/2}$	Illustration C
2.	Any population $\mu < \infty, \sigma < \infty$ σ-known, n-large	$H_0 : \mu = \mu_0$ $H_1 : \mu \neq \mu_0$	$\dfrac{\bar{X} - \mu}{\sigma/\sqrt{n}} \approx N(0, 1)$	$\left| \dfrac{\bar{x} - \mu_0}{\sigma/\sqrt{n}} \right| > z_{\alpha/2}$	"
3.	$N(\mu, \sigma^2)\sigma^2$- unknown $n \geqslant 30$	"	$\dfrac{\bar{X} - \mu}{S/\sqrt{n}} \approx N(0, 1)$	$\left| \dfrac{\bar{x} - \mu_0}{S/\sqrt{n}} \right| > z_{\alpha/2}$	"
4.	$N(\mu, \sigma^2)$ σ^2- unknown	$H_0 : \mu = \mu_0$ $H_1 : \mu \neq \mu_0$	$\left| \dfrac{\bar{X} - \mu_0}{S/\sqrt{n}} \right| : t_{n-1}$	$\left| \dfrac{\bar{x} - \mu_0}{s'/\sqrt{n}} \right| > t_{n-1, \alpha/2}$	Illustration D

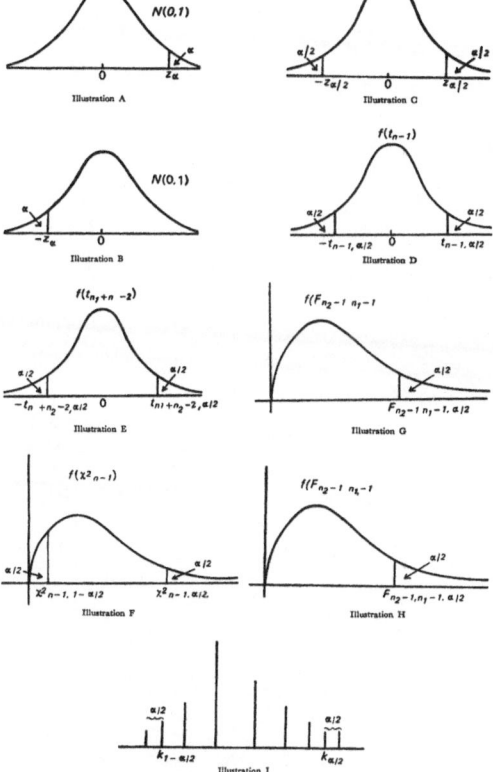

5.	$N(\mu_1, \sigma_1^2)$ and $N(\mu_2, \sigma_2^2)$, σ_1, σ_2-known	$H_0: \mu_1 = \mu_2$ $H_1: \mu_1 \neq \mu_2$	$\dfrac{(\bar{X}-\bar{Y})-(\mu_1-\mu_2)}{\left(\dfrac{\sigma_1^2}{n_1}+\dfrac{\sigma_2^2}{n_2}\right)^{1/2}} : N(0,1)$	$\left\| \dfrac{\bar{x}-\bar{y}}{\left(\dfrac{\sigma_1^2}{n_1}+\dfrac{\sigma_2^2}{n_2}\right)^{\frac{1}{2}}} \right\| \geqslant z_{a/2}$	Illustration C
6.	$N(\mu_1, \sigma_1^2)$ and $N(\mu_2, \sigma_2^2)$, σ_2-unknown $n_1>30, n_2>30$	,,	$\dfrac{(\bar{X}-\bar{Y})-(\mu_1-\mu_2)}{\left(\dfrac{S_1^2}{n_1}+\dfrac{S_2^2}{n_2}\right)^{1/2}} \approx N(0,1)$	$\left\| \dfrac{\bar{x}-\bar{y}}{\left(\dfrac{s_1^2}{n_1}+\dfrac{s_2^2}{n_2}\right)^{1/2}} \right\| \geqslant z_{a/2}$	,,
7.	$N(\mu_1, \sigma_1^2)$ and $N(\mu_2, \sigma_2^2)$, $\sigma_1=\sigma_2=\sigma$ (unknown) $n_1+n_2-2>30$	,,	$\dfrac{(\bar{X}-\bar{Y})-(\mu_1-\mu_2)}{S\left(\dfrac{1}{n_1}+\dfrac{1}{n_2}\right)^{1/2}} \approx N(0,1)$	$\left\| \dfrac{\bar{x}-\bar{y}}{s\left(\dfrac{1}{n_1}+\dfrac{1}{n_2}\right)^{1/2}} \right\| \geqslant z_{a/2}$	,,
8.	$N(\mu_1, \sigma_1^2)$ and $N(\mu_2, \sigma_2^2)$, $\sigma_1=\sigma_2=\sigma$ (unknown)	,,	,, $: t_{n_1+n_2-2}$	,, $\geqslant t_{n_1+n_2-2,\, a/2}$	Illustration E
9.	$N(\mu, \sigma^2)$	$H_0: \sigma=\sigma_0$ $H_1: \sigma \neq \sigma_0$	$\dfrac{\sum(X_i-\bar{X})^2}{\sigma^2} : \chi^2_{n-1}$	$\dfrac{ns^2}{\sigma_0^2} \leqslant \chi^2_{n-1,\,1-a/2}$ or $\geqslant \chi^2_{n-1,\,a/2}$	Illustration F

TABLE 9.1—(continued)

	Population	Hypothesis	Statistic	Criterion Reject H_0 if	Illustration
10.	$N(\mu_1, \sigma_1^2)$ and $N(\mu_2, \sigma_2^2)$	$H_0: \sigma_1 = \sigma_2$ $H_1: \sigma_1 \neq \sigma_2$	$\dfrac{\sum(X_i - \bar{X})^2/(n_1-1)\sigma_1^2}{\sum(Y_i - \bar{Y})^2/(n_2-1)\sigma_2^2}$ $: F_{n_1-1,\, n_2-1}$	$\dfrac{s_1'^2}{s_2'^2} \geq F_{n_1-1,\, n_2-1,\, \alpha/2}$ if $s_1' \geq s_2'$ $\dfrac{s_2'^2}{s_1'^2} \geq F_{n_2-1,\, n_1-1,\, \alpha/2}$ if $s_2' \geq s_1'$	Illustration G Illustration H
11.	Binomial parameters N and P, $NP > 5$, $N(1-P) > 5$, N-large	$H_0: p = p_0$ $H_1: p \neq p_0$	$\dfrac{X - NP}{\sqrt{NP(1-P)}} \approx N(0,1)$	$\dfrac{x - NP_0}{\sqrt{NP_0(1-P_0)}} \geq z_{\alpha/2}$	Illustration C
12.	Binomial	,,	$X : \dbinom{N}{x} P^x(1-P)^{N-x}$	$x \leq k_{1-\alpha/2}$ or $\geq k_{\alpha/2}$	Illustration I
13.	Bivariate Normal with correlation ρ. The sample correlation coefficient r	$H_0: \rho = 0$ $H_1: \rho \neq 0$	$\dfrac{\tfrac{1}{2}\log\dfrac{1+r}{1-r} - \tfrac{1}{2}\log\dfrac{1+\rho}{1-\rho}}{\left(\dfrac{1}{n-3}\right)^{1/2}}$ $\approx N(0,1)$	$\dfrac{\tfrac{1}{2}\log\dfrac{1+r}{1-r}}{\left(\dfrac{1}{n-3}\right)} \geq z_{\alpha/2}$	Illustration C

(5) $\dfrac{\Sigma(X_i-\bar{X})^2/(n_1-1)}{\Sigma(Y_i-\bar{Y})^2/(n_2-1)}$: $F_{n_1-1,\,n_2-1}$ when $\sigma_1=\sigma_2=\sigma$

When $\hat{p}=X/N$ is the proportion of successes from a Binomial population with parameters p and N then we have,

(1) $\quad \sqrt{\dfrac{\hat{p}-p}{\dfrac{p(1-p)}{N}}} \approx N(0,\,1)$ when N is large $(N \geqslant 20,\ Np>5,$

$$N(1-p)>5$$

(2) $\quad \sqrt{\dfrac{\hat{p}-p}{\dfrac{\hat{p}(1-\hat{p})}{N}}} \approx N(0,\,1)$ when N is very large.

Example 9.4.1 A random sample of 9 experimental animals, under a certain diet give the following increase in weights. $\Sigma x_i=45$ lb, $\Sigma x_i^2=279$ lb where x_i denotes the increase in weight of the ith animal.   Assuming that the increase in weight is normally distributed as a $N(\mu,\ \sigma^2)$ test the following hypothesis: (1) $H_0 : \mu=6$, $H_1 : \mu<6$; (2) $H_0 : \mu=4$, $H_1 : \mu>4$; (3) $H_0 : \mu=1$, $H_1 : \mu \neq 1$, at 95% level.

Solution. Here the basic distribution is a $N(\mu,\ \sigma^2)$ where σ^2 is unknown and the sample size n is 9 which is not large either.   The hypothesis to be tested is on μ.   From Table 9.1 we see that the appropriate test is based on a Student-t statistic with $n-1=8$ degrees of freedom.   From the given observations we have,

$$\bar{x}=\frac{\Sigma x_i}{n}=\frac{45}{9}=5 \text{ and } \Sigma(x_i-\bar{x})^2=\Sigma x_i^2-n\bar{x}^2=54$$

$$S'^2=\frac{\Sigma(x_i-\bar{x})^2}{n-1}=\frac{54}{8}$$

But, $\dfrac{\bar{X}-\mu}{S'/\sqrt{n}}$: t_{n-1} (See section 7.33)

(1) $\begin{cases} H_0 : \mu=6 \\ H_1 : \mu<6 \end{cases}$ Under H_0 an observed value of

$$t_{n-1}=\frac{5-6}{\sqrt{\left(\dfrac{54}{8}\right)\cdot\left(\dfrac{1}{9}\right)}}=-1.2$$

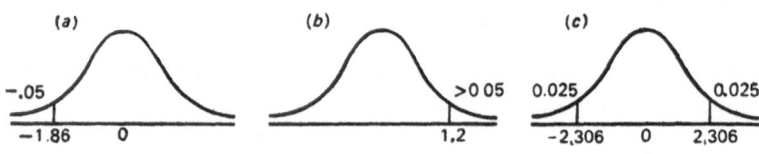

FIGURE 9.12 A Student-t with $n-1=8$ degrees of freedom

From Figure 9.12 (a) we have

$$P\{t_8 \leqslant -1.86\} = 0.05 \quad \text{(from Student-}t\text{ tables)}$$

But if the null hypothesis that $\mu=6$ is true then an observed value of t_8 is $-1.2 > -1.86$ and thus -1.2 is not in the critical region and hence we cannot reject this hypothesis that $\mu=6$.

(2) $\left.\begin{array}{l} H_0 : \mu=4 \\ H_1 : \mu>4 \end{array}\right\}$. Under H_0 an observed value of

$$t_{n-1} = \frac{5-4}{\sqrt{\left(\dfrac{54}{8}\right)\left(\dfrac{1}{9}\right)}} = 1.2$$

From figure 9.12(b) we have

$$P\{t_8 \geqslant 1.2\} > 0.05$$

Hence, 1.2 is not in the critical region and so we cannot reject this hypothesis that $\mu=4$ in favour of the hypothesis that $\mu>4$.

(3) $\left.\begin{array}{l} H_0 : \mu=1 \\ H_1 : \mu \neq 1 \end{array}\right\}$. Under H_0, $t_8 = \dfrac{5-1}{\sqrt{\left(\dfrac{54}{8}\right)\left(\dfrac{1}{9}\right)}} = 2.48$.

But from Student-t tables we see that (also see Figure 9.12(c))

$$P\{|t_{n-1}| \geqslant t_{n-1,\,a/2}\} = \frac{a}{2}.$$

That is, $P\{|t_8| \geqslant 2.306\} = 0.025.$

Therefore, $P\{|t_8| \geqslant 2.48\} < 0.025.$

That is, the observed value of $t_8=2.48>2.306$ lies in the critical region and hence we reject $H_0 : \mu=1$ in favour of the alternative $H_1 : \mu \neq 1$.

Comment. In a practical testing problem, once an appropriate statistic is selected, then the problem is a simple one of evaluating some probabilities. The observed value of the statistic need not be evaluated completely because we are only interested in seeing whether or not the observed value is in the critical region. In the above problem, if the sample size was large (>30) we would have used a normal approximation that

$$\frac{\overline{X}-\mu}{S'/\sqrt{n}} : \mathcal{N}(0, 1)$$

rather than using a Student-t test. The approximation will be good enough, even though a Student-t is a Student-t and it never becomes a standard normal variate. The approximations mentioned in the beginning of section 9.4 follow from the Central Limit Theorem discussed in section 7.23 of Chapter 7.

Example 9.4.2 The I.Q.'s of random samples of 20 boys and 20 girls yield averages 110 and 105 with sample standard deviations 2 and 1 respectively. If we can assume that the boys' and girls' I.Q.'s are normally distributed with the same variance, test the hypothesis that the expected I.Q.'s of boys and girls are different, at 99% level.

Solution. We are given,

$$\bar{x}=110, \bar{y}=105, n_1=n_2=20$$

$$s_1^2 = \frac{\Sigma(x_i-\bar{x})^2}{n_1}=4 \quad \text{and} \quad s_2^2 = \frac{\Sigma(y_i-\bar{y})^2}{n_2}=1$$

The populations are given to be, $\mathcal{N}(\mu_1, \sigma_1^2)$ and $\mathcal{N}(\mu_2, \sigma_2^2)$ where $\sigma_1^2=\sigma_2^2$. The populations can be assumed to be independent. The hypothesis to be tested is $\mu_1 \neq \mu_2$. We will formulate this in the following way $H_0 : \mu_1=\mu_2$ and $H_1 : \mu_1 \neq \mu_2$.

If H_0 is rejected H_1 is accepted and vice versa. Since $\sigma_1^2=\sigma_2^2=\sigma^2$ (unknown) the appropriate statistic is,

$$\frac{(\overline{X}-\overline{Y})-(\mu_1-\mu_2)}{S\left(\frac{1}{n_1} + \frac{1}{n_2}\right)^{1/2}} : t_{n_1+n_2-2} \text{ (See section 7.33)},$$

where $\quad \dfrac{n_1S_1^2+n_2S_2^2}{n_1+n_2-2} = S^2$ and $s^2 = \dfrac{20(4)+20(1)}{38} = \dfrac{100}{38}$

An observed value under $H_0 : \mu_1 = \mu_2$, is

$$t_{n_1+n_2-2} = t_{38} = \frac{(x-\bar{y})-0}{s\left(\dfrac{1}{n_1}+\dfrac{1}{n_2}\right)^{1/2}} = \frac{110-105}{\left(\dfrac{10}{\sqrt{38}}\right)\left(\dfrac{\sqrt{2}}{\sqrt{20}}\right)} = \sqrt{95}$$

According to the criterion (Table 9.1) we reject $H_0 : \mu_1 = \mu_2$ if

$$|t_{n_1+n_2-2}| \geqslant t_{n_1+n_2-2,\,a/2} = t_{38,0.005}.$$

Since $n_1+n_2-2=38>30$ we get a good normal approximation and so instead of a Student-t table we will use the approximation

$$\mathcal{Z} = \frac{(\bar{X}-\bar{Y})-(\mu_1-\mu_2)}{S\left(\dfrac{1}{n_1}+\dfrac{1}{n_2}\right)^{1/2}} : \mathcal{N}(0,\,1)$$

and look into a standard normal table, corresponding to $\dfrac{a}{2}=0.005$

From Normal tables we have,

$$P\{|z| \geqslant z_{a/2}\} = P\{|z| \geqslant 2.58\} = 0.005.$$

But $\sqrt{95}>2.58$ and hence the observed value of z lies in the critical region. Hence we reject $H_0 : \mu_1 = \mu_2$ in favour of $H_1 : \mu_1 \neq \mu_2$

Comment. In this problem we wanted to test the hypothesis $\mu_1 \neq \mu_2$ but we tested $\mu_1 = \mu_2$ against $\mu_1 \neq \mu_2$. That is, we tested the hypothesis that $\mu_1 - \mu_2 = 0$. This is one reason for H_0 to be called the null hypothesis. But in present day statistical analysis ' null hypothesis ' is another unfortunate term and it has no meaning other than ' the hypothesis being tested ' or the region of its rejection when it is true is the critical region with size a.

Example 9.4.3 A random sample of 16 nuts, produced by a machine, has an average diameter of 2 cm with the sample standard deviation of 0.2 cm. Test the hypothesis that the true standard deviation is 0.3 against the alternative that it is less, at the 95% level. Assume a Normal distribution for the diameter measurement.

Solution. Here the population is $\mathcal{N}(\mu,\,\sigma^2)$ and we want to test,

$$H_0 : \sigma=0.3,\ H_1 : \sigma<0.3.$$

We will use the result that when the sample is from a $\mathcal{N}(\mu,\,\sigma^2)$

$$\frac{\Sigma(X_i-\bar{X})^2}{\sigma^2} : \chi_{n-1}^2 \qquad \text{(See section 7.32)}.$$

We have, $\dfrac{\Sigma(x_i-\bar{x})^2}{n} = (0.2)^2 \Longrightarrow \Sigma(x_i-\bar{x})^2 = 16(0.2)^2 = 0.64.$

Under $H_0 : \sigma = 0.3$ an observed value of $X_{n-1}^2 = X_{15}^2$ is,

$$\frac{\Sigma(x_i-\bar{x})^2}{\sigma_0{}^2} = \frac{0.64}{(0.3)^2} = 7.1.$$

From a chi-square table corresponding to the degrees of freedom 15 we have, 7.261 such that,

$$P\{X_{15}^2 \leqslant 7.26\} = 0.05.$$

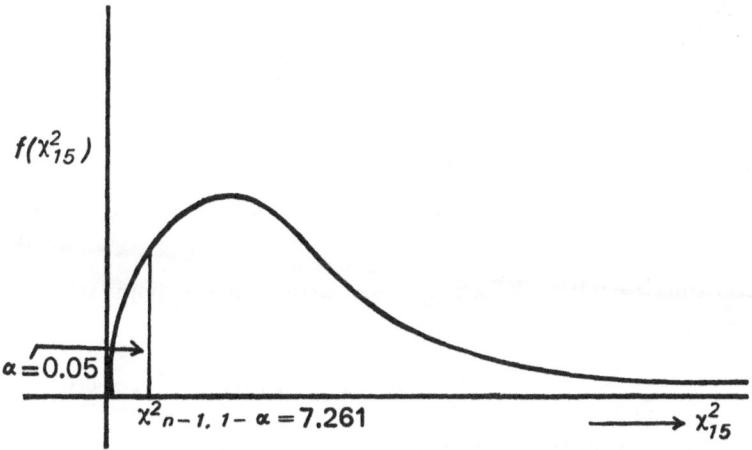

FIGURE 9.13 A one-tail chi-square test

In this problem the alternative is H_1: $\sigma < 0.3$ and hence by the likelihood ratio principle, we can show that it leads to a one tail test as shown in Figure 9.13. Since $7.1 < 7.251$ the observed value of χ_{15}^2 is in the critical region and hence we reject the hypothesis that $\sigma = 0.3$ in favour of the alternative that $\sigma < 0.3$.

Comment. If the alternative was $\sigma > 0.3$ we would have used the right tail and if the alternative was $\sigma \neq 0.3$ we would have used both the tails corresponding to areas $\dfrac{\alpha}{2} = 0.025$ each. Forgetting about the likelihood ratio principle, we may look at this problem in the following way from a layman's point of view. If the hypothesis that $\sigma = 0.3$ is true then,

$$\frac{\Sigma(x_i-\bar{x})^2}{\sigma_0{}^2} = \frac{0.64}{(0.3)^2} = 7.1$$

is an observed value of a chi-square with 15 degrees of freedom. The probability of getting an observed value smaller than or equal to 7.1 is less than 0.05 and hence we reject H_0 at the 95% level. In all these examples, the basic assumption of Normality is necessary. If the underlying distribution is unknown, we may use the Central Limit Theorem to construct some tests if the sample size is large. Otherwise we may use some distribution-free tests (tests where a basic distribution is not assumed) some of which will be discussed in the next chapter.

Example 9.4.4 Pumpkins were grown under two experimental conditions. Two random samples of 11 and 9 pumpkins, show the sample standard deviations of their weights as 0.8 and 0.5 respectively. Assuming that the weight distributions are Normal, test the hypothesis that the true variances are equal, against the alternative that they are not, at the 90% level.

Solution. Here the populations are $\mathcal{N}(\mu_1, \sigma_1^2)$ and $\mathcal{N}(\mu_2, \sigma_2^2)$ and the hypothesis to be tested is, $H_0: \sigma_1^2 = \sigma_2^2$, $H_1: \sigma_1^2 \neq \sigma_2^2$. $a = 0.1$ and thus $\dfrac{a}{2} = 0.05$. When $\sigma_1^2 = \sigma_2^2$, assuming the populations to be independent, we know that,

$$\frac{\Sigma(X_i - \overline{X})^2/(n_1 - 1)}{\Sigma(Y_i - \overline{Y})^2/(n_2 - 1)} : F_{n_1 - 1, \, n_2 - 1} \text{ (See section 7.34)}.$$

We have the observed value of,

$$\frac{\Sigma(x_i - \bar{x})^2}{n_1} = (0.8)^2 \text{ and } \frac{\Sigma(y_i - \bar{y})^2}{n_2} = (0.5)^2$$

and $n_1 = 11$ and $n_2 = 9$. Therefore,

$$\frac{\Sigma(x_i - \bar{x})^2}{n_1 - 1} = \frac{11(0.8)^2}{10} = 0.704 \text{ and } \frac{\Sigma(y_i - \bar{y})^2}{n_2 - 1} = \frac{9(0.5)^2}{8} = 0.28125$$

Hence when H_0 is true an observed value of $F_{n_1 - 1, \, n_2 - 1} = F_{10, \, 8}$ is

$$= \frac{0.704}{0.28125} = 2.5$$

From an F-table corresponding to 10 and 8 degrees of freedom we get a value 3.35 such that,

$$P\{F_{10 \ 8} \geqslant 3.35\} = 0.05.$$

The probability of getting an $F_{10, \, 8}$ larger than or equal to 2.5 is $> 0.05 = \dfrac{a}{2}$, and hence we cannot reject the hypothesis that $\sigma_1 = \sigma_2$ in favour of $H_1 : \sigma_1 \neq \sigma_2$.

Comment. Since there are three numbers, a and the degrees of freedoms m and n in order to tabulate $F_{m,n,a}$, usually the F-table is given only for selected values of a, for example $a=0.05$ and $a=0.01$. For a one-tail test such as,

$$H_0 : \sigma_1 = \sigma_2, H_1 : \sigma_1 > \sigma_2$$

we may use the criterion that, if

$$\frac{\Sigma(x_i-\bar{x})^2/(n_1-1)}{\Sigma(y_i-\bar{y})^2/(n_2-1)} \geq F_{n_1-1,\,n_2-1,\,a}$$

then reject H_0. Similarly for the one-tail test,

$$H_0 : \sigma_1 = \sigma_2, H_1 : \sigma_1 < \sigma_2$$

we may use the criterion, if

$$\frac{\Sigma(x_i-\bar{x})^2/(n_1-1)}{\Sigma(y_i-\bar{y})^2/(n_2-1)} = F_{n_1-1,\,n_2-1} \geq F_{n_1-1,\,n_2-1,\,1-a}$$

then reject H_0. But we may not get a tabulated value for $F_{n_1-1,\,n_2-1,\,1-a}$ due to the fact that the table is available only for selected values of a. But we can use the result that

$$F_{n_1-1,\,n_2-1,\,a} = \frac{1}{F_{n_2-1,\,n_1-1,\,1-a}}$$

and

$$\frac{\Sigma(y_i-\bar{y})^2/(n_2-1)}{\Sigma(x_i-\bar{x})^2/(n_1-1)} = F_{n_2-1,\,n_1-1}$$

and obtain some more points from an F-table. Using this technique a two-tail test,

$$H_0 : \sigma_1 = \sigma_2, H_1 : \sigma_1 \neq \sigma_2$$

is tested by using the following criterion. Reject H_0 if

$$\frac{\Sigma(x_i-\bar{x})^2/(n_1-1)}{\Sigma(y_i-\bar{y})^2/(n_2-1)} = F_{n_1-1,\,n_2-1} \geq F_{n_1-1,\,n_2-1,\,a/2} \text{ for}$$

$$\frac{\Sigma(x_i-\bar{x})^2}{n_1-1} > \frac{\Sigma(y_i-\bar{y})^2}{n_2-1}$$

and in the case when,

$$\frac{\Sigma(x_i-\bar{x})^2}{n_1-1} < \frac{\Sigma(y_i-\bar{y})^2}{n_2-1}, \text{ reject } H_0 \text{ if}$$

$$\frac{\Sigma(y_i-\bar{y})^2/(n_2-1)}{\Sigma(x_i-\bar{x})^2/(n_1-1)} = F_{n_2-1,\,n_1-1} \geq F_{n_2-1,\,n_1-1,\,a/2}$$

In all the tests discussed so far the basic assumption of Normality is there. Some tests are quite sensitive to departure from Normality but some are not. This aspect is called *robustness* of tests which will not be discussed here.

Example 9.4.5 A coin is tossed 10 times and 6 heads are obtained. Test the hypothesis, that the coin is unbiased, against the alternative that it is not at 95% level.

Solution. This is a Binomial probability situation with the parameter p and $N=10$. The hypothesis to be tested is,

$$H_0 : p=\tfrac{1}{2}, H_1 : p\neq\tfrac{1}{2}.$$

Since $a=0.05$, $a/2=0.025$. So we will find two numbers $k_{a/2}$ and $k'_{a/2}$ such that,

$$\sum_{x=0}^{k'_{a/2}} \binom{10}{x} \left(\frac{1}{2}\right)^x \left(\frac{1}{2}\right)^{10-x} = 0.025$$

and

$$\sum_{x=k_{a/2}}^{10} \binom{10}{x} \left(\frac{1}{2}\right)^x \left(\frac{1}{2}\right)^{10-x} = 0.025.$$

Since it is a discrete case we may not get $k_{a/2}$ and $k'_{a/2}$ such that the tail-end probabilities are exactly 0.025 each. Hence we will find two numbers k and k' such that

$$\sum_{x=0}^{k'} \binom{10}{x} \left(\frac{1}{2}\right)^x \left(\frac{1}{2}\right)^{10-x} \leqslant 0.025$$

and

$$\sum_{x=k}^{10} \binom{10}{x} \left(\frac{1}{2}\right)^x \left(\frac{1}{2}\right)^{10-x} \leqslant 0.025.$$

From the Binomial tables we have $k=9$ and $k'=1$.

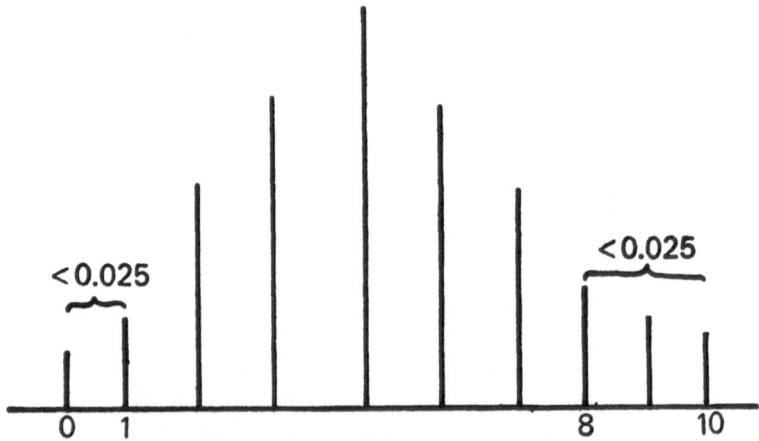

FIGURE 9.14   Critical region in a Binomial case

Hence the probability of getting the number of successes smaller than or equal to or larger than or equal to 9 is <0.05. But the observed number of successes is 6 which is not in the critical region and hence we cannot reject the hypothesis that $p=\frac{1}{2}$ in favour of $H_1 : p \neq \frac{1}{2}$.

Comment. If the alternative was $H_1 : p < \frac{1}{2}$, then we would have used only one tail-end (left tail) and from a Binomial table we find that

$$P\{x \leqslant 1\}=0.0107 \text{ and } P\{x \leqslant 2\}=0.0547$$

and hence we get the number 1 such that

$$P\{x \leqslant 1\} < 0.05 = \alpha$$

Again at 95% level we will not reject the hypothesis $H_0 : p=\frac{1}{2}$. If the alternative $H_1 : p > \frac{1}{2}$ then from the right-tail of the Binomial distribution we find a number 9 such that

$$P\{x \geqslant 9\} < 0.05 = \alpha$$

and again we will not reject H_0. If the number of trials N was very large then we would have used the Normal approximation that

$$\frac{\overline{X} - Np}{\sqrt{Np(1-p)}} \approx \mathcal{N}(0, 1),$$

with $p=\frac{1}{2}$ and under H_0, for constructing the test.

Exercises

9.14 Suppose that the time taken by a certain particle to move from one fixed point to another fixed point is distributed as a $\mathcal{N}(\mu, \sigma^2=4)$. A random sample of 9 readings has mean 50. Test the hypothesis that (1) $H_0 : \mu=52$ against $H_1 : \mu < 52$; (2) $H_0 : \mu=52$, $H_1 : \mu \neq 52$ at 95% level.

9.15 The lifetime of electric bulbs of a certain category is assumed to be exponentially distributed. One bulb selected at random has a life time of 950 hrs. Test the manufacturer's claim that the expected lifetime is 1000 hrs against the alternative that it is less, at 99% level.

9.16 A random sample of 100 turkeys weigh an average of 10 lb with a sample variance of 4 lb. Test the hypothesis that the expected weight of turkeys of that type is 12 lb against the alternative that it is not, at 95% level. *Hint:*. Use a Normal approximation.

9.17 Two diets are compared by conducting an experiment on two sets of 40 and 45 experimental animals. The average increase in weights are 20 and 25 lb with standard deviations 2 and 3 lb respectively. Check the claim that diet 2 increase the weight by 3 lb more than diet 1 on the average.

9.18 Two random samples of 10 experimental plots show average yields of 150 and 140 bushels with standard deviations 5 and 4 bushels respectively. If the two sets of plots are given two different fertilizers, check the claim that the fertilizers are equally effective, at 95% level. Assume that the yields can be considered to be independently Normally distributed as $\mathcal{N}(\mu_1, \sigma^2)$ and $\mathcal{N}(\mu_2, \sigma^2)$.

9.19 Two methods of teaching are compared by teaching two classes of 20 and 25 students. The teacher's I.Q.'s are 110 and 120 respectively. Suppose that the marks obtained by the students can be assumed to be Normal with expected values $T_j + I_j/10$ for $j=1$, 2 and with the same variances, where T_1, T_2 and I_1, I_2 denote the effects of teaching and the I.Q.'s of teachers respectively. Test the hypothesis that the two methods are equally effective, at 95% level, if the average marks are 65 and 75 with standard deviations 2 and 3 respectively.

9.20 Let $\mathcal{N}(\mu_1, \sigma_1^2)$ and $\mathcal{N}(\mu_2, \sigma_2^2)$ be independent. Consider $H_0 : \mu_1 - \mu_2 = 20$ and $H_1 : \mu_1 - \mu_2 = 25$. Obtain n if random samples of the same size n are taken, $\sigma_1^2 = 16$, $\sigma_2^2 = 25$, $\alpha = 0.05$, $\beta = 0.05$ and if the test is based on the right tail of $[(\overline{X} - \overline{Y}) - (\mu_1 - \mu_2)] / (\sigma_1^2/n + \sigma_2^2/n)^{1/2}$.

9.21 A random sample of 40 temperature readings shows a standard deviation of 0.05. Test the hypothesis that the true standard deviation is $H_0 : \sigma = 0.03$ against $H_1 : \sigma \neq 0.03$ at 95% level, assuming that the temperature reading is a $\mathcal{N}(\mu, \sigma^2)$. *Hint:*. $\sqrt{2x^2} - \sqrt{2n-1}$ is approximately a standard normal for $n > 30$.

9.22 Random samples of 5 boys and 4 girls took the following amounts of time to solve a problem.
Girls. 20, 21, 22, 23, 24;   Boys. 18, 22, 25, 23.

Assuming that the distribution of time by girls and boys are $\mathcal{N}(\mu_1, \sigma_1^2)$ and $\mathcal{N}(\mu_2, \sigma_2^2)$ test the hypothesis that $\sigma_1^2 = \sigma_2^2$ against the alternative that $\sigma_1^2 \neq \sigma_2^2$, at 98% level.

9.23   Twenty people were attacked by a disease and only 18 survived. Will you reject the hypothesis that the survival rate, if attacked by this disease, is 85% in favour of the hypothesis that it is more, at 95% level?

9.24   Random samples of 400 men and 600 women were asked whether they would like to have an overpass at a certain railway track. 200 men and 325 women were in favour of that. Test the hypothesis that the true proportions are the same, against the alternative that they are not the same, at 96% level.

9.25   In Binomial trials, if $p=0.3$ is tested against $H_1 : p=0.2, 0.5, 0.7$, obtain the probability of the type II error if $\alpha=0.05$ and $\mathcal{N}=20$.

Additional Exercises

9.1   In an exponential population if the hypothesis $H_0 : \theta=5$ is tested against $H_1 : \theta=7$ based on the criterion; reject H_0 if a single observation is greater than 10, calculate the probabilities of the type I and II errors.

9.2   In a Normal population $\mathcal{N}(\mu, \sigma^2=4)$ if a hypothesis $H_0 : \mu=12$ is tested against $H_1 : \mu \neq 12$ by using the criterion that if the standardized sample mean, under H_0, is greater than 3 in absolute value, reject H_0, (1) calculate the size of the critical region, (2) draw the power curve, (3) draw the OC-curve (sample size 10).

9.3   A box contains 8 marbles out of which θ are red and the rest are green. Consider the hypothesis that $\theta=4$. Consider the tests (a) two marbles are taken at random and the hypothesis is rejected if both are of the same colour, (b) two marbles are taken at random and the hypothesis is rejected if at least one of them is green. Assuming that the marbles are taken without replacement, calculate α and draw the power curves.

9.4   A random sample of size 10 from a $N(\mu, \sigma^2=4)$ gives a sample mean 25. Test $H_0 : \mu=28$ against $H_1 : \mu<28$ at $a=0.025$.

9.5   A random sample of size 15 from a $N(\mu, \sigma^2)$ gives a sample mean of 35 with a standard deviation of 5. Test the hypothesis that $\mu=30$ against the alternative that $\mu \neq 30$ at 90% level.

9.6   A random sample of size 40 from a $N(\mu, \sigma^2)$ gives a sample mean 22 with a standard deviation 4. Test the hypothesis that $\mu=25$ against $H_1 : \mu \neq 25$ at 99% level.

9.7   A random sample of 10 observations from a $N(\mu, \sigma^2)$ gives a sample standard deviation of 0.5. Test the hypothesis that the true variance $\sigma^2=0.04$ against the alternative that it is more at 96% level.

9.8   A random sample of 48 observations from a $N(\mu, \sigma^2)$ gives a sample variance of 4. Test the hypothesis that the true variance is 8, at 99% level.

9.9   A random sample of 100 from some population with variance 2 has a sample mean 28. Test the hypothesis that the mean value of the population is 30 against the alternative that it is not, at 99% level.

9.10 Two independent random samples of sizes 10 and 12 from $N(\mu_1, \sigma_1^2=4)$ and $N(\mu_2, \sigma_2^2=9)$ give the sample means 20 and 22 respectively. Test the hypothesis that $\mu_1-\mu_2=5$ against the alternative that $\mu_1-\mu_2<5$, at 90% level.

9.11 Two independent random samples of sizes 20 and 25 from $N(\mu_1, \sigma^2)$ and $N(\mu_2, \sigma^2)$ give the sample means 10 and 12 with sample variances 4 and 5 respectively. Test the hypothesis that $\mu_1=\mu_2$ against the alternative that $\mu_1 \neq \mu_2$ at 90% level.

9.12 Two independent random samples of sizes 10 and 11 from $N(\mu_1, \sigma^2)$ and $N(\mu_2, \sigma^2)$ have the sample means 5 and 8 with sample variances 1 and 2 respectively. Test the hypothesis that $\mu_1=\mu_2$ against the alternative that $\mu_1 \neq \mu_2$ at 95% level.

9.13 If two random samples are given from $N(\mu_1, \sigma_1^2)$ and $N(\mu_2, \sigma_2^2)$ and if nothing is known about σ_1^2 and σ_2^2 can you test $H_0 : \mu_1=\mu_2$, $H_1 : \mu_1 \neq \mu_2$?

9.14 In a Binomial situation of 12 trials if the observed number of successes is 8, test the hypothesis that $p=0.6$ against the alternative that $p \neq 0.6$, at 95% level.

9.15 In a Binomial situation of 58 trials, if the observed number of successes is 20, test the hypothesis that $p=\frac{1}{2}$ against the alternative that $p<\frac{1}{2}$ at 95% level.

9.16 By using the likelihood ratio principle, obtain a test criterion for testing $\mu_1=\mu_2=\ldots=\mu_k$ in independent $N(\mu_1, \sigma^2)$, $N(\mu_2, \sigma^2),\ldots, N(\mu_k, \sigma^2)$.

9.17 By using the likelihood ratio principle obtain a test criterion for testing $\sigma_1=\sigma_2=\ldots=\sigma_k$ in independent $N(\mu_1, \sigma_1^2),\ldots,$ $N(\mu_k, \sigma_k^2)$.

9.18 By using the likelihood ratio principle obtain a test criterion for testing $\mu_1=\mu_2=\ldots=\mu_k$, $\sigma_1=\sigma_2=\ldots=\sigma_k$ in independent $N(\mu_1, \sigma_1^2),\ldots, N(\mu_k, \sigma_k^2)$.

9.19 If the tests in 9.16, 9.17, and 9.18 are made by using the approximation $-2 \log \lambda=\chi_r^2$, obtain r in these cases.

9.20 By the likelihood ratio principle can you get an exact test for testing $\mu_1=\ldots=\mu_k$ in $N(\mu_1, \sigma_1^2),\ldots N(\mu_k, \sigma_k^2)$?

9.21 In the likelihood ratio principle, $\lambda=\left(\underset{\Omega}{\max} L_0\right) \Big/ (\max L)$, show that $0<\lambda\leqslant A$ for some A is a reasonable critical region from a layman's point of view.

9.22 Derive the criteria for testing the hypotheses in Table 9.1 and show that the tests obtained by the likelihood ratio principle are the same as the tests given in Table 9.1 wherever the standard normal or Student-t is used as a test statistic.

CHAPTER TEN

Some Non-Parametric Tests

In the last chapter we considered a number of parametric hypotheses where a basic distribution was assumed and the hypotheses were on the parameters of this basic distribution. In this chapter we will consider some tests which will not be of this nature.

There are a number of practical situations where one would like to test the association between two attributes such as testing whether there is any association between, the colour of the eyes and the intelligence, habits of wearing tall hats and longevity of life, interest in high altitude jumping and height, and so on. There are other problems where an experimental scientist may have a number of observations and he would like to know whether they came from a given population such as a Normal population or a Poisson population and so on. These problems are known as 'goodness of fit' problems. First, we will consider a statistic which has an approximate chi-square distribution and which can be used for testing 'goodness of fit' as well as the association between attributes.

10.1 A GOODNESS OF FIT PROBLEM

Here we will consider a set of 'goodness of fit' problems. The basic problem is to see whether a given set of data could be considered to be from a given population. A statistic often used for this purpose is Karl Pearson's χ^2 statistic

$$\chi^2 = \sum_{i=1}^{k} \frac{(O_i - e_i)^2}{e_i} \approx \chi_{k-r-1}{}^2, \; k \geqslant 5, \; e_i\text{'s} \geqslant 5,$$

where χ^2 is approximately distributed as a chi-square with $k-r-1$ degrees of freedom where k is the number of classes, r is the number of independent parameters estimated when calculating e_i's the expected frequencies and O_i's are the observed frequencies. A good approximation is available for $k \geqslant 5$ and e_i's $\geqslant 5$.

For example, if the heights of a set of people are classified into a number of different classes then the frequencies in the various classes are the observed frequencies and if we assume that the height measurements are normally distributed then the expected frequencies in the various classes can be estimated by using the assumption of normality. In order to estimate the frequencies we may need the estimates of the 2 parameters μ and σ^2 in a $\mathcal{N}(\mu, \sigma^2)$. Thus, 2 parameters are estimated and $r=2$. In other words the degrees of freedom for Pearson's χ^2 is $k-2-1=k-3$ where k is the number of classes. These aspects will be clear from the following examples.

Example 10.1.1 A bird watcher sitting in a park has spotted a number of birds belonging to 5 categories. The exact classification is given below.

Category	1	2	3	4	5	6
Frequency	6	7	13	17	6	5

Examine whether or not the data is compatible with the assumption that this particular park is visited by birds belonging to these six categories in the proportion $1:1:2:3:1:1$.

Solution. The total number of birds seen by the bird watcher is 54. Under the hypothesis that the proportion of birds in these categories is $1:1:2:3:1:1$, the expected frequencies in the various categories are,

$$54(\tfrac{1}{9})=6, \; 54(\tfrac{2}{9})=12 \text{ and so on.}$$

That is,

Observed frequency	6	7	13	17	6	5
Expected frequency	6	6	12	18	6	6

Here the number of classes is $k=6$ and no parameters are estimated in obtaining the expected frequencies and hence the d.f.$=k-1=5$.

An observed value of Pearson's χ^2 is,

$$\chi^2 = \frac{\Sigma(\text{Observed frequency} - \text{Expected frequency})^2}{\text{Expected frequency}}$$

$$= \frac{(6-6)^2}{6} + \frac{(7-6)^2}{6} + \frac{(13-12)^2}{12} + \frac{(17-18)^2}{18} + \frac{(6-6)^2}{6}$$

$$+ \frac{(5-6)^2}{6} = 0.47.$$

The tabulated value of a X^2 with 5 degrees of freedom at 95% level is 11.07 which is greater than 0.47. That is,

$$P\{\chi_5^2 \geqslant 0.47\} > 0.05.$$

Hence we will accept the hypothesis that the data is compatible with the assumption that the various proportions are as given above. If the data were not compatible then we would have expected large deviations of the expected frequencies from the observed frequencies and thus a large value for Pearson's χ^2.

Comment. In a particular problem if $k \not> 5$ or if any one of the expected frequency $e_i \not> 5$ then the approximation is not good. If some e_i's are <5 then this deficiency can be corrected by combining two or more adjacent classes together. The corresponding expected frequencies are added up and the corresponding observed frequencies are also added up. Now the total number of classes will be reduced and correspondingly the degrees of freedom will also be reduced. The same χ^2 statistic can be used for testing the hypothesis that

$$p_1 = p_2 = \ldots = p_k$$

in a set of k independent Binomial populations.

Example 10.1.2 The number of telephone calls received by a switchboard in successive 5 minute intervals is given in the following table:

Number of calls	0	1	2	3	4	5	6
Number of 5 minute intervals (Frequency)	25	30	20	15	5	3	2

Test the hypothesis that the data is compatible with the assumption that the phone calls there are Poisson distributed.

Solution. Let X denote the number of calls. Then the hypothesis says that X has a Poisson distribution

$$f(x) = \frac{\lambda^x e^{-\lambda}}{x!}$$

where λ is unknown (to be estimated). Since the sample mean satisfies some desirable properties and it can be considered to be a good estimate for λ we will estimate λ by the sample mean,

$$\bar{x} = [0(25) + 1(30) + \ldots + 6(2)]/100 = 1.62 = \hat{\lambda}.$$

That is, now we want to see whether or not

$$f(x) = \frac{(1.62)^x}{x!} e^{-1.62}$$

is a good fit to the given data. The expected frequencies can be evaluated by multiplying the probabilities in the various classes by the total frequency 100. That is, the expected frequencies are,

$$100 \frac{(1.62)^x}{x!} e^{-1.62} \text{ for } x = 0, 1, 2, 3, 4, 5, 6.$$

Number of calls x	Frequencies O_i	Poisson probability $f(x)$ for $\lambda = 1.62$	Expected frequency e_i
0	25	0.2019	20.19
1	30	0.3230	32.30
2	20	0.2584	25.84
3	15	0.1378	13.78
4	15 ⎫	0.0551	5.51 ⎫
5	3 ⎬ 20	0.0176	1.76 ⎬ 7.74
6	2 ⎭	0.0047	0.47 ⎭

The expected frequencies in the last 2 classes are <5 and the total of the last 3 expected frequencies is >5 and, hence, we combine the last 3 classes. Thus, the total number of classes is now 5. But we estimated one parameter λ and hence one degree of freedom is lost and thus we have the degrees of freedom

$$k - r - 1 = 5 - 1 - 1 = 3.$$

An observed value of Pearson's χ^2 is,

$$\chi^2 = \sum_{i=1}^{k} \frac{(O_i - e_i)^2}{e_i} = \frac{(25 - 20.19)^2}{20.19} + \frac{(30 - 32.30)^2}{32.30} + \ldots$$

$$+ \frac{(10 - 7.74)^2}{7.74} = 3.09.$$

The tabulated value of a X_3^2 at 95% level is greater than 3.09. Thus, the observed χ_3^2 does not fall in the critical region and hence we will accept the hypothesis at 95% level.

Comment. In a ' goodness of fit ' problem when a hypothesis is accepted we will say that the data is seen to be compatible with the assumption. This does not mean that the particular distribution is a good fit. If the data is a random sample then we may accept the hypothesis that the particular distribution is a good fit to such data. If the randomness of the data is questionable it is safer to say that there is no evidence to say that the data is not compatible with the assumption.

Example 10.1.3 The yield of corn in 80 experimental plots is given in the following table. The mean and the standard deviation, before the observations are classified, are 35 and 2 respectively. Test whether or not a normal distribution is a good fit to the data.

Yield (bushels)	30 or less	31-32	33-34	35-36	37-38	39 or more
Frequency	8	12	15	20	15	10

Solution. We will make the classes continuous by assuming that the true classes are, less than 30.5, 30.5 to 32.5, and so on. The distribution to be fitted to the data is a continuous distribution namely a $\mathcal{N}(\mu, \sigma^2)$ where μ and σ^2 are estimated to be 35 and 4 respectively. That is,

$$f(x) = \frac{1}{2\sqrt{2\pi}} \exp\left[-\frac{(x-35)^2}{8} \right]$$

The expected frequencies can be evaluated by calculating the probabilities of an observation falling in the various intervals if $f(x)$ is the underlying density and then multiplying these probabilities by the total frequency 80. That is, the frequency in the interval ' less than 30.5 ' is,

$$= 80 \int_{-\infty}^{30.5} \frac{1}{2\sqrt{2\pi}} \exp\left[-\frac{(x-35)^2}{8} \right] dx$$

$$= 80 \int_{-\infty}^{(30.5-35)/2} \frac{1}{2\sqrt{2\pi}} e^{-t^2/2} dt$$

$$= 80(0.0122) = 0.976.$$

The total frequency in the class 32.5 or less

$$= 80 \int_{-\infty}^{(32.5-35)/2} \frac{1}{2\sqrt{2\pi}} e^{-t^2/2} \, dt = 8.448.$$

Hence the frequency in the class 30.5 to 32.5

$$= 8.448 - 0.976 = 7.472.$$

The frequencies in the other classes can be calculated in a similar fashion. These are given below.

Class interval	Observed frequency	Cumulative frequency (expected)	Expected frequency
30.5 or less	8 ⎱ 20	0.976	0.976 ⎱ 8.448
30.5-32.5	12 ⎰	8.448	7.472 ⎰
32.5-34.5	15	32.104	23.656
34.5-36.5	20	61.872	29.768
36.5-38.5	15 ⎱ 25	76.792	14.920 ⎱ 18.128
38.5 or more	10 ⎰		3.208 ⎰

Here $k=4$ and 2 parameters are estimated. Hence the d.f. for for the χ^2 is $4-2-1=1$. The Pearson's χ^2 statistic is,

$$\chi^2 = \Sigma \frac{(O_i - e_i)^2}{e_i} = \frac{(20 - 8.448)^2}{8.448} + \dots + \frac{(25 - 18.128)^2}{18.128} > 6.635,$$

and $\chi^2 : \chi_1^2$. The tabulated value of χ_1^2 at 99% level is 6.635. Hence the hypothesis is rejected. That is, the Normal distribution does not seem to be a good fit.

Comment. Here $k=4<5$ and hence the approximation is not a good one. If in a certain problem the effective value of k is zero or less than it is meaningless which implies that this test cannot be applied at all. When calculating the value of Pearson's χ^2 we need not evaluate the expression completely since we want to see only whether or not it falls in the critical region. In a practical problem, perhaps the evaluation of the largest deviation alone may serve the purpose. Even though the frequencies cannot be fractions, while calculating the expected frequencies all the numbers obtained should be taken as expected frequencies. They should not be rounded up.

Exercises

10.1 A man fishing at a particular place caught fish in the following weight group.

Weight	Less than 1 lb	1-2	2-3	3-4	4-5	5-6	more than 6 lb
Frequency	5	6	13	19	11	7	5

Is the data compatible with the assumption that anybody fishing at that spot will catch fishes in the various weight groups in the proportion 1:1:2:3:2:1:1:?

10.2 At a border crossing point cars containing 1, 2, 3, 4, 5 passengers crossed the border. The exact classification is as follows.

Number of passengers per car	1	2	3	4	5
Number of cars (frequency)	50	400	250	100	200

Is the data compatible with the assumption that the proportion in the various groups are 1:4:2:1:2 respectively?

10.3 The traffic accidents on a high way, per month, are given in the following table

Number of accidents (x)	0	1	2	3	4	5	6	7
Number of months (frequency)	20	40	30	20	15	10	8	7

Test the 'goodness of fit' of a Poisson distribution to this data.

10.4 The following table gives the increase in sales in a particular shop, after advertisements.

Increase	5-10	11-15	16-22	23-27	28-32	33 or more
Number of days (frequency)	200	180	170	140	100	30

Test the ' goodness of fit ' of an exponential distribution to this data. The average increase (before classification) is 18.

10.5 The bust measurements of 100 women are given below. The mean and standard deviation before classification are 36 and 2 respectively. Test whether or not a Normal distribution is a good fit.

Measurements	30 or less	31-32	33-34	35-36	37-38	39 or more
Frequency	10	15	30	30	10	5

Can you generalize your findings? If not why?

10.2 CONTINGENCY TABLES

The following table gives a set of people classified according to their weights and intelligence.

Intelligence Weight (kg)	Intelligent	Average	Below Average
less than 50	10	7	8
51-60	6	7	8
61-70	7	8	7
71-80	5	6	5
81 or more	5	5	5

The numbers in the various cells are the frequencies. If some observations are classified according to two characteristics (quantitative or qualitative) and if the frequencies are given in a two-way table then such a table is called a *contingency table*. Our aim is to test the hypothesis that there is no association between the two characteristics under consideration in a contingency table. This can be achieved by testing the hypothesis with the help of Pearson's χ^2 statistic.

Let the characteristics under consideration be $A_1, A_2,..., A_r$ and $B_1, B_2,..., B_s$ respectively. Then the contingency table can be given as follows,

	B_1	B_2	...	B_s	total
A_1	n_{11}	n_{12}	...	n_{1s}	$n_{1.}$
A_2	n_{21}	n_{22}	...	n_{2s}	$n_{2.}$
$\vdots$	$\vdots$	$\vdots$	$\vdots$	$\vdots$	$\vdots$
A_r	n_{r1}	n_{r2}	...	n_{rs}	$n_{r.}$
Total	$n_{.1}$	$n_{.2}$	...	$n_{.s}$	$n_{..}$

where n_{ij} denotes the frequency (number of observations) in the ith row jth column cell, $n_{i.}$ denotes the sum of the frequencies in the ith row, $n_{.j}$ denotes the sum of the frequencies in the jth column, for $i = 1, 2,..., r$ and $j = 1, 2,..., s$ and $n.. = n$ denotes the total frequency. The above notation is a very convenient one where a summation with respect to a subscript is denoted by a dot. For example,

$$\sum_{j=1}^{s} n_{ij} = n_{i.}; \quad \sum_{i=1}^{r} n_{ij} = n_{.j}; \quad \sum_{i=1}^{r} \sum_{j=1}^{s} n_{ij} = n..$$

Let p_{ij} denote the probability of getting an observation in the (i, j)th cell. Then $p_{i.} = \sum_{j=1}^{s} p_{ij}$ denotes the probability of getting an observation in the 1st or 2nd or...or the sth column of the ith row=probability of getting an observation in the ith row and

$p.j$=the probability of getting an observation in the jth column. We would like to test the hypothesis,

H_0: There is no association between the characteristics of classification.

If H_0 is true then p_{ij}=probability of getting an observation in the (i, j)th cell will be the product of the probabilities of getting an observation in the ith row and that in the jth column. That is,

$p_{ij}=(p_{i.}) (p_{.j})$ under H_0.

Now the hypothesis can be written as,

H_0: $p_{ij}=(p_{i.}) (p_{.j})$ and $H_1 : p_{ij} \neq (p_{.i}) (p_{.j})$

Under H_0, the expected frequency in the (i, j)th cell is,

$$e_{ij}=n(p_{i.}) (p_{.j})$$

But $\sum\limits_{i=1}^{r} p_{i.}=1= \sum\limits_{j=1}^{s} p_{.j}$ and hence only $(r-1)+(s-1)$ independent parameters are estimated when all $p_{.i}$ and $p_{.j}$ are estimated.

Hence, $\qquad \hat{e}_{ij}=n \dfrac{(n_{i.})}{n} \dfrac{(n_{.j})}{n} = \dfrac{n_{i.}}{n} n_{.j}$

where $\dfrac{n_{i.}}{n}$ and $\dfrac{n_{.j}}{n}$ are the maximum likelihood estimates of $P_{i.}$ and $P_{.j}$ respectively. That is, the χ^2 statistic for testing H_0 is

$$\chi^2 = \Sigma \frac{(n_{ij}-\hat{e}_{ij})^2}{\hat{e}_{ij}} \approx \chi^2_{(r-1)\,(s-1)}$$

There are altogether rs frequencies (there are rs cells) and $(r-1)+(s-1)$ parameters are estimated and hence the degrees of freedom is

$$=rs-(r-1)-(s-1)-1=rs-r-s+1=(r-1)\,(s-1)$$

Example 10.2.1 Check whether there is any association between the weight and intelligence, based on the data given in section 10.2.

Solution. The data is reproduced here for convenience. The numbers in the corners of the cells are the expected frequencies calculated under H_0 : that there is no association.

For example the expected frequencies

$$\hat{e}_{11} = \frac{(26) (33)}{100} = 8.58, \, \hat{e}_{12} = \frac{(26) (34)}{100} = 8.84, \text{ etc.}$$

Intelligence Weight (kg)	Intelligent	Average	Below average	Total
less than 50	10 8.58	8 8.84	8 8.58	26
51-60	6 6.93	7 7.14	8 6.93	21
61-70	7 7.26	8 7.48	7 7.26	22
71-80	5 5.28	6 5.44	5 5.28	16
81 or more	5 4.95	5 5.10	5 4.95	15
Total	33	34	33	100

The degrees of freedom $= (r-1)\,(s-1) = (4)\,(2) = 8$.

An observed value of Pearson's X^2 is,

$$\chi^2 = \sum_{i=1}^{5} \sum_{j=1}^{3} \frac{(n_{ij} - \acute{e}_{ij})^2}{\acute{e}_{ij}} = \frac{(10-8.58)^2}{8.58} + \frac{(8-8.84)^2}{8.84} + \cdots$$

$$+ \frac{(5-4.95)^2}{4.95} < 15.507,$$

where 15.507 is the tabulated value of a chi-square with 8 degrees of freedom at the 95% level.   Hence we will accept the null hypothesis that there is no evidence of association between weight and intelligence based on the given data.

Comment.   Our findings here cannot be generalized because the data we have may not be a representative of a classification of weight and intelligence in the population to which generalizations are to be made.   So in short we can only test compatibility of the data with the hypothesis of no association.

Exercises

10.6 The following table gives a categorized data classifying 100 students according to their interest in statistics and their

	Interest in Statistics			
Disposition	Highly interested	Interested	Indifferent	Not interested
Good	15	12	10	5
Tolerable	10	10	5	5
Intolerable	5	10	8	5

disposition. Test whether there is any evidence of association between these characteristics.

10.7 The following categorized data gives the classification of 95 people according to their intelligence and mood upon getting up in the morning.

Mood	Intelligence		
	Intelligent	Average	Below average
Good	10	8	5
Tolerable	8	9	10
Intolerable	5	6	14

Test whether there is any evidence of association between these characteristics.

10.3 Kolmogorov-Smirnov Statistic

In the 'goodness of fit' problems discussed in sections 10.1 and 10.2 we used Pearson's χ^2 statistic. There are some disadvantages of using this particular statistic. This can be used only in a data which is categorized or classified according to some characteristics and if the data is not categorized then we have to form a frequency table by selecting a suitable number of classes. This will induce some arbitrariness in selecting the class intervals and the number of classes. Another disadvantage is that the number of classes k and the expected frequencies e_{ij} are to be greater than or equal to 5 in order to get a good chi-square approximation. This limits the uses of the Pearson's χ^2 statistic in small sample problems. But another statistic known as the Kolmogorov-Smirnov statistic, named after its inventors, is of great help especially in small sample problems.

Let $x_1, x_2,..., x_n$ be an observed random sample of size n. We want to test whether or not this sample can be considered to be from a specified population $f(x, \theta_0)$. For example suppose that we have a sample of size 5 and we would like to know whether or not this could be considered to be from a $N(\mu, \sigma^2)$ with $\mu=10$ and $\sigma^2=4$. In order to construct Kolmogorov-Smirnov statistic we arrange the observations according to their order of magnitude. Let $u_1, u_2, ..., u_n$ be an arrangement of $x_1, x_2, ..., x_n$. That is, $u_1 \leqslant u_2 \leqslant ... \leqslant u_n$. Form the *sample distribution function* $S_n(x)$. That is,

$$S_n(x) = \begin{cases} 0, & x < u_1 \\ \dfrac{r}{n}, & u_r \leqslant x < u_{r+1} \\ 1, & x \geqslant u_n. \end{cases}$$

The hypothesis to be tested is that $x_1,..., x_n$ came from a population with density function $f(x, \theta_0)$ or with distribution function (cummulative density) $F(x, \theta_0)$, where θ_0 indicates that all the parameters are known. Then the following statistics,

$$D_n = \sup_x \ |S_n(x) - F(x, \theta_0)|$$

and $\qquad W^2 = E\,|S_n(x) - F(x, \theta_0)|^2$

are known as the Kolmogorov-Smirnov statistic and the W^2-statistic respectively, where E denotes mathematical expectation and '$\sup_x$' denotes the maximum value over x. In other words, D_n is the maximum absolute difference between $S_n(x)$ and $F(x, \theta_0)$ over the values of x. Figure 10.1 gives an illustration of D_n.

The values of D_n are tabulated for various values of n.

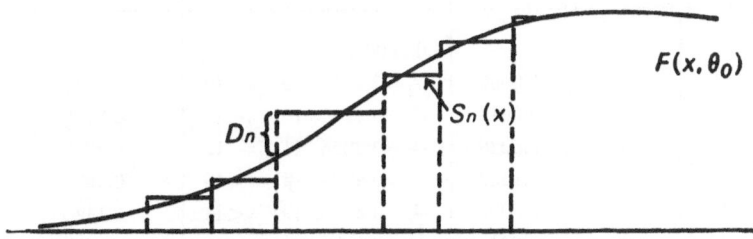

FIGURE 10.1 An illustration of D_n, the Kolmogorov-Smirnov Statistic

Example 10.3.1 Check whether the following data could be considered to have come from a Normal population with parameters $\mu = 13$ and $\sigma^2 = 1$: 16, 15, 15, 9, 10, 10, 12, 12, 11, 13, 13, 13, 14, 14.

Solution. Here $n = 14$ and under H_0 the population is

$$f(x, \theta_0) = \frac{1}{\sqrt{2\pi}} \exp \left[-\frac{(x-13)^2}{2} \right].$$

Thus, $$F(y, \theta_0) = \int_{-\infty}^{y} f(x, \theta_0) \, dx$$

For example, $F(x, \theta_0)$ at $x = 9$ can be found out as follows.

For $x = 9$, $$F(x, \theta_0) = \int_{-\infty}^{9} \frac{1}{\sqrt{2\pi}} \exp \left[-\frac{(x-13)^2}{2} \right] dx$$

$$= \int_{-\infty}^{-4} \frac{1}{\sqrt{2\pi}} e^{-t^2/2} \, dt = 0.0000.$$

For $x = 10$, $$F(x, \theta_0) = \int_{-\infty}^{10} \frac{1}{\sqrt{2\pi}} \exp \left[-\frac{(x-13)^2}{2} \right] dx = 0.0013.$$

For $9 \leqslant x < 10$, $S_n(x) = S_{14}(x) = \frac{1}{14} = 0.0714$.

The various values of x, $S_{14}(x)$ and $F(x, \theta_0)$ are given below.

x	Frequency	$F(x, \theta_0)$	$S_{14}(x)$	$\lvert S_{14}(x) - F(x, \theta_0) \rvert$
			0.0000 for $-\infty < x < 9$	
9	1	0.0000	$\frac{1}{14} = 0.0714$, $9 \leqslant x < 10$	0.0714
10	2	0.0013	$\frac{3}{14} = 0.2142$, $10 \leqslant x < 11$	0.2129
11	1	0.0228	$\frac{4}{14} = 0.2856$, $11 \leqslant x < 12$	0.2628
12	2	0.1587	$\frac{6}{14} = 0.4284$, $12 \leqslant x < 13$	0.2697
13	3	0.5000	$\frac{9}{14} = 0.6426$, $13 \leqslant x < 14$	0.1426
14	2	0.8413	$\frac{11}{14} = 0.7854$, $14 \leqslant x < 15$	0.1987
15	2	0.9772	$\frac{13}{14} = 0.9282$, $15 \leqslant x < 16$	0.1918
16	1	0.9987	$\frac{14}{14} = 1.0000$, $16 \leqslant x < \infty$	0.0705

Hence the largest of the absolute deviations between $S_n(x)$ and $F(x, \theta_0)$ is,

$$D_n = D_{14} = 0.2697$$

The tabulated value of D_{14} at 95% level is 0.35. (Tables are available from the references at the end of the book). The observed $D_n = 0.2697 < 0.35$ and hence we cannot reject H_0. Therefore we may consider a $\mathcal{N}(\mu = 13, \sigma^2 = 1)$ as a good fit to the given data at the 95% level, assuming that the data is a random sample from some population, since the observed D_n is not in the critical region. (Notice that if H_0 is correct, large values of D_n are less likely).

Exercises

10.8 The following table gives the telephone calls received by a switch-board. Test whether or not a Poisson distribution with parameter $\lambda = 2$ is a good fit. (The tabled value of $D_{256} = 0.085$ at 95% level).

x	0	1	2	3	4	5	6	7	8
Frequency	52	68	60	40	22	10	3	1	0

10.9   The following table gives the waist measurements of 35 girls. Test whether or not a $\mathcal{N}(\mu,\ \sigma^2)$ is a good fit to the data. (Tabled value of $D_{35}=0.27$ at 99% level).

x	10	12	13	14	15	16	17
Frequency	2	4	6	10	7	4	2

Hint.: Estimate the parameters. The test is not exact due to estimation. Use a 99% level.

10.4 SOME DISTRIBUTION-FREE TESTS

Here we will consider a few more distribution-free tests which can be used for testing, independence, randomness, normality and so on.   We will consider the sign test, the rank test, the run test and the H-test briefly.   For further reading on these tests see the Bibliography at the end of this book.

10.41 *The Sign Test*

This is a distribution-free test which is quite appropriate when the population is known to be *continuous* and *symmetrical*.   Suppose that we want to test the hypothesis $\theta=\theta_0$ (a specified value) in a population known to be symmetrical about the ordinate at $x=\theta$. A single sample sign test can be formulated as follows.   Let the observed sample be $x_1, x_2,..., x_n$.   Give a plus sign to all the observations which exceed θ_0 and a minus sign to all the observations which are less than θ_0.   Without any loss of generality we can assume that the probability of getting an observation greater than θ_0 is the same as that of less than θ_0.   Hence an equivalent test is that of testing a Binomial probability $p=\frac{1}{2}$.   That is, the following are equivalent tests in this case

$$\begin{cases} H_0 : \theta=\theta_0 \quad p=\frac{1}{2} \\ H_1 : \theta<\theta_0 \quad p>\frac{1}{2}; \end{cases} \Rightarrow \begin{cases} H_0 : \theta=\theta_0 \quad p=\frac{1}{2} \\ H_1 : \theta>\theta_0 \quad p<\frac{1}{2}; \end{cases} \Rightarrow \begin{cases} H_0 : \theta=\theta_0 \quad p=\frac{1}{2} \\ H_1 : \theta\neq\theta_1 \quad p\neq\frac{1}{2}. \end{cases}$$

Example 10.41.1 Use a sign test for testing the hypothesis that the mean yield of a particular variety of wheat is 50 at the 95%

level.   The following are the yields of wheat from 18 experimental plots: 49, 49, 48, 48.5, 52, 50, 57, 54, 53.5, 52, 49.5, 48.5, 53, 51, 47, 51.5, 52, 53.

Solution. We want to test the expected yield $\theta = 50$.   That is,
$$H_0 : \theta = 50, \ H_1 : \theta \neq 50.$$
When we consider the sign test, this is equivalent to testing
$$H_0 : p = \tfrac{1}{2}, \ H_1 : p \neq \tfrac{1}{2}$$
in a Binomial situation of 18 trials.   Giving $+$ and $-$ signs we have the following.

$-$	$-$	$-$	$-$	$+$	$+$	$+$	$+$	$+$	$-$	$-$	$+$	$+$
49,	49,	48,	48.5,	52,	50,	57,	54,	53.5,	52,	49.5,	48.5,	53,	51,

$-$	$+$	$+$	$+$
47,	51.5,	52,	53.

We have one observation equal to 50 and hence we omit this. Thus, out of 17 numbers there are 10 plus signs.   That is, in 17 Binomial trials we have 10 successes.   At 95% level the critical region is $x \leqslant 4$ or $x \geqslant 13$.   That is,
$$P\{x \leqslant 4\} \leqslant 0.025 \text{ and } P\{x \geqslant 13\} \leqslant 0.025.$$
The observed number of successes equals 10 which is not in the critical region and hence we accept the hypothesis.

Comment. When the sample size is large we can use the Normal approximation to the Binomial; see Chapter 9, section 9.4.   The same technique can be used for testing the equality of two populations if we have paired samples.   Suppose that we have n pairs of numbers of which the first numbers are from one population and the second numbers are from the other population.   Then a plus sign can be assigned to the pair where the first number is greater than the second.   Again ties may be omitted.   The assumption of symmetry can be omitted if θ_0 is the median point since then
$$P\{x \geqslant \theta_0\} = P\{x \leqslant \theta_0\} = \tfrac{1}{2}$$
when X is continuous.

10.42   *The Rank Test*

Suppose that we have two samples and we want to test the hypothesis that the samples have come from the same population.   This is a test of equality of two populations.   Let the sample sizes be n_1

and n_2. Now pool the samples, and then arrange the numbers according to the order of their magnitudes and assign ranks 1 to n_1+n_2. If a number of numbers have the same magnitudes then assign the mean rank to each of them. For example, if the first two numbers have the same magnitudes then assign the rank $(1+2)/2=1.5$ each. The third number has the rank 3 and so on. A commonly used test based on the rank sums is the *Mann-Whitney U-test* where,

$$U = n_1 n_2 + \frac{n_1(n_1+1)}{2} - R_1$$

where n_1 and n_2 are the sample sizes and R_1 is the sum of the ranks occupied by the first sample of size n_1. It can be shown that,

$$E(U) = n_1 n_2/2 \text{ and Var } (U) = \sigma_u{}^2 = n_1 n_2 \frac{(n_1+n_2+1)}{12}$$

and the standardized score,

$$T = \frac{U - E(U)}{\sigma_u}$$

is approximately normally distributed, under the null hypothesis, when n_1 and n_2 are large. A good approximation is available for n_1, n_2 as small as 8. Exact tables of U are available from D. B. Owen, *Handbook of Statistical Tables* (Addison Wesley, 1962).

Example 10.42.1 Corn is planted in 10 and 9 experimental plots and two different fertilizers A and B are applied to these 10 and 9 plots respectively. The yields are given in the following table.

Fertilizer A	15	14	12	13	12	10	11	16	17	18
Fertilizer B	9	8	10	14	19	21	20	22	22	

Test the hypothesis that the fertilizers are equally effective, at the 95% level.

Solution. We will test this hypothesis by testing that the corresponding populations are identical. This can be achieved by applying a rank test. The following table gives the pooled samples and the corresponding ranks, where A denotes that the corresponding observation came from the first sample (fertilizer A)

Pooled sample	8	9	10	10_A	11_A	12_A	12_A	13_A	14_A
Ranks	1	2	3.5	3.5	5	6.5	6.5	8	9.5

14	15_A	16_A	17_A	18_A	19	20	21	22	22
9.5	11	12	13	14	15	16	17	18.5	18.5

For example, the two 10's were supposed to occupy the ranks 3 and 4 and hence they are given $(3+4)/2=3.5$ each.

Now R_1=sum of the ranks occupied by the first sample
$$=3.5+5+6.5+6.5+8+9.5+11+12+13+14=89,$$
and $n_1 = 10$ and $n_2=9$.

An observed value of U is,
$$U = n_1 n_2 + n_1(n_1+1)/2 - R_1 = (10)(9) + (10)(11)/2 - 89 = 56.$$
$E(U) = n_1 n_2/2 = 45$ and $\sigma_u{}^2 = n_1 n_2(n_1+n_2+1)/12 = 150.$

An observed value of the standardized score is,
$$t = \frac{U-E(U)}{\sigma_u} = \frac{(56-45)}{\sqrt{150}} < 1.96.$$

Hence the hypothesis that the populations are identical cannot be rejected, since the standardized score is approximately normally distributed and the critical region is $t \geqslant 1.96$ at the 95% level.

Comment. The same rank test can be modified to test the hypothesis that the populations are identical against the alternative that the population variances are different.

10.43 *The Run Test*

This is a very convenient test for testing randomness. That is, suppose that we have a set of numbers and we would like to test and see whether or not we could consider the data to be a random sample from some population. The run test is based on the runs in a sample. If there is a sequence of two symbols then a succession of identical symbols is called a· run. For example, in the sequence, $GDDDGGDG$ there are 5 runs. Suppose that there is a sequence of n_1+n_2 symbols of two types where n_1 are of one type and the remaining n_2 are of the second type. Then it can be shown that when n_1 and n_2 are large the standardized run, (R denotes the number of runs in the sequence),

$$T = \frac{R-E(R)}{\sigma_R}$$

is approximately a standard normal under the null hypothesis that the symbols appear at random in the sequence, where, (it can be shown),

$$E(R) = \frac{2n_1n_2}{n_1+n_2} + 1 \text{ and Var } (R) = \sigma_R^2 = \frac{2n_1n_2 \, (2n_1n_2-n_1-n_2)}{(n_1+n_2)^2 \, (n_1+n_2-1)}.$$

A good approximation is available when n_1 and n_2 are greater than 10.

Example 10.43.1 A machine produces a particular item which can be classified as either defective (D) or good (G). One item at every 5 minutes is tested and the following sequence is obtained $GGGDDGGDGGGDDDGGDGDDGDGGDDGDG$. Test at 95% level whether or not the occurrence of a defective can be considered to be random.

Solution. The sequence is $GGG\ DD\ GG\ D\ GGG\ DDD\ GG\ D\ G$ $DD\ G\ D\ GG\ DD\ G\ D\ G$. There are 17 runs, 16 G's and 13 D's. Hence,

$$n_1=16, \ n_2=13, \ R=17.$$

$$E(R) = \frac{2n_1n_2}{n_1+n_2} + 1 = \frac{2(16)\,(13)}{16+13} + 1 = 15.4,$$

$$\sigma_R^2 = \frac{2n_1n_2\,(2n_1n_2-n_1-n_2)}{(n_1+n_2)^2\,(n_1+n_2-1)} = 6.8.$$

Thus an observed value of the standardized run

$$= \frac{17-15.4}{\sqrt{6.8}} < 1.96.$$

Hence we accept the hypothesis.

10.44 *Kruskal-Wallis H-Test*

For testing whether k independent samples have come from k identical populations, a usual test used is the H-test. The statistic

$$H = \frac{12}{n(n+1)} \sum_{i=1}^{k} \frac{R_i^2}{n_i} - 3(n+1)$$

where $n = \sum_{i=1}^{k} n_i$, n_i is the size of the ith sample and R_i is the sum of the ranks occupied by the ith sample when the samples are pooled and ranked. When $n_i > 5$ for all i, H has an approximate chi-square distribution with $k-1$ degrees of freedom under the null hypothesis.

Exercises

10.10 The following data gives the increase in weight of 15 experimental animals under a certain diet. Test the hypothesis that the expected increase is 25, at 95% level, against the alternative that is more. 22, 23, 25, 26, 27, 25, 28, 29, 30, 21, 22, 24, 24, 29, 31.

10.11 A set of 20 animals of a particular type were exposed to a poisonous gas till they died. The time interval is recorded. Use a sign test to test the hypothesis that the expected time interval is 2 minutes against the alternative that it is not. 1.8, 1.9, 2.0, 2.1, 2.2, 2.7, 2.5, 2.1, 1.9, 1.7, 1.8, 2.2, 2.3, 2.5, 1.6, 1.7, 2.1, 2.3, 2.5, 2.4. (Time being measured in minutes). Test at 95% level.

10.12 Two types of missiles are test fired. The flight distances are given below:

Type 1	2000	2025	2010	2020	2019	1990	1985	2012	2020	2010
Type 2	2010	2015	2022	2020	1980	2000	2013	2014	2100	

Test the hypothesis that both are equally efficient, at 95% level.

10.13 Two random samples of 12 boys and 14 girls have the following I.Q.'s.

Boys	110 112 108 97 99 98 102 105 120 118 113 114
Girls	100 102 108 110 104 109 121 118 116 117 119 120 114 110

Test the hypothesis that the boys and girls are equally intelligent, at the 95% level.

10.14 In a production process the following sequence of good (G) and defective (D) items are observed. Test at 99% level to see whether or not the defectvies are occurring at random. *GGGG D GG DD G DDDDDD GG D GGGGG DDD G D.*

10.15 The following is a sequence of numbers taken from a table of random numbers. Test at 95% level to see whether even and odd numbers are appearing at random. 04433806742452018222106100579437515.

10.16 The following observations are made on three random samples of girls of a particular age-group. The numbers are the percentage reduction in skin rash by applying three brands of beauty creams. Test the hypothesis that the three brands are equally good, at 99% level.

Brand A	10	15	25	30	22	27	28	31	40	33
Brand B	20	18	17	20	21	23	22	24	26	
Brand C	18	19	20	21	20	18	29			

10.17 The following table gives the yields of corn from a number of test plots under 4 different fertilizers. Test the hypothesis that the fertilizers are equally effective, at 95% level.

Fertilizer 1	10	8	11	12	16	9	14
Fertilizer 2	8	10	12	13	15	8	
Fertilizer 3	9	12	11	14	14	10	
Fertilizer 4	8	10	10	15	10	12	

Additional Exercises

10.1   A historical monument is visited by groups of 1, 2, 3, 4, 5, 6 people and the exact classification is given below.

Numbers of people per set	1	2	3	4	5	6
Number of sets (frequency)	10	41	42	27	12	8

Test the hypothesis that the true proportions are 1:4:4:3:1:1.

10.2   A die is rolled 50 times and the following table gives the outcomes.

Face numbers	1	2	3	4	5	6
Frequencies	7	8	9	8	8	10

Test the hypothesis that the die is balanced.

10.3   Emission of gamma particles by a radio-active source in 5 minute intervals is given below.

Number of emissions (x)	0	1	2	3	4	5	6	7
Number of 5 minute intervals	20	65	50	45	30	20	15	5

Test whether or not a Poisson distribution is a good fit.

10.4   The consumption of electricity in a township per day is given below. The average consumption before classification is 2,000 k.w.

Consumption (x)	0-2000	2000-2020	2020-2030	2030-2040	2040-2050	2050-2060
Number of days	60	50	40	15	10	5

Test the 'goodness of fit' of an exponential population by (1) Pearson's χ^2 statistic, (2) Kolmogorov-Smirnov Statistic.

10.5   The marks obtained by the students in a class have an average of 60 with a standard deviation of 5. The exact classification is given below.

Marks	less than 40	41-50	51-60	61-70	71-80	81-90	91-100
Number of students	5	10	35	22	15	8	5

Test the 'goodness of fit' of a normal distribution to this data.

10.6   Test whether the following data is compatible with the assumption that there is no association between the weight and colour of the eyes.

Weight	Colour of the eyes		
	Black	Brown	Blue
Less than 100	10	20	20
100-150	12	22	30
150-175	15	35	40
175-200	8	10	8

10.7   *Coefficients of Contingency.* Similar to the coefficient of correlation between quantitative characteristics we can have some measures of association between either two qualitative characteristics or between a qualitative and a quantitative characteristics. Some of the measures usually used are (1) Square contingency, (2) Pearson's coefficient of contingency, which are defined as follows

$$\text{Square Contingency } S = n \left[\sum_{i=1}^{r} \sum_{j=1}^{s} \frac{n_{ij}^2}{n_i \cdot n_{\cdot j}} - 1 \right]$$

where $\sum_{i=1}^{r} n_{ij} = n_{\cdot j}, \ \sum_{j=1}^{s} n_{ij} = n_{i\cdot}$ and $n = \sum_{i=1}^{r} \sum_{j=1}^{s} n_{ij}$

K. Pearson's Coefficient of Contingency $P = \left[\dfrac{\chi^2}{n + \chi^2} \right]^{\frac{1}{2}}$

where χ^2 is Pearson's χ^2 statistic.

Test for association and if there is any evidence of association calculate S and P from the following table

	B_1	B_2	B_3	B_4
A_1	8	12	18	24
A_2	10	13	20	28
A_3	10	15	22	30

10.8   20 tea-tasters tasted two brands A and B of tea. Some preferred A to B. The following sequence gives the exact result. *ABAABBAAABAAABAABBAA.* Use a sign test to test the hypothesis that the true proportions are the same.

10.9   The marks obtained by a sample of 15 students are, 85, 80, 78, 70, 65, 68, 69, 82, 83, 87, 88, 90, 80, 81, 82. Test the hypothesis that the true average marks of such students is 80, by using a sign test.

10.10 The efficiencies of two machines are compared by considering the time taken by the machines to do a certain job. The data is given below. Test the hypothesis that both the machines are equally effective.

A	2.5	2.8	2.6	2.8	3.1	3.3	3.5	3.0	3.1	3.2	2.1
B	2.0	2.6	3.1	2.9	3.2	3.2	2.9	3.2	2.9	3.3	2.2

10.11 The marks obtained by two sets of students, of the same background knowledge who are taught by two different methods of teaching, are given in the following table:

Method 1	65	82	81	80	70	75	71	55	78	82	85
Method 2	70	80	85	82	75	70	69	58	62	71	75

Test the hypothesis that both the methods are equally effective.

10.12 The following sequence of numbers are from a table of random numbers. Check and see whether even numbers are occurring at random 68921081417922705748512765714 331927.

10.13 The following is a sequence of good (G) and defective (D) items produced by a machine. Test and see whether the defectives are appearing at random. *GGG DD GG DD GGG DDDDD GG D G D G D GGG DDD G.*

10.14 The yields of a number of test plots under 3 different fertilizers are given below. Test the hypothesis that all the fertilizers are equally effective as far as the yields are concerned.

A	31	28	32	30	31	28	31
B	32	29	35	28	29	30	
C	35	30	29	27	22		

10.15   The percentage reduction in cavities due to 3 different toothpastes is given in the following table:

Brand A	50	29	28	29	45	42	30	32	35	38
Brand B	28	35	31	30	41	41	28			
Brand C	30	32	31	28	47	31				

Test the hypothesis that all the toothpastes are equally effective in reducing cavities.

Choose an appropriate size of the critical region in all the above problems.

10.16   Show that the variance expression $\dfrac{12}{n(n+1)} \displaystyle\sum_{i=1}^{k} n_i(\overline{R}_i - \overline{R})^2$

reduces to Kruskal-Wallis H-statistic, namely,

$$\frac{12}{n(n+1)} \sum_{i=1}^{k} \frac{R_i^2}{n_i} - 3(n+1) \text{ where } \overline{R}_i = \frac{R_i}{n_i},$$

$$\overline{R} = \frac{1}{n} \sum_{i=1}^{k} R_i = \frac{n+1}{2}, \; n = \sum_{i=1}^{k} n_i, \; R_i \text{ is the sum of ranks}$$

occupied by the ith sample and n_i is the ith sample size.

CHAPTER ELEVEN

Statistical Regression

Regression and correlation analyses are often used in problems in Econometrics, Biometry, Social Sciences and so on. Regression analysis is a widely used tool in evaluating the relationship among a number of variables and then using this relationship to predict the value of a variable or to forecast the value that will be taken by a variable at a particular time. The word regression means ' going back '. The theory of regression analysis started when statisticians investigated the genetical problem of making inference regarding the parents by observing the offsprings. But in the present day statistical discussion this branch is the study of structural relationship among observable (on which numerical observations can be made) variables. For example, the price of a commodity may be related to the demand for that commodity. Both the price and demand are observable, but the exact nature of the relationship is unknown. This is evidently not a mathematical relationship such as $y=2x+3$. In the mathematical relationship $y=2x+3$ there is one and only one value for y, which is determinate, corresponding to a given value for x. For example if $x=2$ then $y=7$. But in the case of demand and price suppose that we estimate the relationship between the demand and price based on a given data but for a given price we won't be able to say that the demand will be exactly a particular number. Here we can only predict the demand based on a model for the relationship between the price and the demand. Unless the particular nature of the relationship (a particular model) is assumed we won't be able to predict the value of a variable by observing another. In order to obtain an idea regarding the nature of the relationship between two observable variables x and y the experimental scientists often use a scatter diagram.

11.1 THE SCATTER DIAGRAM

This is a diagrammatic representation of a set of paired values. Suppose that x_1, x_2, ..., x_n are the observations on one variable x and y_1, y_2, ..., y_n are the corresponding observations on another variable y then by plotting the points $(x_1, y_1), ..., (x_n, y_n)$ we get a scatter diagram or *scattergram*.

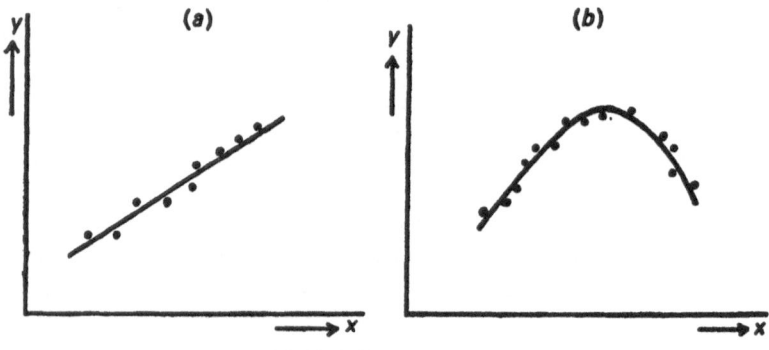

FIGURE 11.1 A scattergram

In Figure 11.1 there are two scattergrams. In (a) the points seem to concentrate around a straight line and in (b) the best fitting curve seems to be a parabola. An equation $y = a + bx$ where a and b are some constants denotes a straight line and the particular straight line in (a) can be estimated by using some methods of estimation. An equation of the form $y = \alpha + \beta x + \gamma x^2$ will in general define a family of parabolas where α, β, γ are some constants. From a given data we can estimate the model by using some techniques. The usual technique used to estimate a particular model is the method of least squares.

11.2 THE METHOD OF LEAST SQUARES

Suppose that x and y are two observable variates and suppose that from a scattergram or otherwise we think that $y = f(a_1, a_2, ..., a_k, x)$ is an appropriate model to describe the observations on x and y, where $a_1, a_2, ..., a_k$ are some constants. For example in Figure 11.1 (a) we thought that a straight line could be a good model to describe the data. But a general equation for a straight line contains 2 constants which can be estimated from a given set of observations on the variables. Thus, once we assume that model then evidently $e = y - f(a_1, a_2,..., a_k, x)$ can be called the error in taking $f(a_1, a_2, ..., a_k, x)$ for y. If the unknown constants $a_1, a_2,...,a_k$ are estimated by minimizing the error sum of squares, that is,

$$\Sigma e^2 = \Sigma \{y - f(a_1, a_2,..., a_k, x)\}^2$$

then the method of estimating the model is called the *method of least squares* and the estimates of the constants $a_1, a_2,..., a_k$ from a given data are called the least square estimates. In order to minimize Σe^2 (the error sum of squares) we can use the method of Calculus. That is, we differentiate Σe^2 partially with respect to $a_1, a_2, ..., a_k$, equate them to zero, and solve for the unknowns and choose those values which will minimize Σe^2. That is, we obtain the following equations:

$$\frac{\partial}{\partial a_j} (\Sigma e^2) = 0 \quad \text{for} \quad j = 1, 2, ..., k.$$

That is,

$$-2 \Sigma [y - f(a_1, ..., a_k, x)] \frac{\partial}{\partial a_j} f(a_1, ..., a_k, x) = 0,$$

$$j = 1, 2, ..., k$$

$$\Longrightarrow \quad \Sigma [y - f(a_1, ..., a_k, x)] \frac{\partial}{\partial a_j} f(a_1, ..., a_k, x) = 0,$$

$$j = 1, 2, ..., k.$$

These equations are known as the *least square normal equations* (nothing to do with normality). By solving these normal equations we obtain the least square estimates for the unknowns.

11.21 *Curve Fitting*

The selection of a particular model for a given situation and the evaluation of the unknowns based on a given set of observations, can be called the fitting of the appropriate curves (models) to the data because once the unknowns are estimated the selection of a particular model is complete.

Example 11.2.1 For the following data fit a straight line of the form $y = a + bx$, by the method of least squares:

x	0	1	-1	2
y	3	4	2	8

Solution. We will obtain a general formula by taking the observations as $x_1,\ldots, x_n$ and $y_1, y_2,\ldots, y_n$ and then obtain the particular straight line by substituting the above values. The model to be fitted is,

$$a + bx \text{ for } y$$

where a and b are unknowns. Let e be the error in taking $a+bx$ for y. Then we may write the model as,

$$y = a + bx + e.$$

The error sum of squares, $\Sigma\ e^2 = L(\text{say})$, is

$$L = \sum_{i=1}^{n} e_i{}^2 = \sum_{i=1}^{n} (y_i - a - bx_i)^2$$

The least square normal equations are,

$$\frac{\partial L}{\partial a} = 0 \quad \text{and} \quad \frac{\partial L}{\partial b} = 0$$

That is,

$$-2 \sum_{i=1}^{n} (y_i - a - bx_i) = 0 \text{ and } -2 \sum_{i=1}^{n} x_i(y - a - bx_i) = 0$$

That is,

$$\sum_{i=1}^{n} (y_i - a - bx_i) = 0 \qquad\qquad (1)$$

and

$$\sum_{i=1}^{n} x_i (y_i - a - bx_i) = 0 \qquad\qquad (2)$$

From (1) $\Sigma y_i - \Sigma\ a - b\ \Sigma x_i = 0 \Longrightarrow \hat{a} = \dfrac{\Sigma y_i}{n} - \hat{b}\dfrac{\Sigma x_i}{n}$ (3)

where $\hat{a}$ and $\hat{b}$ denote the estimated values of a and b respectively. From (2) $\Sigma x_i y_i - a \Sigma x_i - b \Sigma x_i{}^2 = 0$ (4)

Substituting the value of $\hat{a}$ from (3) in (4) we get,

$$\hat{b} = \frac{\Sigma x_i\, y_i/n - \bar{x}\,\bar{y}}{\Sigma x_i{}^2/n - \bar{x}^2} = \frac{\Sigma(x_i - \bar{x})\,(y_i - \bar{y})/n}{\Sigma(x_i - \bar{x})^2/n} = \frac{\text{Cov}\,(x, y)}{\text{Var}\,(x)}$$

and $\hat{a} = \bar{y} - \hat{b}\, \bar{x},$

where Cov (x, y) and Var (x) denote the covariance between the set of numbers $(x_1,\ldots,x_n)$ and $(y_1,\ldots,y_n)$ and the sample variance in the set of numbers $x_1, \ldots, x_n$ respectively (see Chapter 7). But for our problem here, $x_1=0$, $x_2=1$, $x_3=-1$, $x_4=2$ and $y_1=3$, $y_2=4$, $y_3=2$ and $y_4=8$, and $n=4$. Therefore

$$\frac{\Sigma x_i y_i}{n} - \bar{x}\bar{y} = \frac{0(3)+1(4)-1(2)+2(8)}{4} -$$

$$\frac{(0+1-1+2)}{4}\frac{(3+4+2+8)}{4} = 2.375$$

and $\quad \frac{\Sigma x_i^2}{n} - \bar{x}^2 = \frac{0^2+1^2+(-1)^2+2^2}{4} - \left(\frac{1}{2}\right)^2 = 1.25.$

Hence, $\qquad \hat{b} = 2.375/1.25 = 1.9$ and $\hat{a} = 3.3$

Therefore, the estimated relationship is,

$$y = 3.3+1.9x \text{ (estimated)}.$$

Comment. This relationship is not a mathematical relationship in the sense that the value of y is available for any given value of x but we may use this to predict y. If we had more observations on x and y, for example if we had one more observation $x=3$ and $y=9$ and if we had estimated a and b by the principle of least squares we would have got different estimates for a and b and thus the estimated equation would have been different. Thus, in a curve fitting problem, we are only fitting the best fitting (in the least square sense) curve to the given data. In this problem we obtained the best fitting curve of the form $a+bx$ as,

$$y = 3.3+1.9x \text{ (estimated)}.$$

From this we cannot conclude that ' therefore ',

$$x = \frac{y}{1.9} - \frac{3.3}{1.9}$$

because in estimating the equation $y=a+bx$ we assumed that x is observable without any error and there is an error e in predicting y, in the sense, in using $a+bx$ as a predictor for y. If we had the assumption that y is observable without error and a model $c+dy$ is used to predict x then we would have started with the model,

$$x = c + dy + e$$

and we would have arrived at the estimates for c and d as

$$\hat{c} = \bar{x} - \hat{d}\bar{y} \text{ and } \hat{d} = \frac{\text{Cov}(x, y)}{\text{Var}(y)}$$

which need not be the same as the values for $\hat{c}$ and $\hat{d}$ obtained by solving, the equation $y = \hat{a} + \hat{b}x$. These points are to be remembered in selecting a particular model in a practical situation.

Example 11.2.2. Fit a curve of the form $y = ab^x$ to the following data:

x	0	1	2
y	2	7	17

Solution. Here the model is,

$$y = ab^x + e$$

where e is the error in taking ab^x for y and a and b are to be estimated. For convenience we may write the model in the form,

$$y' = a + b'x + e'$$

where $y' = \log y$, $a' = \log a$, $b' = \log b$ and e' is an error in the model. Now we get a linear model and hence the estimates of a' and b' are,

$$\hat{a}' = \bar{y}' - \hat{b}' \bar{x} \text{ and } \hat{b}' = \frac{\text{Cov}(x, y')}{\text{Var}(x)}$$

We get $\hat{a}$ and $\hat{b}$ by taking the antilogarithms of $\hat{a}'$ and $\hat{b}'$.

In our problem, the values of y are 2, 7 and 17 and hence the values of $y' = \log y$ are, 0.3010, 0.8451 and 1.2304 (common logarithms) respectively. Thus,

$$\bar{x} = \frac{0+1+2}{3} = 1, \quad \bar{y}' = \frac{0.3010+0.8451+1.2304}{3}$$

$$= 0.7952$$

$$\text{Cov}(x, y') = \frac{\Sigma xy'}{n} - \bar{x}\bar{y}' = 0.3068; \text{Var}(x) = \frac{\Sigma x^2}{n} - \bar{x}^2 = 0.6667.$$

Hence,

$$\hat{b}' = \frac{0.3068}{0.6667} = 0.46 \text{ and } \hat{a}' = 0.3352.$$

Now taking the antilogarithms we get the estimates of a and b as,

$$\hat{a} = 2.16 \text{ and } \hat{b} = 2.89$$

Hence, the estimated model is,

$$y = 2.16 \ (2.89)^x \text{ (estimated)}$$

Comment. By using the above techniques complicated expressions can be brought to simpler ones.

Exercises

11.1 Assuming a linear model $y=a+bx+e$ fit a straight line to the following data.

x	64	65	63	66	67
y	125	130	120	140	150

11.2 Fit a curve of the form $y=ab^x$ to the data in problem 11.1.

11.3 Fit the following curves (1) $y=a+bx+cx^2$, (2) $y=ax+\dfrac{b}{x}$, (3) $x=aye^{-by}$, to the following data.

x	10	12	13	15	17	18	19	21	22	24
y	8	9	10	11	13	15	16	18	19	20

11.4 Draw the scattergram and select an appropriate model for the following data.

x	0	1	2	3	4	−1	−2
y	5	8	10	11	12	3	2

11.5 Obtain the estimates of c and d if a model $x=c+dy+e$ is fitted to the data (x_1, y_1), (x_2, y_2),..., (x_n, y_n).

11.3 REGRESSION

In sections 11.1 and 11.2 we considered the problem of curve fitting by the principle of least squares. In a practical situation the observable variables may be stochastic variables and hence the practical problem is to select an appropriate model $f(a_1, a_2, ..., a_k, x)$ for a stochastic variable Y where x is an observed value of a s.v. X and $a_1, a_1,...,a_k$ are some constants to be determined. For example if X denotes the price of a particular commodity and if Y denotes its demand then both X and Y are stochastic variables. We can observe the price x and our aim may be to predict demand y corresponding to a given value x of X. Since we do not know the ' best predictor ' for Y based on X, we can construct an infinite number of ' predictors ' (functions of X) for Y. But we can select the best among them, in the minimum mean square sense, without much difficulty. That is, we will call a predictor $f(a_1, ..., a_k, x)$ the ' best ' for predicting the value of Y if,

$$E\left(\{Y - f(a_1, ..., a_k, x)\}^2\right)$$

is a minimum for every given value x of X, where E denotes mathematical expectation. It can be shown that the minimum value of $E\left(\{Y - f(a_1, ..., a_k, x)\}^2\right)$ is attained when,

$$f(a_1, ..., a_k, x) = E(Y \mid X)$$

where $E(Y \mid X)$ denotes the conditional expectation of Y given any value x of X. Conditional distributions and conditional expectations are discussed in Chapter 7. Here we will not go into the details of conditional expectations. Due to this fact the regression of Y on X is defined as follows.

Definition. The regression of Y on X is defined as the conditional expectation of Y given $X=x$.

Similarly the regression of X on Y can be defined as the conditional expectation of X given $Y=y$. If $Y, X_1, X_2,...,X_n$ are a set of stochastic variables then the regression of Y on $X_1, X_2,...,X_n$ is the conditional expectation of Y given $X_1=x_1, X_2=x_2,...,X_n=x_n$. The functional form of $E(Y \mid X)$ depends upon the nature of the joint distribution of X and Y. If $E(Y \mid X)$ is linear in x then we say that there is a linear regression of Y on X. If $E(Y \mid X)$ is quadratic in x we say that there is a quadratic regression of Y on X and so on. In this section we will consider mainly linear regressions. That is,

$$E(Y \mid X) = a+bx \text{ (linear regression of } Y \text{ on } X),$$

or

$$E(X \mid Y) = c+dy \text{ (linear regression of } X \text{ on } Y).$$

If there are a number of variates then,

$$E(Y \mid X_1, X_2, ..., X_k)=a_0+a_1x_1+...+a_kx_k \text{ (linear regression of } Y \text{ on } X_1, X_2, ..., X_k);$$

$$E(Y \mid X) = a_0+a_1x+a_2x^2 \text{ (quadratic regression of } Y \text{ on } X)$$
and so on.

By using some properties of conditional expectations, it can be shown that if,

$$E(Y \mid X) = a + bx$$

then, $a = E(Y) - bE(X)$ and $b = \dfrac{\text{Cov }(X, Y)}{\text{Var }(X)}$

where Cov (X, Y) = covariance between the s.v.s X and Y, and Var (X)=variance of the s.v. X. (*Hint.* Use the result $E(E(U \mid V)) = E(U)$). But here we will consider only the problem of estimating a regression equation by the principle of least squares, based on a given set of observations on the underlying stochastic variables. That is, if we know that the regression of Y on X is linear then we consider the model,

$$Y = a + bx + e$$
$$\text{or} \quad E(Y \mid X) = a + bx$$

where e is a random error. In other words,

$$e = Y - a - bx$$

is a stochastic variable since Y is a s.v. and x is a value assumed by a s.v. X. If we are given observations on Y and X then we can estimate a and b by considering

$$\Sigma e^2 = \Sigma (y - a - bx)^2$$

and minimizing this error sum of squares by the principle of least squares, where y denotes an observed value of Y. From section 11.2 we have the estimates

$$\hat{a} = \bar{y} - \hat{b}\bar{x} \text{ and } \hat{b} = \dfrac{\text{Cov}(x, y)}{\text{Var }(x)}$$

where Cov (x, y) is the sample covariance between the sets of observations on X and Y and Var (x) denotes the sample variance in the observations on X.

It can be easily seen that if e has a Normal distribution then the maximum likelihood estimates of a and b, under a linear regression of Y on X, are also the same as the least square estimates. That is, when

$$E(Y \mid X) = a + bx \quad \text{or} \quad Y = a + bx + e$$

then $Ee = 0$ and if $\text{Var}(e) = \sigma^2$ (some unknown quantity) then evidently Y has a Normal distribution with mean value $a + bx$ and with variance σ^2 corresponding to every given value x of X, if e has a $N(0, \sigma^2)$. If $x_k, \ldots, x_n$ are the given values of X and if $y_1, \ldots, y_n$ are the corresponding observations on Y then under the normality assumption for e we have the likelihood function

$$L = \prod_{i=1}^{n} f(y_i) = \frac{1}{(\sigma\sqrt{2\pi})^n} \exp\left\{- \sum_{i=1}^{n} (y_i-a-bx_i)^2/2\sigma^2\right\}$$

and by maximizing L we can easily see that the maximum likelihood estimates of a and b are also,

$$\hat{a}=\bar{y}-\hat{b}\bar{x} \text{ and } \hat{b} = \frac{\text{Cov }(x,y)}{\text{Var }(x)}$$

Example 11.3.1 If Y has a linear regression on X, estimate the regression equation by the method of least squares, based on the following data.

x	0	1	2
y	1	5	7

Solution. We are given, $E(Y|X)=a+bx$

For estimating a and b we consider the model

$$y=a+bx+e$$

By minimizing Σe^2 we get the estimates as,

$$\hat{a} = \bar{y} - \hat{b}\bar{x} = 1.33$$

$$\text{and} \quad \hat{b} = \frac{\text{Cov }(x,y)}{\text{Var }(x)} = 3.$$

Hence the estimated regression line is,

$$y = 1.33+3x$$

Comment. In a linear regression $E(Y|X)=a+bx$, a and b are often called the linear regression coefficients. If $b=0$ then evidently there is no linear regression of Y on X. We can test a hypothesis H_0: that there is no linear regression of Y on X by testing $H_0 : b=0$. We will consider a test in the next section. If there is a linear regression of Y on $X_1, X_2, ..., X_k$, that is,

$$E(Y|X_1, ..., X_k) = a_0+a_1x_1+...+a_kx_k$$

then $a_0, a_1, a_2, ..., a_k$ are called the linear regression coefficients of Y on $X_1, ..., X_k$.

11.31 *A Test for Regression*

Consider a linear regression of Y on X. That is,
$$E(Y \mid X) = a + bx.$$
In section 11.3 we have noticed that the value of b is
$$b = \frac{\text{Cov }(X,\ Y)}{\text{Var }(X)} = \rho\,\frac{\sigma_y}{\sigma_x}$$
where ρ denotes the linear correlation between X and Y and σ_x and σ_y denote the standard deviations of X and Y respectively. Since σ_x and σ_y are non-zero, the test,
$$H_0 : b = 0 \iff \rho = 0 \quad (\iff \text{means equivalent}).$$
A test for testing $H_0 : \rho = 0$ is given in Table 9.1 in Chapter 9. Here we will consider a test for the hypothesis $b = 0$. The estimated value of b is $\hat{b}$ and it can be shown that the expected value of $\hat{b}$ is b and an unbiased estimator for σ^2, is
$$\Sigma\,(y - \hat{a} - \hat{b}x)^2/(n-2)$$
where n is the number of given values for X, and σ^2 is the variance of Y. It can also be shown that
$$\text{Var }(\hat{b}) = \sigma^2/[\Sigma(x-\bar{x})^2].$$
Hence when e's are independently normally distributed as $\mathcal{N}(0, \sigma^2)$ then we can show that $\hat{b}$ and $\hat{\sigma}^2/[\Sigma(x-\bar{x}^2]$ where
$$\hat{\sigma}^2 = \Sigma\,y(y - \hat{a} - \hat{b}x)/(n-2)$$
are independently distributed and further,
$$\frac{(\hat{b}-b)\ \sqrt{\Sigma x^2 - n\bar{x}^2}}{\hat{\sigma}} : t_{n-2} \quad (\text{Student-}t \text{ with } n-2 \text{ d.f.})$$
where,
$$\hat{\sigma}^2 = \frac{\Sigma\,y(y - \hat{a} - \hat{b}x)}{n-2} \quad (\text{Least square minimum}/(n-2))$$
$$= [\Sigma\,y^2 - \hat{a}\,\Sigma\,y - \hat{b}\,\Sigma\,xy]/(n-2).$$
Hence a $100(1-a)\%$ confidence interval for b is,
$$\hat{b} \pm t_{a/2}\,\hat{\sigma}/\sqrt{\Sigma x^2 - n\bar{x}^2}.$$

Example 11.31.1 Assuming that there is a linear regression of Y on X, estimate the regression equation. Test for the linear regression by assuming a Normal linear regression [that is, e's are independent and $N(0, \sigma^2)$] and by using the following data:

x	0	1	2
y	4	7	12

Solution. $\hat{a} = \bar{y} - \hat{b}\bar{x}$ and $\hat{b} = \text{Cov } (x, y)/\text{Var } (x)$
$$= 3.67 \qquad\qquad = 4.$$

Hence the estimated regression equation is,
$$y = 3.67 + 4x.$$

In order to test the hypothesis that $b = 0$ we need,
$$\hat{\sigma}^2 = \frac{\Sigma y(y - \hat{a} - \hat{b}x)}{n - 2}$$
$$= 4(0.33) + 7(-0.67) + 12(0.33) = 0.59.$$

Under $H_0: b = 0$, we have an observed value of a Student-t with $n - 2 = 1$ d.f. as,
$$t_1 = \frac{(\hat{b} - 0)}{\hat{\sigma}} \sqrt{\Sigma x^2 - n \bar{x}^2}$$
$$= \frac{4\sqrt{2}}{0.59} = 7.4.$$

The tabulated value of a Student-t with one degree of freedom at 95% level is $t_{0.025} = 12.706$.

Hence $|t_1| < 12.706$

and therefore the hypothesis cannot be rejected at 95% level. A 95% confidence interval for b is,
$$\hat{b} \pm t_{a/2} \hat{\sigma}/\sqrt{[\Sigma x^2 - n \bar{x}^2]}$$

That is,
$$4 \pm 12.706 \ (0.54) \text{ or } (-2.86, 10.86).$$

Comment. Since the hypothesis $H_0: b = 0$ cannot be rejected, we conclude that the difference between $\hat{b}$ and b that is, $\hat{b} - 0 = 4$ is due to chance variation. In other words $\hat{b}$ cannot be considered to be *significantly* different from zero.

Exercises

11.6 Assuming a model $y-\bar{y}=a+b(x-\bar{x})+e$ estimate a and b by the method of least squares, where $\bar{x}$ and $\bar{y}$ are the arithmetic means of the observations on x and y respectively.

11.7 If the joint density of X and Y is given to be

$$f(x, y) = \begin{cases} (x+y)/3, & 0 < x < 1, \ 0 < y < 2 \\ 0, & \text{elsewhere,} \end{cases}$$

obtain regression of X on Y.

11.8 Assuming a linear regression of Y on X estimate the regression equation from the following data

x	0	1	2	3
y	3	6	8	11

11.9 Assuming Normality for Y, test the hypothesis that $b=2$ if $a+bx = E(Y \mid X)$ is the regression equation and obtain a 95% confidence interval, for the data in problem 11.8.

11.10 By using the following data, test the hypothesis that there is no evidence of a linear regression X on Y, at 95% level.

x	1	3	8	10	13
y	0	1	2	3	4

11.11 If $E(Y \mid X)=a+bx$ and if $\hat{a}$ and $\hat{b}$ denote the least square estimation show that $E(\hat{b})=b$, $E(\hat{a})=a$, var $(\hat{b})=\sigma^2/\Sigma(x_i-\bar{x})^2$, Var $(\hat{a}) = \sigma^2 \left[\dfrac{1}{n} + \dfrac{\bar{x}^2}{\Sigma(x_i - \bar{x})^2} \right]$. Construct a $100(1-a)\%$ confidence interval for a, where $\sigma^2 = $ Var $(Y \mid X)$.

Additional Exercises

11.1 If the joint density function of X and Y is

$$f(x, y) = \begin{cases} 2, & 0 < x < 1, \ 0 < x < y < 1 \\ 0, & \text{elsewhere,} \end{cases}$$

evaluate the regression of (1) Y on X, (2) X on Y.

11.2 If the regression of Y on X is linear, that is,

$$E(Y \mid X) = a + bx$$

show that $a = E(Y) - bE(X)$ and $b = \text{Cov}(X, Y)/\text{Var}(X)$.

(*Hint.* Use the properties of the conditional expectations.)

11.3 Estimate a_0, a_1, a_2 by using the model,

$$y = a_0 + a_1 x_1 + a_2 x_2 + e,$$

and from the data,

$$(y_1, x_{11}, x_{21}), \; (y_2, x_{12}, x_{22}), \; \ldots, \; (y_n, x_{1n}, x_{2n}).$$

11.4 From the results in problem 11.3, write down a_0, a_1 and a_2 if there is a linear regression of Y on X_1 and X_2, of the form,

$$E(Y \mid X_1, X_2) = a_0 + a_1 x_1 + a_2 x_2.$$

11.5 If $E(Y \mid X) = a + bx$ and if $\hat{a}$ and $\hat{b}$ denote the least square estimates, show that

$$\Sigma(y - \hat{a} - \hat{b}x)^2/(n-2),$$

that is, the least square minimum divided by $(n-2)$ is an unbiased estimator of σ^2, where $\sigma^2 = \text{Var}(Y \mid X)$.

11.6 Assuming a linear regression of Y on X estimate the regression equation and test for linear regression under normality and evaluate 95% confidence intervals for b and a, based on the following observations.

x	0	1	2	3	4	5
y	4	7	8	12	14	15

11.7 Assuming a linear regression of X on Y and Z estimate the regression equation based on the following observations

z	0	1	1	2	3	5
y	0	1	0	1	2	2
x	3	7	6	7	10	15

11.8   By using the method of least squares fit the following curves

$$(1) \ y=ab^x, \quad (2) \ y=a+bx+cx^2, \quad (3) \ y=a+\frac{b}{x},$$

to the following data.[For (3) delete the point $(x, y)=(0, 2)$].

x	0	1	2	3	4	5
y	2	5	8	10	12	14

CHAPTER TWELVE

Analysis of Variance

Often an experimental scientist consults a statistician when he cannot get the inference that he had expected to get from his experimental data. Usually in such cases the statistician is unable to analyse the data property due to the fact that the experiment was not planned properly, in the sense, that the experiment was not planned keeping in mind the particular analysis to be done later. We had seen that for applying any particular statistical test a number of conditions are to be satisfied by the data. For example suppose that the department of education wants to compare two methods of teaching. Suppose that two classes of students are exposed to the two methods of teaching. The effect of teaching is a hypothetical quantity which cannot be measured directly. Hence the marks obtained by the students may be taken as an indication of the effect of teaching. But the difference in the average marks in the two classes cannot be taken as a measure of the difference in the effects of teaching, because there are a number of other factors, such as the intelligence of the students, their previous acquaintance with the methods of teaching, etc., which contributed towards the average marks. But if the experiment is planned in such a way that all the extraneous factors (factors other than the methods of teaching) are controlled, then the difference in the average marks can be taken as an indication of the difference in the effects of the two methods of teaching. Here we will not consider the problem of designing an experiment (which comes under the topic Design of Experiments), but we will consider the problem of analysing the data collected from a properly designed experiment.

12.1 ONE-WAY CLASSIFICATION

This is the data collected from an experiment which is designed to control all the factors of variation except one factor under consideration. For example, consider an agricultural experiment to compare the effects of k different fertilizers on the yields of wheat. Suppose that the experimenter controls all other factors except the effects of the fertilizers. That is, if a particular plot is given fertilizer F_1 then the yield from that plot can be written as a function of the effect of F_1. In other words, the yield is a response to the effect of F_1. If x_{ij} denotes the yield of the jth plot which received fertilizer i then x_{ij} can be written as,

$$x_{ij} = \phi(\mu_i)$$

where μ_i is the effect of the ith fertilizer and the functional form of ϕ is unknown. The yields from two different plots of the same size which received fertilizer i may not be the same. This means, that there is an effect due to the uncontrolled factor or factors whose existence was/were not known to the experimenter. Hence we may write

$$x_{ij} = \phi(\mu_i) + e_{ij}$$

where e_{ij} is called the *random error* which is the sum total effect of the uncontrolled factors and hence e_{ij} is a stochastic variable. Unless we know something about the nature of $\phi(\mu_i)$ we cannot make any valid inference regarding μ_i. In practice we use a linear model for simplicity. That is,

$$x_{ij} = \mu_i + e_{ji}.$$

Our aim may be to test a hypothesis that $\mu_1 = \mu_2 = ... = \mu_k$, that is, the effect of the k fertilizers are the same. Again for convenience, we may write as,

$$x_{ij} = \mu + a_{\cdot} + e_{ij} \qquad (1)$$

where μ can be called some general effect and a_i can be considered to be the deviation from the general effect due to fertilizer F_i. (In experimental language we call F_i the *treatment* T_i, that is, the factor whose effect is under investigation). Since we have taken a_i as deviations from some general effect $\sum\limits_{i=1}^{k} a_i = 0$ because we can take

$$\mu = \frac{\Sigma \mu_i}{k} \text{ and } a_i = \mu_i - \mu \text{ so that } \Sigma a_i = 0$$

Model (1) is called a *one-way classification* (involving only one type of treatments) *fixed effect* (μ, a_1, ..., a_k are assumed to be some unknown constants and not as stochastic variables) *additive* (x_{ij} is assumed to have been obtained by adding up the various effects involved. In this case μ, a_i and e_{ij}), *linear* (of first degree in a_i) model. Now we consider the analysis of this very simple model.

12.2 ESTIMATION OF THE PARAMETERS

In the one-way classification, linear, fixed effect, additive model,
$$x_{ij} = \mu + a_i + e_{ij}, \; i = 1, 2, \ldots, k; j = 1, 2, \ldots, n,$$
we will estimate μ, a_1, a_2, $\ldots$, a_k by the method of least squares. The data collected from such a design will look like the data in Table 12.1.

TABLE 12.1

A ONE-WAY CLASSIFICATION DATA

	Treatments					
	T_1	T_2	T_3	...	T_k	Total
	x_{11}	x_{21}	x_{31}	...	x_{k1}	$x_{.1}$
	x_{12}	x_{22}	x_{32}	...	x_{k2}	$x_{.2}$
	.	.	.	...	.	.
	.	.	.	...	.	.
	.	.	.	...	.	.
	x_{1n}	x_{2n}	x_{3n}	...	x_{kn}	$x_{.n}$
Total	$x_1.$	$x_2.$	$x_3.$	...	$x_{k.}$	$x_{..}$

The summation with respect to a subscript is denoted by a dot.

That is, $\displaystyle\sum_{j=1}^{n} x_{ij} = x_{i.}; \; \sum_{i=1}^{k} x_{ij} = x_{.j}; \; \sum_{i=1}^{k}\sum_{j=1}^{n} x_{ij} = x_{..}$

For convenience we assumed that there are n observations on each treatment. If the numbers of observations are different then also the procedure of analysis remains the same. The error sum of squares,

$$M = \sum_{i=1}^{k}\sum_{j=1}^{n} e_{ij}^2 = \sum_{i=1}^{k}\sum_{j=1}^{n} (x_{ij} - \mu - a_i)^2$$

Hence the least square normal equations are,

$$\frac{\partial M}{\partial \mu} = 0; \; \frac{\partial M}{\partial a_i} = 0. \; i = 1, 2, \ldots, k$$

There are $k+1$ normal equations. Since μ is contained in every term of the summation, that is,

$$(x_{11}-\mu-a_1)^2+\ldots+(x_{1n}-\mu-a_1)^2+\ldots+(x_{k1}-\mu-a_k)^2+$$
$$+\ldots+(x_{kn}-\mu-a_k)^2$$

$$\frac{\partial M}{\partial \mu}=0 \implies \sum_{ij}(x_{ij}-\hat{\mu}-\hat{a}_i)=0 \qquad (2)$$

Only n terms contain a_i and hence,

$$\frac{\partial M}{\partial a_i}=0 \implies \sum_{j}(x_{ij}-\hat{\mu}-\hat{a}_i)=0 \qquad (3)$$

This can be seen from the complete expansion of $\sum\limits_{ij} e_{ij}^2$

Equation (2) gives, $x_{..}-nk\,\hat{\mu}-na_.=0$

$$\implies \hat{\mu}=\frac{x_{..}}{nk}\text{ (the grand mean)} \qquad (4)$$

where a $(\wedge)$ denotes a least square estimate.

Equation (3) gives, $x_{i.}-n\hat{\mu}-n\,\hat{a}_i=0$

$$\implies \hat{a}_i=\frac{x_{i.}}{n}-\frac{x_{..}}{nk}, \; i=1,2,\ldots,k \quad (5)$$

Now the least square minimum,

$$S^2=\sum_{i,j}(x_{ij}-\hat{\mu}-\hat{a}_i)^2$$

$$=\sum_{i,j} x_{ij}(x_{ij}-\hat{\mu}-\hat{a}_i)+0 \text{ (due to the normal equations)}$$

$$=\sum_{i,j} x_{ij}\left(x_{ij}-\frac{x_{i.}}{n}\right)=\sum_{i,j} x_{ij}^2-\sum_{j}\frac{x_{i.}^2}{n}$$

$$=\left(\sum_{j} x_{ij}^2-\frac{x_{..}^2}{nk}\right)-\left(\sum_{i}\frac{x_{i.}^2}{n}-\frac{x_{..}^2}{nk}\right)$$

Now consider the hypothesis,

$$H_0:\mu_1=\mu_2=\ldots=\mu_k \iff a_1=a_2=\ldots=a_k=0$$

against the alternative that not all the effects are the same.

Under H_0 the model is,

$$x_{ij}=\mu+e_{ij}$$

and the least square minimum under H_0 can be easily seen to be,

$$S_0{}^2 = \left(\sum_{i,j} x_{ij}{}^2 - \frac{x_{..}{}^2}{nk} \right)$$

and hence

$$S_0{}^2 - S^2 = \left(\sum_i \frac{x_{i.}{}^2}{n} - \frac{x_{..}{}^2}{nk} \right)$$

can be taken as the variation due to the hypothesis H_0.

That is,

$$S_0{}^2 = (S_0{}^2 - S^2) + S^2$$

where, by examining the structure, we may call

$\qquad S_0{}^2 - S^2$—as the ' between treatment ' variation

$\qquad\qquad S^2$—' within treatment ' variation and

$\qquad\qquad S_0{}^2$—the total variation.

This aspect of splitting the total variation in a data into different components of variation and the resulting analysis (testing hypotheses) is called the *analysis of variance* or more appropriately called the *analysis of variation*.

If we assume that e_{ij}'s are independently distributed as a $\mathcal{N}(0, \sigma^2)$ then we can show that,

$$\frac{(S_0{}^2 - S^2)}{\sigma^2} : \chi_{k-1}{}^2$$

$$\frac{S^2}{\sigma^2} : \chi_{nk-k}{}^2$$

and further S^2 and $S_0{}^2 - S^2$ are independently distributed and thus under H_0,

$$\frac{(S_0{}^2 - S^2)/(k-1)}{S^2/(nk-k)} : F_{k-1,\ nk-k}$$

By using this F-statistic with $k-1$ and $nk-k=k(n-1)$ d.f. we may test the hypothesis $H_0 : a_1 = a_2 = \ldots = a_n = 0$.

These results can be given in a neat table known as the *analysis of variance table* (ANOVA Table) which is illustrated in Table 12.2.

TABLE 12.2

AN ANALYSIS OF VARIANCE TABLE FOR A ONE-WAY CLASSIFICATION

(1)	(2)	(3)	(4) = (3)/(2)	(5)	(6)
Variation due to	d. f.	Sum of squares (S.S.)	Mean squares (M.S.)	F-ratio	Inference
Between treatments	$k-1$	$\sum\limits_{i} \dfrac{x_{i.}^{2}}{n} - \dfrac{x_{..}^{2}}{nk}$	T	T/E $= F_{k-1}$ $k(n-1)$	Significant or not Significant
Within treatments	$k(n-1)$	by subtraction	E		
Total	$nk-1$	$\sum\limits_{i,j} x_{ij}^{2} - \dfrac{x_{..}^{2}}{nk}$			

For example, $T = \left(\sum\limits_{i} \dfrac{x_{i.}^{2}}{n} - \dfrac{x_{..}^{2}}{nk} \right) / (k-1)$ and since $nk-1 = (k-1)$ $+ k(n-1)$ and $S_0^2 = (S_0^2 - S^2) + S^2$ we obtain the entries in columns (2) and (3) of 'within treatment variation' by subtraction. If the observed value of the $F_{k-1, k(n-1)}$, falls in the critical region (that is, greater than the tabulated value of $F_{k-1, k(n-1)}$ at a chosen level) then we reject $H_0 : a_1 = \ldots = a_k = 0$ and we infer that there is a significant deviation from the null hypothesis and we write in the column (6) 'significant'; otherwise we write 'not significant'. In order to simplify the computations the following computational procedure can be used.

1. Compute $\sum\limits_{i,j} x_{ij}^2$ and $x_{..}$

2. Compute the correction factor (C.F.) $= x_{..}^2/nk$

3. Compute $\sum\limits_{i} x_{i..}^2$

Example 12.2.1. The following table gives the yields of wheat in 30 test plots which were given 3 different fertilizers. Test the hypothesis that the three fertilizers are equally effective. Assume that a one-way classification, linear, additive, fixed effect, Normal model is appropriate.

											Total
Fertilizer *A*	50	60	60	65	70	80	75	80	85	75	700
Fertilizer *B*	60	60	65	70	75	80	70	75	85	80	720
Fertilizer *C*	40	50	50	60	60	60	65	75	70	70	600

Solution. Let x_{ij} denote the jth observation receiving ith fertilizer. Then $j = 1, 2, ..., 10$ and $i = 1, 2, 3$.

$$\sum_{i,j} x_{ij} = x.. = 600 + 720 + 700 = 2020.$$

$$\text{C.F.} = \frac{x..^2}{nk} = \frac{(2020)^2}{30} = 136013.33$$

$$\sum_i \frac{x_{i.}^2}{n} = \frac{1}{10} [600^2 + 720^2 + 700^2] = 136840$$

$$\sum_{i,j} x_{ij}^2 - \text{C.F.} = 50^2 + ... + 70^2 - \text{C.F.} = 3636.67$$

$$\sum_i \frac{x_{i.}^2}{n} - \text{C.F.} = 826.67$$

Now we can set up the analysis of variance table

Variation due to	d.f.	S.S.	M.S.	F-ratio	Inference
Between fertilizers	$k-1=2$	826.67	413.33	$\dfrac{413.33}{104.08} > 4$	significant
Within fertilizers	27	2810.00	104.08		
Total	$nk-1=29$	3636.67			

The d.f. for the 'within' variation $=(nk-1)-(k-1)=29-2=27$.

The within sum of squares $=3636.67-826.67=2810.0$.

The tabled value of an F with 2 and 27 d.f. at 95% level is $F_{2,27,0.025}$ and is between 3.32 and 3.37 (from F-tables). Hence the observed $F_{2,27}$ falls in the critical region and hence we reject the hypothesis that the effects of the fertilizers are all the same.

Comment. The 'within' variation can also be called the residual variation or the error variation and in fact this is contributed by all the unaccounted factors of variation. In a particular analysis if the F-ratio is found to be significant then we can test individual hypotheses such as $H_0 : a_i = a_j$ by using a Student-t test. This we will not consider here. In a 'test of significance' we are testing to see whether or not an observed value of a random quantity can be considered to be significantly different from a fixed number (usually zero) in the sense that whether or not the probability of setting a difference as small as the observed difference is large at a chosen level of significance or at a chosen size of the critical region. Also by evaluating $E(\hat{a_i} - \hat{a_j})$ and var $(\hat{a_i} - \hat{a_j})$ one can construct confidence interval for $a_i - a_j$.

12.3 Two-way Classification

Suppose that we want to test the effect of r different fertilizers on t different varieties of wheat then an experiment can be properly planned so that the yield in a particular plot, for example the plot with the ith fertilizer and the jth variety of wheat, will be a function of the effects of the corresponding fertilizer and variety of wheat. Again a simple *linear, fixed effect, additive, model* can be given as follows

$$x_{ij} = \mu + a_i + \beta_j + e_{ij} \quad i = 1, 2, ..., r; \ j = 1, ..., t$$

where x_{ij} is the observation corresponding to the ith treatment of the first kind say A_i, (here fertilizers) and the jth treatment of the second kind, say B_j, (here variety of wheat), μ is a general effect, a_i is the deviation from the general effect due to A_i, β_j is the deviation from the general effect due to B_j and e_{ij} is the random error. Here we have taken only one observation corresponding to A_i and B_j

combination. This particular model is called a model without *interaction*. Interaction is an effect due to the combination A_iB_j. In other words, if the effect of A_i varies with the effects of $B_1, \ldots B_t$ and vice versa then there is interaction. If there is interaction we would have taken one more effect γ_{ij} which is the deviation from the general effect due to interaction. Again, without any loss of generality we may assume that,

$$\sum_{i=1}^{r} a_i = 0 = \sum_{j=1}^{t} \beta_j$$

Proceeding as in the case of a one-way classification we can set up an analysis of variance table for this two-way classified data with one observation per cell and with a linear, fixed effect, additive model without interaction as in Table 12.4. A two-way classified data with one observation per cell will look like the data in Table 12.3.

TABLE 12.3

A TWO-WAY CLASSIFIED DATA WITH ONE OBSERVATION PER CELL

	B_1	B_2	$\cdots$	B_t	Total
A_1	x_{11}	x_{12}	$\cdots$	x_{1t}	$x_{1.}$
A_2	x_{21}	x_{22}	$\cdots$	x_{2t}	$x_{2.}$
$\cdot$	$\cdot$	$\cdot$		$\cdot$	$\cdot$
$\cdot$	$\cdot$	$\cdot$		$\cdot$	$\cdot$
$\cdot$	$\cdot$	$\cdot$		$\cdot$	$\cdot$
A_r	x_{r1}	x_{r2}	$\cdots$	x_{rt}	$x_{r.}$
Total	$x_{.1}$	$x_{.2}$	$\cdots$	$x_{.t}$	$x_{..}$

TABLE 12.4

ANALYSIS OF VARIANCE TABLE FOR A TWO-WAY CLASSIFICATION

Variation due to (1)	d.f. (2)	S.S. (3)	M.S. (4)=(3)/(2)	F-ratio	Inference
A-treatments	$r-1$	$\sum_i \dfrac{x_i.^2}{t} - \dfrac{x..^2}{rt}$	T_1	T_1/E	
B-treatments	$t-1$	$\sum_j \dfrac{x._j^2}{r} - \dfrac{x..^2}{rt}$	T_2	T_2/E	
error	$(r-1)(t-1)$	(subtract)	E		
Total	$rt-1$	$\sum_{i,j} x_{ij}^2 - \dfrac{x..^2}{rt}$			

T_1/E is an F with $r-1$ and $(r-1)(t-1)$ d.f. under normality for the error distributions. Similarly T_2/E is an F with $t-1$ and $(r-1)(t-1)$ d.f. under normality. By using these F-ratios the hypotheses that,

$$H_0 : a_1 = a_2 = ... = a_r = 0 \quad \text{and}$$
$$H_{00} : \beta_1 = \beta_2 = ... = \beta_t = 0$$

can be tested respectively. For some of the experimental designs from which one can get one-way and two-way classified data, the reader may see the bibliography at the end of this book. It should also be noticed that a two-way classified data is different from a contingency table. In a contingency table we have the frequencies in the various cells whereas in a two-way classified data we have observations on some variables.

Exercises

12.1   The results of a completely randomized experiment (where a one-way classification model is appropriate) to compare the effects of 3 different toothpastes are given below. Test the hypothesis that the toothpastes are equally effective, at 95% level.

	Reduction in cavities				
Toothpaste 1	14	15	17	20	16
Toothpaste 2	10	12	11	13	15
Toothpaste 3	12	14	15	13	11

12.2   The results of a randomized block experiment (where a two-way classification model without interaction is appropriate) to study the effects of 2 fertilizers on 3 varieties of corn, are given below. Test the hypothesis that there is no significant difference in the effects of the fertilizers, at 95% level. The observations are the yields.

	Variety 1	2	3
Fertilizer 1	8	16	26
Fertilizer 2	10	15	22

12.3   In a one-way classification with $n_1, n_2, \ldots, n_k$ observations in the different sets, show that the sum of squares due to the treatment is $\sum\limits_{i} \dfrac{x_{i\cdot}^2}{n_i} - \dfrac{x_{\cdot\cdot}^2}{n_{\cdot}}$ where $n_{\cdot} = n_1 + n_2 + \ldots + n_k$

12.4 For the observations in problem 12.3. show that,

$$\sum_{i,j}\left(x_{ij} - \frac{x_{..}}{n_{.}}\right)^2 = \sum_{i,j}\left(x_{ij} - \frac{x_{i.}}{n_j}\right)^2 + \sum_i n_i\left(\frac{x_{i.}}{n_i} - \frac{x_{..}}{n_{.}}\right)^2$$

Additional Exercises

12.1 Analyse the following one-way classification

(a)

A_1	20	21	22	25	21	23	24
A_2	19	20	23	24	25	27	26
A_3	21	20	22	24	23	24	25

(b)

A_1	10	12	11	10	8	12	10
A_2	8	10	12	13			
A_3	11	12	14	15	16	18	
A_4	19	18	16	14	15		

(c)

A_1	5	4	8	6	7	10	12	11	9	8
A_2	6	4	6	7	8	9	9	10		
A_3	12	8	7	6	5	4	9	12	11	12

12.2 Analyse the following two-way classifications

(a)

	B_1	B_2	B_3	B_4
A_1	12	14	16	12
A_2	10	8	15	11
A_3	8	9	11	10

(b)

	B_1	B_2	B_3	B_4	B_5
A_1	20	22	25	21	20
A_2	12	16	11	10	8

12.3 From a model $x_{ij} = \mu + a_i + \beta_j + \gamma_{ij} + e_{ij}$, can you estimate γ_{ij} if there is only one observation per cell?

12.4 For a two-way classification model with interaction, that is,

$$x_{ijk} = \mu + a_i + \beta_j + \gamma_{ij} + e_{ijk}, \quad i = 1, 2, \ldots, r, \, j = 1, \ldots, s;$$
$$k = 1, 2, \ldots, n$$

set up an analysis of variance table.

12.5 In problem 12.4 if $k = 1, 2, \ldots, n_{ij}$ set up an analysis of variance table.

Bibliography

Alexander, H. W. *Elements of Mathematical Statistics*, John Wiley, 1961.

Bernstein, A. L. *A Handbook of Statistics Solutions for Behavioral Sciences*, Holt, Rinehart and Winston, 1964.

Brunk, H. D. *An Introduction to Mathematical Statistics*, Blaisdell, 1965.

Cohen, L. S. and Hanson, D. N. *Applications and Problems in Elementary Statistics*, Wadsworth, 1963.

Chakravarti, I. M., Laha, R. G. and Roy, J. *Handbook of Methods of Applied Statistics*, Vol. I and II, John Wiley, 1967.

Clark, V. A. and Tarter, M. E. *Preparations for Basic Statistics*, McGraw-Hill, 1968.

Cochran, W. G. *Sampling Techniques*, John Wiley, 1963.

Elzey, F. F. *A programmed Introduction to Statistics*, Wadsworth, 1966.

Feller, W. *An Introduction to Probability Theory*, John Wiley, Vol. I and II, 1950, 1966.

Fraser, D. A. S. *Statistics, An Introduction*, John Wiley, 1958.

Fraser, D. A. S. *Non-parametric Methods in Statistics*, John Wiley, 1957.

Freund, J. E. *Modern Elementary Statistics*, Prentice-Hall, 1967.

Freund, J. E. *Mathematical Statistics*, Prentice-Hall, 1964.

Freund, J. E. Livermore, P. and Millar, I. *A Manual of Experimental Statistics*, Prentice-Hall, 1960.

Fryer, H. C. *Concepts of Methods of Experimental Statistics*, Allyn and Bacon, 1966.

Guttman, I. and Wilks, S. S. *Introduction Engineering Statistics*, John Wiley, 1965.

Hald, A. *Statistical Theory with Engineering Applications*, John Wiley, 1952.

Hodges, J. L. and Lehmann, E. L. *Basic Concepts of Probability and Statistics*, Holden-Day, 1964.

Huntsberger, D. V. *Elements of Statistical Inference*, Allyn and Bacon, 1965.

Hogg, R. V. and Craig, A. T. *Introduction to Mathematical Statistics*, Macmillan, 1965.

Kraft, C. and Van Elden, C. *A Non-parametric Introduction to Statistics*, Macmillan, 1968.

Kendall, M. G. and Stuart, A. *Advanced Theory of Statistics*, Vol. I and II, Griffin, 1958, 1961.

Lehman, E. L. *Testing Statistical Hypotheses*, John Wiley, 1959.

Lev, J. and Walker, H. M. *Statistical Inference*, Holt, Rinehart and Winston, 1953.

Li, C. C. *Introduction to Experimental Statistics*, McGraw-Hill, 1964.

Lindgran, B. W. and McElrath, G. W. *Introduction to Probability and Statistics*, Macmillan, 1966.

Mathai, A. M. *Introduction to Statistical Mathematics*, S. Chand, 1967.

Mathai, A. M. and Rathie, P. N. *Basic Concepts in Information Theory and Statistics; Axiomatic Foundations and Applications*, Wiley Eastern, 1975.

Mathai, A. M. and Saxena, R. K. *Generalized Hypergeometric Functions with Applications in Statistics and Physical Sciences.* Springer-Verlag, No. 348, 1973.

Mendenhall, W. *Introduction to Statistics*, Wadsworth, 1965.

Meyer, P. L. *Introduction to Probability and Statistical Application*, Addison-Wesley, 1965.

Miller, I. and Freund, J. E. *Probability and Statistics for Engineers*, Prentice-Hall, 1965.

Miller, R. G. *Simultaneous Statistical Inference*, McGraw-Hill, 1966.

Mode, E. B. *Elements of Probability and Statistics*, Prentice-Hall, 1966.

Mood, A. M. and Graybill, F. A. *Introduction to the Theory of Statistics*, McGraw-Hill, 1963.

Peng, K. C. *The Design and Analysis of Scientific Experiments*, Addison-Wesley, 1967.

Pfeiffer, P. E. *Concepts of Probability Theory*, McGraw-Hill, 1965.

Rao, C. R. *Linear Statistical Inference and Its Applications*, John Wiley, 1965.

Richmond, S. B. *Statistical Analysis*, Ronald Press, 1964.

Rickmers, A. D. and Todd, H. N. *Statistics, An Introduction*, McGraw-Hill, 1967.

Weinberg, G. H. and Schumaker, J. A. *Statistics, An Intuitive Approach*, Wadsworth, 1965.

Whitney, D. R. *Elements of Mathematical Statistics*, Holt, Rinehart and Winston, 1961.

Wilks, S. S. *Mathematical Statistics*, John Wiley, 1962.

Wine, R. L. *Statistics for Scientists and Engineers*, Prentice-Hall, 1964.

Answers

CHAPTER ONE

1.1. (a) 4; (b) 4; (c) 1; (d) Infinite; (e) Infinite; (f) 16; (g) Nil. **1.3.** (a) Yes, finite; (b) Yes, finite; (c) Yes, continuous in the interval $(-1, \infty)$; (d) No, two different populations; (e) Yes, bivariate; (f) No unless more information is given; (g) No. **1.4.** (a) Yes, three-variate; (b) Yes, $a+1$—variate where a is the number of measurements in 'beauty'; (c) No (no joint measurements). **1.5.** (a) ϕ, $\{1, 2, -1\}$, $\{1\}$, $\{2\}$, $\{-1\}$, $\{1, 2\}$, $\{1, -1\}$, $\{2, -1\}$; (b) ϕ, $\{(0, 0), (0, 1), (1, 0), (1, 1)\}$, $\{(0, 0)\}$, $\{(0, 1)\}$, $\{(1, 0)\}$, $\{(1, 1)\}$, $\{(0, 0), (0, 1)\}$, $\{(0, 0), (1, 0)\}$, $\{(0, 0), (1, 1)\}$, $\{(0, 1), (1, 0)\}$, $\{(0, 1), (1, 1)\}$, $\{(1, 0), (1, 1)\}$, $\{(0, 0), (0, 1), (1, 0)\}$, $\{(0, 0), (0, 1), (1, 1)\}$, $\{(0, 0), (1, 0), (1, 1)\}$, $\{(0, 1), (1, 0), (1, 1)\}$; (c) ϕ, $\{a_1, a_2, a_3\}$, $\{a_1\}$, $\{a_2\}$, $\{a_3\}$, $\{a_1, a_2\}$, $\{a_1, a_3\}$, $\{a_2, a_3\}$. **1.6.** (a) Yes; (b) Yes; (c) No; (d) No. **1.7.** (a) Need not be; (b) Need not be; (c) Yes; (d) Yes; (e) need not be. **1.8.** (a) No; (b) desirable; (c) Yes; (d) Yes; (e) No; (f) desirable. **1.10.** (a) Yes; (b) Yes; (c) No, unless 1 and 7 are the random numbers obtained. **1.11.** (a) Fix a number of yards, say 100. Select 5 random numbers from 00 to 99, and select the yards corresponding to these numbers. Repeat the process if the quality is to be checked continuously. (b) Depending upon the reliability of the machine, consider a set of bulbs as a population, say for example, the first 100. Select the sample as it is in (a); (c) Adopt a procedure similar to the ones in (a) and (b); (d) Mark them by numbers from 00 to 34. By using a table of random numbers select 5 by omitting repetitions. **1.13.** (a) 7.5 to 12.5, 12.5 to 17.5 and so on; (b) 7.5, 12.5,...; (c) 12.5, 17.5,...**1.14.** (a) $4.995-9.995$, $9.995-14.995$, $14.995-19.995$; (b) 7.495, 12.495, 17.495. **1.17.** Bar diagrams. **1.18.** *Hint*: Pie diagrams. **1.20.** No. Represent by two \$ \$. **1.21.** (a) $x_1 f_1 + x_2 f_2 + ... + x_n f_n$; (b) $(x_1 f_1 + ... + x_n f_n) - (y_1 f_1 + ... + y_n f_n)$; (c) $x_1 y_1 f_1 + ... + x_n y_n f_n$. **1.24.** (a) 6; (b) 4; (c) 8.

1.25. (a) $\displaystyle\sum_{i=1}^{2} \sum_{j=1}^{3} x_{ij}$; (b) $\displaystyle\sum_{i=1}^{3} x_{ii} y_{ii}$;

(c) $\displaystyle\sum_{i=1}^{n} f_i x_i^2$; (d) $\displaystyle\sum_{i=1}^{n} (x_i - \bar{x})^2$.

1.27. $1/(n+1)^2$; **1.28.** $a^{\sum_{i=1}^{n} x_i}$; **1.29.** $c^n e^{-(x_1 + ... + x_n)/\beta}$.

1.30. $c^n \, e^{-[(x_1-a)^2+(x_2-a)^2+\ldots+(x_n-a)^2]/\beta}$.

1.31. 3.96, not much use. **1.32.** 30.8, 32.46. **1.33.** Correct, not correct, not correct. **1.34.** No. **1.35.** (a) 16.83; (b) 32.25. **1.37.** (c) 25, 14; In other cases the quantities cannot be calculated unless more information about the data is available (d) and (e) can be done if an assumption of uniform or even distribution of the frequencies in the various classes can be assumed. This will be discussed in the next section. **1.38.** (a) 2.34, 1.99;

(b) $\left\{ \displaystyle\prod_{i=1}^{n} x_i^{f_i} \right\}^{1/f}$; $\dfrac{f}{\displaystyle\sum_{i=1}^{n} \dfrac{f_i}{x_i}}$ where $f = f_1 + f_2 + \ldots + f_n$.

1.39. 125.5. **1.40.** 126. **1.41.** (a) 1; (b) 1. **1.42.** (a) (1) 9.75, (2) 12.5, (3) 4.8; (b) (1) 14, (2) 19.79, (3) 4.9. **1.43.** (a) 9; (b) 8.5. **1.44.** (a) 5.5; (b) 6.2. **1.45.** 21. **1.47.** (a) No mode; (b) 2; (c) 4. **1.48.** (a) 14.34; (b) 22.16 **1.49.** Mean-Mode $= 14.53 - 14.34 = 0s19$,   Mean-Median $= 14.53 - 14.45 = 0.08$. **1.50.** Mean-Mode $= 4.55 - 5 = -0.45$,   Mean-Median $= 4.55 - 5 = -0.45$. **1.51.** (a) 2.8; (b) 3.04; (c) 3. **1.52.** (a) 1.33; (b) 3.85; (c) 4.999; (d) not possible. **1.56.** When and only when all the numbers are the same. **1.57.** No. **1.58.** (a)1.88;(b) 2.936. **1.59.** (1) 6, 16; (2) 2, 2.8; (3) 2.33, 0.3; (4) 2.09, 0.26. **1.62.** Mean deviation: 1.88, 2.936; standard deviation: 2.1, 3.52; Semi-interquartile range: 2, 2.8 (Calculate). **1.63.** $m_3' = -0.2$, $m_4 = 0.8272$, $M_2' = 0.76$.

CHAPTER TWO

2.1. (1) No; (2) can be; (3) No, if he is aware of what he is doing; (4) Yes; (5) (a) No, (b) Yes. **2.2.** (1) $\{(1, 1), (1, 2), \ldots, (1, 6), \ldots, (6, 6)\}$, 36; (2) $\{(a_1, a_2), (a_1, a_3), \ldots, (a_4, a_5)\}$, 10; (3) $\{1, 2, \ldots, 52\}$, 52; (4) $\{x \mid 0 \leqslant x \leqslant 10\}$, infinite; (5) $\{(x, y) \mid 0 \leqslant x, y \leqslant 10\}$, infinite. **2.3.** (1) $\{(6, 4), (5, 5), (4, 6)\}$; (2) $\{(6, 5), (5, 6), (6, 6)\}$; (3) ϕ. **2.4.** (1) $\{(7, 3), (7, 5), (7, 7), (5, 5), (5, 7), (3, 7)\}$; (2) $\{(2, 3), (2, 5), (2, 7), (3, 5), (3, 7), (5, 7)\}$. **2.5.** A-the event of getting a person whose weight is between 20 and 80 (both inclusive). $A \cup B$—the event of getting a person whose weight is either between 20 and 80 or between 60 and 140 (all 4 numbers inclusive). $A \cap B$—the event of getting a person whose weight is in the interval $20 \leqslant x \leqslant 80$ as well as in the interval $60 \leqslant x \leqslant 140$. That is, the event of getting a person whose weight is in the interval $60 \leqslant x \leqslant 80$. **2.6.** A—the event of getting an odd number. $A \cup B$—the event of getting an odd number or an even number (which is a sure event). $A \cap B$—the event of getting an even number and at the same time an odd number (which is impossible and hence $A \cap B = \phi$). **2.8.** (a) 2^n; (b) 6^n. **2.9.** $A \cup B$—that he will see at least one of the animals, a deer and a bear. $A \cap B$—that he will see a deer as well as bear. **2.10.** $A \cup B$ —the event of getting either 2 dimes or a dime and a nickel. $A \cap B$—the event of getting 2 dimes and a dime and a nickel (which is impossible and hence $A \cap B = \phi$).

2.11. (a) Yes; (b) No; (c) No. **2.13.** The event that the business executive selected at random does not have stomach ulcers; (2) either has or does not have ulcer (sure event); (3) has and does not have ulcer (impossible); (4) has at least one of the diseases, heart disease or stomach ulcer; (5) has both; (6) has stomach ulcer or has no heart disease; (7) no stomach ulcer but has heart disease; (8) no ulcer, no heart disease; (9) does not have both. **2.14.** (1) will see a new immigrant, a hippie and a tourist; (2) a new immigrant, and no hippies no tourist; (3) a new immigrant or a hippie and no tourist; (4) at least one of them, an immigrant, a hippie and a tourist. **2.15.** (1) height is in the interval 4-7; (2) in the interval 5-6; (3) in the interval 4-6 or not in the interval 5-7. **2.16.** 34650. **2.17.** $25 \times (25)_3$. **2.18.** $(26)_2 \times (10)_5$. **2.19.** $\binom{13}{4}\binom{13}{2}\binom{13}{2}\binom{13}{0}$.

2.20. (1) 24; (2) 6; (3) 360. **2.21.** (1) 30; (2) 100. **2.22.** $\binom{400}{7}\binom{300}{2}\binom{100}{1}$.

2.23. 9. **2.24.** 56, 4. **2.25.** $\{1, 0\}$, $\{1, -1\}$, $\{1, 3\}$, $\{0, -1\}$, $\{0, 3\}$, $\{-1, 3\}$; $\{1, 0, -1\}$, $\{1, 0, 3\}$, $\{1, -1, 3\}$, $\{0, -1, 3\}$. **2.28.** 40320, 1·5%.
2.29. 3/8; $-45/2048$. **2.30.** (1) $(1 - 0.02)^{-1/3} = 1 + \frac{1}{3}(0.02) + \frac{1.4}{3.3}\frac{(0.02)^2}{2!}$

$+ \frac{1.4.7}{3.3.3}\frac{(0.02)^3}{3!} + ...$; (2) $2^{2/3}(1 + 0.02)^{2/3} = 2^{2/3}\left[1 + \frac{2}{3}(0.02) - \frac{2.1}{3.3}\frac{(0.02)^2}{2!}\right.$

$\left. + \frac{2.1.4}{3.3.3}\frac{(0.02)^3}{3!} - ...\right]$

CHAPTER THREE

3.1. No. **3.2.** (1) No; (2) No; (3) Yes; (4) No. **3.4.** (1) at least one of them, a teenage girl and a migratory bird, 0.7; (2) either teenage girl or no migratory bird or both, 0.8; (3) either no teenage girl or migratory bird, or both, 0.7; (4) not both, teenage girl and a migratory bird, 0.8; (5) no teenage girl and a migratory bird, 0.2; (6) not any of, a teenage girl or a migratory bird, 0.3. **3.5.** (1) No; (2) No. **3.6.** Yes. **3.7.** (1) 57/203; (2) 8/27; (3) 57/203. **3.8.** $\binom{13}{2}\binom{13}{3}\binom{13}{3}/\binom{52}{8}$. **3.9.** 7/18. **3.10.** $\frac{1}{4}$. **3.11.** $\frac{2}{3}$. **3.12.** 53/144. **3.13.** 15/128. **3.14.** (1) $\frac{2}{5}$; (2) 0. **3.15.** 2/7735. **3.16.** (1) 6/77; (2) 54/133; (3) $\frac{1}{4}$. **3.17.** (9.11.19.39) / (2.4.17.46.47.49). **3.18.** (1) 1/10; (2) 1/10; (3) $\frac{1}{5}$. **3.19.** (1) $\frac{2}{5}$; (2) $\frac{2}{5}$. **3.21.** 0.25. **3.22.** $\frac{2}{3}$. **3.23.** 0.705. **3.24.** $2P/(1+P)$ where P is the probability of getting a man from the community. **3.25.** 6/13. **3.26.** (1) 1.01; (2) 1.09.

CHAPTER FOUR

4.3. $\binom{4}{x}\binom{26}{5-x}/\binom{30}{5}$, $x = 0, 1, 2, 3, 4, 5$. **4.4.** $\binom{20}{x}(\frac{1}{2})^{20}$, $x = 0, 1, 2, ..., 20$.
4.5. (1) Yes; (2) No; (3) Yes; (4) No; (5) Yes; (6) No; (7) No; (8) Yes. **4.6.** (1) $\frac{1}{2}$; (2) $\frac{4}{3}$; (3) 1; (4) $\frac{1}{4}$. **4.7.** (1) $\frac{1}{8}$; (2) $\frac{1}{4}$; (3) $\frac{3}{4}$; (illustrate). **4.8.** (1) 0; (2) $\frac{1}{2}$; (3) $\frac{1}{4}$. **4.9.** Yes.

4.11. (1)
$$F(x) = \begin{cases} 0, & x < 0 \\ \frac{1}{2}, & 0 \leqslant x < 1 \\ \frac{3}{4}, & 1 \leqslant x < 5 \\ 1, & x \geqslant 5 \end{cases}$$

(2)
$$F(x) = \begin{cases} 0, & x < -2 \\ \frac{1}{5}, & -2 \leqslant x < -1 \\ \frac{9}{5}, & -1 \leqslant x < 1 \\ \frac{4}{5}, & 1 \leqslant x < 2 \\ 1, & x \geqslant 2. \end{cases}$$

4.12. (a) $F(x) = \dfrac{x}{5}$, $0 < x < 5$; $F(2) = \frac{2}{5}$; $F(3) - F(2) = \frac{1}{5}$; $F(-1) = 0$;

(b) $F(x) = (1 - e^{-x/2})$; $F(2) = (1 - e^{-1})$;
$F(3) - F(2) = e^{-1} - e^{-3/2}$; $F(-1) = 0$.

4.13. (a)
$$F(x) = \begin{cases} 0, & x < 0 \\ \dfrac{x^2}{2}, & 0 < x \leqslant 1; \ F(2) = \frac{7}{8}; \ F(1) = \frac{1}{2}. \\ -\frac{1}{8} + \dfrac{3x}{4} - \dfrac{x^2}{8}, & 1 < x \leqslant 3 \\ 1, & x \geqslant 3. \end{cases}$$

(b)
$$F(x) = \begin{cases} 0, & x < 0 \\ \dfrac{x^2}{4}, & 0 < x \leqslant 1; \ F(2) = \frac{5}{8}; \ F(1) = \frac{1}{4} \\ -\frac{3}{8} + \dfrac{3x}{4} - \dfrac{x^2}{8}, & 1 < x \leqslant 2 \\ \frac{1}{8} + \dfrac{x}{4}, & 2 < x \leqslant 3 \\ -1 + x - \dfrac{x^2}{8}, & 3 < x \leqslant 4 \\ 1, & x \geqslant 4. \end{cases}$$

4.14. (1)
$$f(x) = \begin{cases} \frac{4}{5}x, & 0 < x \leqslant 1 \\ \frac{2}{5}(3-x), & 1 < x < 2; \\ 0 \text{ elsewhere} \end{cases}$$

(2)
$$f(x) = \begin{cases} x, & 0 < x \leqslant 1 \\ \frac{1}{4}(3-x), & 1 < x < 3 \\ 0 \text{ elsewhere}. \end{cases}$$

4.15.
$$f(x) = \begin{cases} \frac{1}{4}, & x = 1 \\ \frac{1}{4}, & x = 2 \\ \frac{2}{4}, & x = 3 \text{ and } 0 \text{ elsewhere.} \end{cases}$$

4.16. (1) θ, θ; (2) 0, 1; (3) $-4/3$, 2. **4.17.** (1) 17/15; (2) 7/4. **4.18.** \$ 12.
4.19. \$ 3/2 (loss). **4.20.** \$ ($\frac{1}{2}$). **4.21.** (1) $\theta/2$; (2) $\theta/2$; (3) No mode; (4) $\theta/4$.
4.22. 11/16. **4.23.** 0. **4.26.** (1) 500; (2) 500. **4.27.** (1) don't accept;
(2) accept; (3) don't accept; (4) 1. **4.28.** Yes, first. **4.29.** (1) party 3; (2) 5/8.
4.30. 5. **4.31.** (1) $(1-t)^{-1}$; (2) does not exist. **4.32.** $M(t) = (e^{bt} - e^{at})/t(b-a)$;
$\mu_1' = (a+b)/2$; $\mu_2' = (b^3 - a^3)/3(b-a)$. **4.38.** Beta distribution with parameters
a and β.

CHAPTER FIVE

5.1. 0.2 363. **5.2.** 15/4^5. **5.3.** (1) 0.086; (2) 0.7361. **5.4.** (1) 7/8; (2) 3/8.
5.5. (1) 0.5490; (2) 0.1710. **5.9.** $N(N-1)(N-2)p^3 + 3N(N-1)p^2 + Np$.
5.10. (1) Binomial $N=5$, $p=\frac{1}{2}$; (2) Binomial $N=4$, $p=\frac{1}{4}$; (3) Binomial $N=3$,

$p=2/3$. **5.11.** (1) 0.1008; (2) 0.9502. **5.12.** 0.1901, 0.1804. **5.13.** (1) 0.0164; (2) 0.0361 **5.14.** 0.9862. **5.16.** λ, λ. **5.19.** (1) Poisson, $\lambda = 2$; (2) Poisson. $\lambda=\frac{1}{3}$; (3) Binomial, $p=\frac{1}{3}$, $N=3$. **5.21.** (1) 27/95; (2) 68/95. **5.22.** 0.0645, **5.23.** 0.4752. **5.24.** $0.01(0.99)^7$. **5.25.** 0.01806. **5.30.** (1) Geometric with $p=\frac{1}{2}$. (2) Negative Binomial with $k=2$, $p=\frac{2}{3}$.

5.31. $p(\theta) = \sum\limits_{x=0}^{2} \dfrac{\dbinom{20\ \theta}{x}\dbinom{20-20\ \theta}{10-x}}{\dbinom{20}{10}}$; $\sum\limits_{x=0}^{2}\dbinom{10}{x}\theta^x\ (1-\theta)^{10-x}$ (draw).

5.32. $\theta p(\theta)$. (Draw). **5.33.** (1) 0.044; (2) 0.10. **5.34.** (1) 2, 2; (2) 0.6769, 0.6767. **5.35.** 0.07892, 0.2.

CHAPTER SIX

6.1. (1) $(2\pi)^{-1/2} (0.2)^{-1} \exp(-x^2/0.08)$; (2) $(2\pi)^{-1/2} 4^{-1} \exp\{-(x+5)^2/32\}$; (3) $(2\pi)^{-1/2} 2^{-1} \exp\{-(x-1)^2/8\}$. **6.2.** (1) 0.2266; (2) 0.6586; (3) 0.6279; (4) 0.5; (5) 0.6388. **6.3.** (1) Yes; (2) Yes; (3) Yes; (4) Yes; (5) Yes; (6) No; (7) No. **6.6.** (1) 3 (approximately); (2) 1.96. **6.7.** $t_0=1.96$, $t_1=1.96$, No. **6.8.** (1) 0.383; (2) 0.0668. **6.10.** (1) Normal, $\mu=0$, $\sigma^2=2$; (2) Normal $\mu=-2$, $\sigma^2=8$; (3) Normal $\mu=3$, $\sigma^2=2$. **6.12.** (1) 0.4370 (approximately), (2) 0.21186; (3) 0. **6.14.** (1) 0.0214, 0.25, (2) 0.2023, $\frac{4}{9}$. **6.15.** (1) 0.0124; (2) 0.9938. **6.16.** 0.04. **6.17.** $1-e^{-6}$. **6.18.** (1) $e^{-2/5}$; (2) $1-e^{-6/5}$. **6.19.** (1) $\frac{7}{2}e^{-5/2}$; (2) 0. **6.20.** $e^{-7.5}$. **6.23.** (1) $1-e^{-1/5}$; (2) $1-e^{-2}$; (3) $e^{-1/5}$; (4) $1-e^{-1/5}$. **6.25.** (1) Binomial, $N=2$, $p=\frac{3}{4}$; (2) Poisson, $\lambda=2$; (3) Normal, $\mu=0$, $\sigma^2=6$; (4) Gamma, $\beta=2$, $\alpha=3$; (5) Exponential, $\beta=5$.

CHAPTER SEVEN

7.1. $h(x,y) = \begin{cases} \frac{1}{4} \text{ for } x=2, y=0 \\ \frac{1}{2}, \quad x=1, y=1 \\ \frac{1}{4}, \quad x=0, y=2 \\ 0, \quad \text{elsewhere.} \end{cases}$ $\qquad f(x) = \begin{cases} \frac{1}{4}, x=0 \\ \frac{1}{2}, x=1 \\ \frac{1}{4}, x=2 \\ 0, \text{ elsewhere.} \end{cases}$

$g(y) = \begin{cases} \frac{1}{4}, y=0 \\ \frac{1}{2}, y=1 \\ \frac{1}{4}, y=2 \\ 0, \text{ elsewhere.} \end{cases}$

7.2.

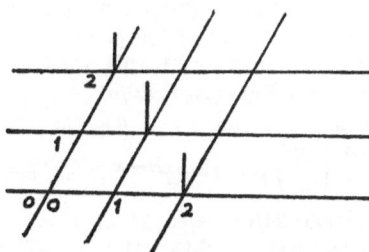

7.3. For example $P\{x=1, y=1\}=\frac{1}{2}\neq P\{x=1\}P\{y=1\}=\frac{1}{2}\cdot\frac{1}{2}=\frac{1}{4}$.

7.4. $f(x) = \begin{cases} \frac{1}{4}, & x=1 \\ \frac{1}{4}, & x=2 \\ \frac{1}{2}, & x=3 \\ 0, & \text{elsewhere.} \end{cases}$ $\qquad g(y) = \begin{cases} \frac{1}{4}, & y=0 \\ \frac{1}{4}, & y=3 \\ \frac{1}{2}, & y=5 \\ 0, & \text{elsewhere.} \end{cases}$

7.5. $f(x) = \begin{cases} \frac{2}{8}, & x=0 \\ \frac{6}{8}, & x=1 \\ 0, & \text{elsewhere.} \end{cases}$ $\qquad g(y) = \begin{cases} \frac{2}{8}, & y=0 \\ \frac{6}{8}, & y=1 \\ 0, & \text{elsewhere.} \end{cases}$

7.7. No. **7.9.** Yes; independent. **7.10.** Yes.
7.11. 1; 10.67. **7.14.** 0. **7.21.** 1 approx. **7.22.** 0.9266. **7.23.** 0.0047. **7.24.**
0.15. **7.25.** 0.05. **7.26.** 0.90. **7.27.** Yes at 99%. **7.28.** Yes at 95%. **7.29.** 0.95.

7.30. 0·4 approx. **7.33.** $\dfrac{u}{(1+v)^2}\, e^{-u}$, u, $v>0$ and 0 elsewhere. **7.34.** $\dfrac{1}{6}\left(\dfrac{1}{\sqrt{2\pi}}\right)^3$

$\exp\left[\dfrac{-\left(\dfrac{u^2}{3}+\dfrac{v^2}{6}+\dfrac{w^2}{2}\right)}{2}\right]$. **7.35.** 0·064. **7.36.** 0·0034. **7.37.** 0·09.

7.39. (a) $[(1-t_1)(1-t_2)]^{-1}$; (b) $\dfrac{2}{t_1}\left\{\dfrac{e^{t_1+t_2}-1}{t_1+t_2}-\dfrac{e^{t_2}-1}{t_2}\right\}$.

7.41. $f(y) = \dfrac{e^{-(y+3)^2/52}}{\sqrt{26}\,\sqrt{2\pi}}$, $-\infty < y < \infty$.

Chapter Eight

8.1. (1) 3, 10/3; (2) 27/10, 10/9. **8.2.** (1) 3, 10/3. **8.3.** M.L.E. is the largest of the observations and the method of moments gives $2\bar{x}$. **8.4.** 0.1. **8.5.** \$ 5000.
8.6. 10. **8.7.** \$17.5. **8.8.** $\hat{a}=$ smallest of the observations and $\hat{\beta}=$ the largest of the observations. **8.16.** (41.432, 44.568). **8.17.** (9930·13, 10069·87). **8.18.** (46·61, 53·39). **8.19.** (4·9883, 5·0317); (1) No; (2) Yes. **8.20.** (\$ 9490, \$ 10510). **8.21.** (−0·426, 2·426). **8.22.** (−8·5346, −1·4654). **8.23.** (18·9265, 21·0735). **8.24.** (8266, 9774). **8.25.** (0, 2599). **8.26.** (−14·51%, 4·51%). **8.27.** (0·05, 0·45) approx. **8.28.** (10·6, 144·5). **8.29.** (2·86, 10·54). **8.30.** (0·155, 0·299).

Chapter Nine

9.1. (1) 0.1209; (2) all points in the sample space for which the number of heads is greater than or equal to 7. (3) 0·1719. **9.2.** (1) all the points in the sample space of the form $(R\,R)$ out of 100 points; (2) 1/4; (3) 16/25. **9.3.** $C=\{x\mid x \geqslant 15,000\}$; $a=e^{-3}$; $\beta=1-e^{-3/2}$. **9.4.** $C=\{x_1,\ldots,x_{16}\}\mid 4(\bar{x}-25)\geqslant 3\}$; $a=0\cdot0014$. **9.5.** 5. **9.6.** Reject H_0 if $ns^2/\sigma^2 \geqslant \chi^2_{a,\,n-1}$. **9.7.** Reject H_0 if $\bar{x} \geqslant g_a$ where $P\{\text{Gamma variate, } a=n, \beta=\theta/n \text{ is} \geqslant g_a\}=a$. **9.8.** Reject H_0 if $x \geqslant k_a$ where $\sum\limits_{k_a}^{n}\binom{N}{x} p_0^x (1-p_0)^{N-x} \geqslant a$. **9.9.** Any size a test.
9.10. 0; 1/5; 16/45; 21/45; 24/45; 24/45; 21/45; 16/45; 1/5; 0. **9.12.** Same as the ones given in section 9·31. **9.13.** (1) $\chi_1^2 \geqslant \chi_{1,\,a}^2$; (2) $\chi_1^2 \leqslant \chi_{1,\,1-a}^2$; (3) $\chi_1^2 \geqslant \chi_{1,\,a/2}^2$ or $\leqslant \chi_{1,\,1-a/2}^2$. **9.14.** (1) reject; (2) reject. **9.15.** Accept.

9.16. Reject. **9.17.** Reject at 95%. **9.18.** Reject. **9.19.** Reject. **9.20.** 18. **9.21.** Reject. **9.22.** Accept. **9.23.** No. **9.24.** Accept. **9.25.** 0·7939; 0·7483; 0·0171.

CHAPTER TEN

10.1. Yes. **10.2.** No. **10.3.** Not good. **10.4.** Not a good fit. **10.5.** Not good; No; the sample may not be random. **10.6.** No. **10.7.** No. **10.8.** Not good. **10.9.** Good. **10.10.** Accept. **10.11.** Accept. **10.12.** Accept. **10.13·** Accept. **10.14.** Accept. **10.15.** Reject. **10.16.** Accept. **10.17.** Accept.

CHAPTER ELEVEN

11.1. $\hat{a}=-354.5$; $\hat{b}=7.5$. **11.2.** $\hat{a}=3.49$; $\hat{b}=1.055$. **11.3.** (1) $\hat{a}=-0.007$, $\hat{b}=-0.475$, $\hat{c}=-0.05$; (2) $\hat{a}=1.10$, $\hat{b}=21.18$; (3) $\hat{a}=4.86$, $\hat{b}=0.106$. **11.4.** straight line. **11.5.** See ex: 11.2.1. **11.6.** $\hat{a}=0$, $\hat{b}=$Cov $(x, y)/$Var (x). **11.7.** $(2+3y) / 9(1+2y)$. **11.8.** $y=3.1+2.6x$. **11.9.** Accept; (2.17,3.03). **11.10.** reject.

CHAPTER TWELVE

12.1. Significant. **12.2.** Not significant.

Answers to 'Additional Exercises'

1.1. (1) (a) Yes, (b) Yes, set of marks of which the 10 marks is a subset (could be the marks of all students in the class). (2) (a) Yes, (b) Yes (could be the tax returns of all the people in the province). (3) (a) Yes, (b) Yes, (could be the prices of round steak in all shops having round steak). (4) (a) Need not be, (b) Need not be. (5) (a) Yes, (b) Yes (could be such numbers for a year). (6) (a) Yes, (b) Yes (could be demand over the years).

1.2. (1) need not be; (2) need not be; (3) need not be; (4) need not be; (5) need not be; (6) yes, the set of possible numbers that can be obtained in 5 rolls.

1.4. (1) The classification is not unique; only one classification is given here. (2) 20-21, 22-23, 24-25, 26-27, 28-29. (3) 19.5-21.5, 21.5-23.5, etc. (4) bar diagrams. (5) no (can not be assumed to be continuous). (6) 21.

1.5. (1) The classification is not unique; only one such classification is given here. (2) 100-108, 109-117, 118-126, 127-135. (3) 99.5-108·5, 108·5-117·5, etc. (4) bar diagrams or histograms. (5) o.k.

1.6. The following results are obtained from the classification as reported in the answer to 1.4. (1) 24.7, 24.7; (2) 25, 25; (3) 21, 22, 25, 26; 22.17, 25.61; (4) 2.58, 2.6; (5) 2.16, 2.09; (6) 22, 22.6; (7) 2.25, 2; (8) 29.84, 22.84.

1.9. 1.17, 0.546.   **1.10.** 27, 5.6.

CHAPTER TWO

2.5. (1) $-63/2^8$; (2) -4; (3) 10.   **2.6.** (1) 3573000, 1.5%; (2) 39300000, 1.6%; (3) 472000000, 1.5%. **2.7.** (1) 210; (2) 120; (3) 1. **2.8.** (1) 210 (2) 12600; (3) 12600. **2.9.** $\binom{20}{5}$. **2.10.** $\binom{13}{3}\binom{13}{3}\binom{13}{2}\binom{13}{0}$. **2.11.** (1) 24; (2) 24. **2.12.** (1) 84; (2) $\binom{16}{6}$.   **2.13.** 14832.   **2.14.** (1) 16; (2) 216. **2.15.**

(1) $\bigcup\limits_{i=1}^{\infty} A_i$, where $A_1 = \{(H), (T)\}$, $A_2 = \{(H, H), (H, T), (T, H), (T, T)\}$,

$A_3 = \{(H,H,H),...\}$, ... (2) $\bigcup\limits_{i=1}^{\infty} B_i$ where $B_1 = \{1,..., 6\}$, $B_2 = \{(1, 1), (1, 2),...,$ $(6, 6)\}$, $B_3 = \{(1, 1, 1),..., (6, 6, 6)\}$,... **2.17.** no. **2.18.** no. **2.19.** no (depends upon $A \cup \bar{A}$). **2.20.** (1) $0 \leqslant x \leqslant 7$; (2) ϕ; (3) $\{0 \leqslant x \leqslant 2, \ 3 \leqslant x \leqslant 5\}$; (4) ϕ; (5) $0 \leqslant x \leqslant 2$; (6) $0 \leqslant x \leqslant 2$; (7) $\{-\infty < x < 0, \ 7 < x < \infty\}$. **2.21.** 18. **2.22.** 6. **2.23.** 8.

Chapter Three

3.1. (1) yes; (2) no; (3) no; (4) yes. **3.2.** (1) yes; (2) no; (3) yes; (4) yes; (5) no; (6) no. **3.5.** (1) 507/16660; (2) 33/66640; **3.6.** (1) $\frac{1}{3}$; (2) $\frac{1}{6}$; (3) $\frac{1}{2}$; (4) $\frac{1}{6}$. **3.7.** (1) 48/125; (2) 12/25. **3.8.** (1) 11/30; (2) $\frac{3}{5}$. **3.9.** (1) $\frac{1}{5}$; (2) $\frac{1}{5}$; (3) 4/25. **3.10.** (1) 7/29; (2) 15/29; (3) 15/58; (4) 1/58. **3.11.** 0·42. **3.12.** (1) 1/36; (2) $\frac{1}{2}$; (3) 1/12; (4) $\frac{1}{6}$. **3.13.** (16) (39) (39)/(33) (49) (97). **3.14.** $\frac{1}{8}$. **3.15.** $\frac{9}{7}$. **3.16.** 0.64. **3.17.** (1) 0.1; (2) 0·006. **3.18.** 0·275. **3.19.** $\frac{1}{2}$. **3.20.** 2/19. **3.21.** $\frac{1}{2}$.

Chapter Four

4.2. (1) no; (2) no; (3) no, if $\theta \neq 1$; (4) yes. **4.3.** (1) no; (2) no; (3) no; (4) yes.

4.4. (1) $F(x) = \begin{cases} 0, & -\infty \leqslant x < -1 \\ \frac{1}{8}, & -1 \leqslant x < 0 \\ \frac{2}{8}, & 0 \leqslant x < 2 \\ \frac{5}{8}, & 2 \leqslant x < 4 \\ 1, & x \geqslant 4; \end{cases}$ (2) $F(x) = \begin{cases} 0, & -\infty < x < -2 \\ \frac{1}{2}, & -2 \leqslant x < 0 \\ 1, & x \geqslant 0. \end{cases}$

(3) $F(x) = \begin{cases} 0, & -\infty < x \leqslant 0 \\ 1 - e^{-x/\theta}, & x > 0; \end{cases}$ (4) $F(x) = \begin{cases} 0, & -\infty < x \leqslant 0 \\ 1 - (x+1)e^{-x}, & x > 0. \end{cases}$

4.5. (1) 1; (2) 12/7; (3) 2/25; (4) 1. **4.6.** (1) $f(0) = f(2) = \frac{1}{8}$, $f(3) = f(4) = \frac{3}{8}$ and $f(x) = 0$, elsewhere; (2) $f(1) = f(6) = \frac{1}{8}$, $f(4) = \frac{2}{4}$ and $f(x) = 0$, elsewhere; (3) $f(x) = 1 - x/2$, $0 < x < 2$ and $f(x) = 0$, elsewhere; (4) $f(x) = x$, $0 < x \leqslant 1$, $f(x) = \frac{1}{2}$, $1 < x \leqslant 2$ and $f(x) = 0$, elsewhere. **4.7.** (1) 7/4; (2) 11/4; (3) 27/16; (4) 2. **4.8.** 2; 2. **4.13.** (1) 0; (2) 0. **4.14.** (1) 0; (2) $-6/5$. **4.15.** (1) $1 - e^{-2/\theta}$; (2) $e^{-1/\theta} (1 - e^{-4/\theta})$. **4.16.** 1/4; 2/4; 0. **4.17.** X is degenerate at $E(X)$. **4.18.** (1) not; (2) not. **4.19.** 5. **4.20.** 50%. **4.21.** $\frac{1}{4}$; $\frac{3}{4}$. **4.22.** (1) $(1 - 2t)^{-3}$; (2) $(1 + e^t)/2$. **4.23.** (1) $f(x) = x^3 e^{-x/2}/3^4 \ 3!$, $x > 0$ and $f(x) = 0$, elsewhere; (2) $f(x) = 1$, $x = 5$ and $f(x) = 0$, elsewhere. **4.24.** (1) 12, 36; (2) 5, 0. **4.26.** (1) no; (2) no; (3) no.

Chapter Five

5.1. $\frac{1}{8}$. **5.2.** 26/27. **5.3.** 0.9274. **5.4.** $\binom{20}{0} \binom{30}{5} \Big/ \binom{50}{5}$. **5.5.** $(30/50)^5$.

5.7. see page 167. **5.8.** $\bar{x} = \Sigma x_i/n$; $\Sigma (x_i - \bar{x})^2/n$. **5.9.** 0.0183. **5.10.** (1) 0.4481; (2) 0.8009. **5.11.** 0.5370. **5.12.** 0.0053. **5.13.** $\Sigma e^{tx_i}/n$. **5.14.** (1) Geometric, $p = \frac{1}{5}$; (2) Geometric, $p = \frac{1}{2}$. **5.15.** $-(1-p)/p \log p$. **5.16.** (1)−(4) see table 5.3;

(5) $f(x+1)=(1-p)$ $[x/(x+1)]$ $f(x)$; (6) $f(x+2)=[(N-x-2)/(x+N)]$ (p/q) $f(x)$.

5.17. (1) 0, $\sum_{x=2}^{4} \binom{x-1}{1} p^2 (1-p)^{x-2}$, 0, 1; (2) $-(1-p)/\log p$, $-\sum_{x=2}^{4} (1-p)^x/x$

$\log p$, 0, $1+(1-p)/\log p$. **5.18.** (1) 21/38; (2) 21/1292. **5.19.** 0.3679.
5.20. (1) 0.02; (2) 0.247. **5.22.** (1) 0.2852; 0.0755. (2) 0.2707; 0.0803.

CHAPTER SIX

6.3. (1) symmetric of the same shape, the line of symmetry changes. (2)
symmetric about the same line but the curves become more peaked when σ^2
decreases. **6.5.** (1) 0.0668; (2) 0.6687; (3) 0; (4) 0.8664; (5) 0.0000; (6) 0.0013.
6.7. 0.1036. **6.8.** (1) 0.4971; (2) 0.8186; (3) 0.9759. **6.9.** $(e^{t\beta}-e^{t\alpha})/t(\beta-\alpha)$.
6.10. (1) $(\alpha+\beta)/2$; (2) $(\beta-\alpha)^2/12$. **6.13.** (1) 0.9599; (2) 0.0329. **6.14.** e^{-1}.
6.15. (1) $2e^{-1}-e^{-2}$; (2) $(1-e^{-2})$. **6.16.** $1-4e^{-3}$. **6.17.** (8/5) $e^{-1.5}$. **6.18.** (1)
$2e^{-1}-3e^{-2}$; (2) 0. **6.19.** (1) $g(y)=(y-3)/2$, $3 < y < 5$ and $g(y)=0$, elsewhere;
(2) $g(t)=2$ $(t+2)/9$, $-2 < t < 1$ and 0, elsewhere. **6.20.** $g(y)=-1+y/4$ for

$4 < y < 6$, $(5-y/2)/8$ for $6 \leqslant y < 10$ and 0, elsewhere. **6.27.** $\dfrac{1}{na}$ $y^{(1/n)-1}$.

6.28. $(\sqrt{y}+1)/(16\sqrt{y})$, $y>1$; $1/(8\sqrt{y})$, $y\leqslant1$.

CHAPTER SEVEN

7.11. 0.5. **7.12.** 0. **7.14.** 0.99. **7.15.** 0.84. **7.17.** (1) e^{-x}, $x > 0$; (2) e^{-y},
$y>0$. **7.18.** (1) $h(x \mid y=\frac{1}{2})=2$, $0 < x < \frac{1}{2}$ and 0, elsewhere; $k(y \mid x = \frac{1}{4}) = \frac{4}{3}$,
$\frac{1}{4} <y< 1$ and 0, elsewhere. **7.19.** $h(x \mid y=2)=\frac{1}{6}$ for $x=0$ and $x=1$ and $\frac{3}{6}$ for
$x=2$ and 0, elsewhere. **7.20.** 1. **7.25.** $\frac{4}{3}$. **7.26.** -0.4. **7.28.** 0.01 (approx.).
7.29. yes (90−95%). **7.30.** 0.5 (approx.). **7.31.** no (at 95%). **7.34.** 0.99.
7.35. 0.05.

CHAPTER EIGHT

8.1. $\hat{\beta}=(\Sigma x_i)/n\hat{a}$, $\log \hat{a} - d/d\hat{a} \log (\hat{a}) = \log (\Sigma x_i) - (1/n) \log (\pi x_i) - \log n$.
8.4. $[X(X-1) / N(N-1) + 2]$. **8.5.** λ/n. **8.11.** $[2N\hat{p} + z_{a/2}^2 \pm z_{a/2}$

$\sqrt{4N\hat{p}+z_{a/2}^2-4N\hat{p}^2} / 2(N+z_{a/2}^2)]$. **8.12.** $\left[s' / \left(1 + \dfrac{z_{a/2}}{\sqrt{2n}}\right), s' / \left(1 - \dfrac{z_{a/2}}{\sqrt{2n}}\right)\right]$

8.13. (0, 0.4). **8.14.** (1) (0.014, 0.086); (2) (0.05, 0.28). **8.15.** (1)

$\left(\bar{x} + \dfrac{z_{a/2}^2}{2n}\right) \pm \dfrac{z_{a/2}}{2\sqrt{n}} \sqrt{\left(\dfrac{z_{a/2}}{n} + 4\bar{x}\right)}$; (2) Read about general procedures

(Mood and Graybill, *Introduction to Theory of Statistics*, pp. 256-262).
8.16. $[(0.01) / \sqrt{1.929}, (0.01) / \sqrt{0.342}]$. **8.17.** (1) (1.0542, 1.1458);
(2) $(\sqrt{0.037}, 0.043)$. **8.18.** (1) (98.77, 101.63); (2) $(3/\sqrt{9.49}, 3/\sqrt{0.71})$.
8.19. (1.45, 2.55). **8.20.** $(-5344.19, -4655.81)$.

CHAPTER NINE

9.1. e^{-2}; $1-e^{-10/7}$. **9.2.** (1) 0.0027. **9.3.** 3/7; 11/14. **9.4.** reject. **9.5.** reject.
9.6. reject. **9.7.** reject. **9.8.** accept. **9.9.** reject. **9.10.** reject. **9.11.** reject.
9.12. reject. **9.13.** no exact test unless $\sigma_1^2=\sigma_2^2$. **9.14.** accept. **9.15.** reject.

9.16. based on $F_{k-1,\ n-k} = \dfrac{\Sigma(\bar{x}_i-\bar{x})/(k-1)}{\Sigma n_i s_i^2/(\Sigma n_i-k)}$

where $\bar{x}_i$ and s_i^2 are the ith sample mean and sample variance and $\bar{x}$ is the grand
mean $=\Sigma n_i\bar{x}_i\ /\ \Sigma n_i$ where n_i is the ith sample size, $i=1,...,k$. **9.19.** (1) $k-1$;
(2) $k-2$; (3) $2k-2$. **9.20.** no, unless the σ's are equal.

CHAPTER TEN

10.1. Accept. **10.2.** Accept. **10.3.** Reject. **10.4.** (1) not good; (2) not good.
10.5. Not good. **10.6.** Compatible. **10.7.** No evidence of association. **10.8.**
Accept. **10.9.** Accept. **10.10.** Accept. **10.11.** Accept. **10.12.** Accept. **10.13.**
Accept. **10.14.** Accept. **10.15.** Accept.

CHAPTER ELEVEN

11.1. (1) $\dfrac{1}{2(1-x)}$; (2) $\dfrac{1}{2y}$. **11.3.** $\hat{a}_0=\bar{y}-\hat{a}_1\bar{x}_1-\hat{a}_2\bar{x}_2$,

$\hat{a}_1 = \dfrac{\text{cov}(x,y)}{\text{var}(x_1)} - \hat{a}_2\dfrac{\text{cov}(x_1,x_2)}{\text{var}(x_1)}$ and $\hat{a}_2 = \dfrac{\text{cov}(x_1,x_2)}{\text{var}(x_1)}$ and

$\hat{a}_2 = [\text{cov}(x_1,x_2)\,\text{cov}(x_1,y)-\text{Var}(x_1)\,\text{cov}(x_2,y)]/$
$\qquad [\text{cov}^2(x_1,x_2)-\text{var}(x_1)\,\text{var}(x_2)]$

where the variances and covariances are the sample variances and covariances
respectively. **11.4.** Replace all the sample values by the population values in
the estimates in 11.3 then the exact values of a_0, a_1 and a_2 are obtained. **11.6.**
$y=0.25+2.3x$; no evidence of linear regression; $(-4.7, 7.3)$. **11.7.** $x=1.33$
$-14.67\,y+10.67\,z$. **11.8.** (1) $\hat{a}=2.93$, $\hat{b}=1.43$; (2) $\hat{a}=2.24$, $\hat{b}=2\cdot5$, $\hat{c}=-0.02$;
(3) $\hat{a}=14.6$, $\hat{b}=-8.33$.

CHAPTER TWELVE

12.1. (a) not significant; (b) significant; (c) not significant. **12.2.** (a) A-
effects (not significant), B-effects (not significant); (b) A-effects (significant),
B-effects (not significant). **12.3.** No, unless more assumptions are made about
the model. **12.4.** and **12.5.** See O. Kempthorne, *Design and Analysis of
Experiments*, Wiley, New York, 1962.

Statistical Tables

TABLE 1: BINOMIAL COEFFICIENTS

Entry: $\left(\begin{array}{c} n \\ x \end{array}\right) = \left(\begin{array}{c} n \\ n-x \end{array}\right) = \dfrac{n!}{x!(n-x)!}$

n	$x=0$	1	2	3	4	5	6	7	8	9	10	11
5	1	5	10									
6	1	6	15	20								
7	1	7	21	35								
8	1	8	28	56	70							
9	1	9	36	84	126							
10	1	10	45	120	210	252						
11	1	11	55	165	330	462						
12	1	12	66	220	495	792	924					
13	1	13	78	286	715	1287	1716					
14	1	14	91	364	1001	2002	3003	3432				
15	1	15	105	455	1365	3003	5005	6435				
16	1	16	120	560	1820	4368	8008	11440	12870			
17	1	17	136	680	2380	6188	12376	19448	24310			
18	1	18	153	816	3060	8568	18564	31824	43758	48620		
19	1	19	171	969	3876	11628	27132	50388	75582	92378		
20	1	20	190	1140	4845	15504	38760	77520	125970	167960	184756	
21	1	21	210	1330	5985	20349	54264	116280	203499	293930	352716	
22	1	22	231	1540	7315	26334	74613	170544	319770	497420	646646	705432
23	1	23	253	1771	8855	33649	100947	245157	490314	817190	1144066	1352078
24	1	24	276	2024	10626	42504	134596	346104	735471	1307504	1961256	2496144
25	1	25	300	2300	12650	53130	177100	480700	1081575	2042975	3268760	4457400
26	1	26	325	2600	14950	65780	230230	657800	1562275	3124550	5311735	7726160
27	1	27	351	2925	17550	80730	296010	888030	2220075	4686825	8436285	13037895
28	1	28	378	3276	20475	98280	376740	1184040	3108105	6906900	13123110	21474180
29	1	29	406	3654	23751	118755	475020	1560780	4292145	10015005	20030010	34597290
30	1	30	435	4060	27405	142506	593775	2035800	5852925	14307150	30045015	54627300

n	$x=12$	13	14	15
24	2704156			
25	5200300			
26	9657700	10400600		
27	17383860	20058300		
28	30421755	37442160	40116600	
29	51895935	67863915	77558760	
30	86493225	119759850	145422675	155117520

TABLE 2: CUMULATIVE BINOMIAL PROBABILITIES

Entry: $$\sum_{r=0}^{x} \binom{n}{r} p^r (1-p)^{n-r} = \sum_{r=n-x}^{n} \binom{n}{r} (1-p)^r p^{n-r}$$

n	x	p=.05	.10	.15	.20	.25	.30	.35	.40	.45	.50
1	0	.9500	.9000	.8500	.8000	.7500	.7000	.6500	.6000	.5500	.5000
	1	1.0000	1.0000	1.0000	1.0000	1.0000	1.0000	1.0000	1.0000	1.0000	1.0000
2	0	.9025	.8100	.7225	.6400	.5625	.4900	.4225	.3600	.3025	.2500
	1	.9975	.9900	.9775	.9600	.9375	.9100	.8775	.8400	.7975	.7500
	2	1.0000	1.0000	1.0000	1.0000	1.0000	1.0000	1.0000	1.0000	1.0000	1.0000
3	0	.8574	.7290	.6141	.5120	.4219	.3430	.2746	.2160	.1664	.1250
	1	.9927	.9720	.9392	.8960	.8437	.7840	.7182	.6480	.5747	.5000
	2	.9999	.9990	.9966	.9920	.9844	.9730	.9571	.9360	.9089	.8750
	3	1.0000	1.0000	1.0000	1.0000	1.0000	1.0000	1.0000	1.0000	1.0000	1.0000
4	0	.8145	.6561	.5220	.4096	.3164	.2401	.1785	.1296	.0915	.0625
	1	.9860	.9477	.8905	.8912	.7383	.6517	.5630	.4752	.3910	.3125
	2	.9995	.9963	.9880	.9728	.9492	.9163	.8735	.8208	.7585	.6875
	3	1.0000	.9999	.9995	.9984	.9961	.9919	.9850	.9744	.9590	.9375
	4	1.0000	1.0000	1.0000	1.0000	1.0000	1.0000	1.0000	1.0000	1.0000	1.0000
5	0	.7738	.5905	.4437	.3277	.2373	.1681	.1160	.0778	.0503	.0313
	1	.9774	.9185	.8352	.7373	.6328	.5282	.4284	.3370	.2562	.1875
	2	.9988	.9914	.9734	.9421	.8905	.8369	.7648	.6826	.5931	.5000
	3	1.0000	.9995	.9978	.9933	.9844	.9692	.9460	.9130	.8688	.8125
	4	1.0000	1.0000	.9999	.9997	.9990	.9976	.9947	.9898	.9815	.9688
	5	1.0000	1.0000	1.0000	1.0000	1.0000	1.0000	1.0000	1.0000	1.0000	1.0000

n	r										
6	0	·7351	·5314	·3771	·2621	·1780	·1176	·0754	·0467	·0277	·0156
	1	·9672	·8857	·7765	·6554	·5339	·4202	·3191	·2333	·1636	·1094
	2	·9978	·9841	·9527	·9011	·8306	·7443	·6471	·5443	·4415	·3438
	3	·9999	·9987	·9941	·9830	·9624	·9295	·8826	·8208	·7447	·6562
	4	1·0000	·9999	·9996	·9984	·9954	·9891	·9777	·9590	·9308	·8906
	5	1·0000	1·0000	1·0000	·9999	·9998	·9993	·9982	·9959	·9917	·9844
	6	1·0000	1·0000	1·0000	1·0000	1·0000	1·0000	1·0000	1·0000	1·0000	1·0000
7	0	·6983	·4783	·3206	·2097	·1335	·0824	·0490	·0280	·0152	·0078
	1	·9556	·8503	·7166	·5767	·4449	·3294	·2338	·1586	·1024	·0625
	2	·9962	·9743	·9262	·8520	·7564	·6471	·5323	·4199	·3164	·2266
	3	·9998	·9973	·9879	·9667	·9294	·8740	·8002	·7102	·6083	·5000
	4	1·0000	·9998	·9988	·9953	·9871	·9712	·9444	·9037	·8471	·7734
	5	1·0000	1·0000	·9999	·9996	·9987	·9962	·9910	·9812	·9643	·9375
	6	1·0000	1·0000	1·0000	1·0000	·9999	·9998	·9994	·9984	·9963	·9922
	7	1·0000	1·0000	1·0000	1·0000	1·0000	1·0000	1·0000	1·0000	1·0000	1·0000
8	0	·6634	·4305	·2725	·1678	·1001	·0576	·0319	·0168	·0084	·0039
	1	·9428	·8131	·6572	·5033	·3671	·2553	·1691	·1064	·0632	·0352
	2	·9942	·9619	·8948	·7969	·6786	·5518	·4278	·3154	·2201	·1445
	3	·9996	·9950	·9786	·9437	·8862	·8059	·7064	·5941	·4770	·3633
	4	1·0000	·9996	·9971	·9896	·9727	·9420	·8939	·8263	·7396	·6367
	5	1·0000	1·0000	·9998	·9988	·9958	·9887	·9747	·9502	·9115	·8555
	6	1·0000	1·0000	1·0000	·9999	·9996	·9987	·9964	·9915	·9819	·9648
	7	1·0000	1·0000	1·0000	1·0000	1·0000	·9999	·9998	·9993	·9983	·9961
	8	1·0000	1·0000	1·0000	1·0000	1·0000	1·0000	1·0000	1·0000	1·0000	1·0000
9	0	·6302	·3874	·2316	·1342	·0751	·0404	·0207	·0101	·0046	·0020
	1	·9288	·7748	·5995	·4362	·3003	·1960	·1211	·0705	·0385	·0195
	2	·9916	·9470	·8591	·7382	·6007	·4628	·3373	·2318	·1495	·0898
	3	·9994	·9917	·9661	·9144	·8343	·7297	·6089	·4826	·3614	·2539
	4	1·0000	·9991	·9944	·9804	·9511	·9012	·8283	·7334	·6214	·5000

CUMULATIVE BINOMIAL PROBABILITIES—(*Continued*)

n	x	p=·05	·10	·15	·20	·25	·30	·35	·40	·45	·50
	5	1·0000	·9999	·9994	·9969	·9900	·9747	·9464	·9006	·8342	·7461
	6	1·0000	1·0000	1·0000	·9997	·9987	·9957	·9888	·9750	·9502	·9102
	7	1·0000	1·0000	1·0000	1·0000	·9999	·9996	·9986	·9962	·9909	·9805
	8	1·0000	1·0000	1·0000	1·0000	1·0000	1·0000	·9999	·9997	·9992	·9980
	9	1·0000	1·0000	1·0000	1·0000	1·0000	1·0000	1·0000	1·0000	1·0000	1·0000
10	0	·5987	·3487	·1969	·1074	·0563	·0282	·0135	·0060	·0025	·0010
	1	·9139	·7361	·5443	·3758	·2440	·1493	·0860	·0464	·0233	·0107
	2	·9885	·9298	·8202	·6778	·5256	·3828	·2616	·1673	·0996	·0547
	3	·9990	·9872	·9500	·8791	·7759	·6496	·5138	·3823	·2660	·1719
	4	·9999	·9984	·9901	·9672	·9219	·8497	·7515	·6331	·5044	·3770
	5	1·0000	·9999	·9986	·9936	·9803	·9527	·9051	·8338	·7384	·6230
	6	1·0000	1·0000	·9999	·9991	·9965	·9894	·9740	·9452	·8980	·8281
	7	1·0000	1·0000	1·0000	·9999	·9996	·9984	·9952	·9877	·9726	·9453
	8	1·0000	1·0000	1·0000	1·0000	1·0000	·9999	·9995	·9983	·9955	·9893
	9	1·0000	1·0000	1·0000	1·0000	1·0000	1·0000	1·0000	·9999	·9997	·9990
	10	1·0000	1·0000	1·0000	1·0000	1·0000	1·0000	1·0000	1·0000	1·0000	1·0000
11	0	·5688	·3138	·1673	·0859	·0422	·0198	·0088	·0036	·0014	·0005
	1	·8981	·6974	·4922	·3221	·1971	·1130	·0606	·0302	·0139	·0059
	2	·9848	·9104	·7788	·6174	·4552	·3127	·2001	·1189	·0652	·0327
	3	·9984	·9815	·9306	·8389	·7133	·5696	·4256	·2963	·1911	·1133
	4	·9999	·9972	·9841	·9496	·8854	·7897	·6683	·5328	·3971	·2744
	5	1·0000	·9997	·9973	·9883	·9657	·9218	·8513	·7535	·6331	·5000
	6	1·0000	1·0000	·9997	·9980	·9924	·9784	·9499	·9006	·8262	·7256
	7	1·0000	1·0000	1·0000	·9998	·9988	·9957	·9878	·9707	·9390	·8867
	8	1·0000	1·0000	1·0000	1·0000	·9999	·9994	·9980	·9941	·9852	·9673

n	x										
	9	1·0000	1·0000	1·0000	1·0000	1·0000	1·0000	·9998	·9993	·9978	·9941
	10	1·0000	1·0000	1·0000	1·0000	1·0000	1·0000	1·0000	1·0000	·9998	·9995
	11	1·0000	1·0000	1·0000	1·0000	1·0000	1·0000	1·0000	1·0000	1·0000	1·0000
12	0	·5404	·2824	·1422	·0687	·0317	·0138	·0057	·0022	·0008	·0002
	1	·8816	·6590	·4435	·2749	·1584	·0850	·0424	·0196	·0083	·0032
	2	·9804	·8891	·7358	·5583	·3907	·2528	·1513	·0834	·0421	·0193
	3	·9978	·9744	·9078	·7946	·6488	·4925	·3467	·2253	·1345	·0730
	4	·9998	·9957	·9761	·9274	·8424	·7237	·5833	·4382	·3044	·1938
	5	1·0000	·9995	·9954	·9806	·9456	·8822	·7873	·6652	·5269	·3872
	6	1·0000	·9999	·9993	·9961	·9857	·9614	·9154	·8418	·7393	·6128
	7	1·0000	1·0000	·9999	·9994	·9972	·9905	·9745	·9427	·8883	·8062
	8	1·0000	1·0000	1·0000	·9999	·9996	·9983	·9944	·9847	·9644	·9270
	9	1·0000	1·0000	1·0000	1·0000	1·0000	·9998	·9992	·9972	·9921	·9807
	10	1·0000	1·0000	1·0000	1·0000	1·0000	1·0000	·9999	·9997	·9989	·9968
	11	1·0000	1·0000	1·0000	1·0000	1·0000	1·0000	1·0000	1·0000	·9999	·9998
	12	1·0000	1·0000	1·0000	1·0000	1·0000	1·0000	1·0000	1·0000	1·0000	1·0000
13	0	·5133	·2543	·1209	·0550	·0238	·0097	·0037	·0013	·0004	·0001
	1	·8646	·6213	·3983	·2336	·1267	·0637	·0296	·0126	·0049	·0017
	2	·9755	·8661	·6920	·5017	·3326	·2025	·1132	·0572	·0269	·0112
	3	·9969	·9658	·8820	·7473	·5843	·4206	·2783	·1686	·0929	·0461
	4	·9997	·9935	·9658	·9009	·7940	·6543	·5005	·3530	·2279	·1334
	5	1·0000	·9991	·9925	·9700	·9198	·8346	·7159	·5744	·4268	·2905
	6	1·0000	·9999	·9987	·9930	·9757	·9376	·8705	·7712	·6437	·5000
	7	1·0000	1·0000	·9998	·9988	·9944	·9818	·9538	·9023	·8212	·7095
	8	1·0000	1·0000	1·0000	·9998	·9990	·9960	·9874	·9679	·9302	·8666
	9	1·0000	1·0000	1·0000	1·0000	·9999	·9993	·9975	·9922	·9797	·9539
	10	1·0000	1·0000	1·0000	1·0000	1·0000	·9999	·9997	·9987	·9959	·9888
	11	1·0000	1·0000	1·0000	1·0000	1·0000	1·0000	1·0000	·9999	·9995	·9983
	12	1·0000	1·0000	1·0000	1·0000	1·0000	1·0000	1·0000	1·0000	1·0000	·9999
	13	1·0000	1·0000	1·0000	1·0000	1·0000	1·0000	1·0000	1·0000	1·0000	1·0000

CUMULATIVE BINOMIAL PROBABILITIES—(Continued)

n	x	p=·05	·10	·15	·20	·25	·30	·35	·40	·45	·50
14	0	·4877	·2288	·1028	·0440	·0178	·0068	·0024	·0008	·0002	·0001
	1	·8470	·5846	·3567	·1979	·1010	·0475	·0205	·0081	·0029	·0009
	2	·9699	·8416	·6479	·4481	·2811	·1608	·0839	·0398	·0170	·0065
	3	·9958	·9559	·8535	·6982	·5213	·3552	·2205	·1243	·0632	·0287
	4	·9996	·9908	·9533	·8702	·7415	·5842	·4227	·2793	·1672	·0898
	5	1·0000	·9985	·9885	·9561	·8883	·7805	·6405	·4859	·3373	·2120
	6	1·0000	·9998	·9978	·9884	·9617	·9064	·8164	·6925	·5461	·3953
	7	1·0000	1·0000	·9997	·9976	·9897	·9685	·9247	·8499	·7414	·6047
	8	1·0000	1·0000	1·0000	·9996	·9978	·9917	·9757	·9417	·8811	·7880
	9	1·0000	1·0000	1·0000	1·0000	·9997	·9983	·9940	·9825	·9574	·9102
	10	1·0000	1·0000	1·0000	1·0000	1·0000	·9998	·9989	·9961	·9886	·9713
	11	1·0000	1·0000	1·0000	1·0000	1·0000	1·0000	·9999	·9994	·9978	·9935
	12	1·0000	1·0000	1·0000	1·0000	1·0000	1·0000	1·0000	·9999	·9997	·9991
	13	1·0000	1·0000	1·0000	1·0000	1·0000	1·0000	1·0000	1·0000	1·0000	·9999
	14	1·0000	1·0000	1·0000	1·0000	1·0000	1·0000	1·0000	1·0000	1·0000	1·0000
15	0	·4633	·2059	·0874	·0352	·0134	·0047	·0016	·0005	·0001	·0000
	1	·8290	·5490	·3186	·1671	·0802	·0353	·0142	·0052	·0017	·0005
	2	·9638	·8159	·6042	·3980	·2361	·1268	·0617	·0271	·0107	·0037
	3	·9945	·9444	·8227	·6482	·4613	·2969	·1727	·0905	·0424	·0176
	4	·9994	·9873	·9383	·8358	·6865	·5155	·3519	·2173	·1204	·0592
	5	·9999	·9978	·9832	·9389	·8516	·7216	·5643	·4032	·2608	·1509
	6	1·0000	·9997	·9964	·9819	·9434	·8689	·7548	·6098	·4522	·3036
	7	1·0000	1·0000	·9994	·9958	·9827	·9500	·8868	·7869	·6535	·5000
	8	1·0000	1·0000	·9999	·9992	·9958	·9848	·9578	·9050	·8182	·6964
	9	1·0000	1·0000	1·0000	·9999	·9992	·9963	·9876	·9662	·9231	·8491
	10	1·0000	1·0000	1·0000	1·0000	·9999	·9993	·9972	·9907	·9745	·9408

n	r										
	11	·9824	·9937	·9981	·9995	·9999	1·0000	1·0000	1·0000	1·0000	1·0000
	12	·9963	·9989	·9997	·9999	1·0000	1·0000	1·0000	1·0000	1·0000	1·0000
	13	·9995	·9999	1·0000	1·0000	1·0000	1·0000	1·0000	1·0000	1·0000	1·0000
	14	1·0000	1·0000	1·0000	1·0000	1·0000	1·0000	1·0000	1·0000	1·0000	1·0000
16	0	·0000	·0001	·0003	·0010	·0033	·0100	·0281	·0743	·1853	·4401
	1	·0003	·0010	·0033	·0098	·0261	·0635	·1407	·2839	·5147	·8108
	2	·0021	·0066	·0183	·0451	·0994	·1971	·3518	·5614	·7892	·9571
	3	·0106	·0281	·0651	·1339	·2459	·4050	·5981	·7899	·9316	·9930
	4	·0384	·0853	·1666	·2892	·4499	·6302	·7982	·9209	·9830	·9991
	5	·1051	·1976	·3288	·4900	·6598	·8103	·9183	·9765	·9967	·9999
	6	·2272	·3660	·5272	·6881	·8247	·9204	·9733	·9944	·9995	1·0000
	7	·4018	·5629	·7161	·8406	·9256	·9729	·9930	·9989	·9999	1·0000
	8	·5982	·7441	·8577	·9329	·9743	·9925	·9985	·9998	1·0000	1·0000
	9	·7728	·8750	·9417	·9771	·9929	·9984	·9998	1·0000	1·0000	1·0000
	10	·8949	·9514	·9809	·9938	·9984	·9997	1·0000	1·0000	1·0000	1·0000
	11	·9616	·9851	·9951	·9987	·9997	1·0000	1·0000	1·0000	1·0000	1·0000
	12	·9894	·9965	·9991	·9998	1·0000	1·0000	1·0000	1·0000	1·0000	1·0000
	13	·9979	·9994	·9999	1·0000	1·0000	1·0000	1·0000	1·0000	1·0000	1·0000
	14	·9997	·9999	1·0000	1·0000	1·0000	1·0000	1·0000	1·0000	1·0000	1·0000
	15	1·0000	1·0000	1·0000	1·0000	1·0000	1·0000	1·0000	1·0000	1·0000	1·0000
17	0	·0000	·0000	·0002	·0007	·0023	·0075	·0225	·0631	·1668	·4181
	1	·0001	·0006	·0021	·0067	·0193	·0501	·1182	·2525	·4818	·7922
	2	·0012	·0041	·0123	·0327	·0774	·1637	·3096	·5198	·7618	·9497
	3	·0064	·0184	·0464	·1028	·2019	·3530	·5489	·7556	·9174	·9912
	4	·0245	·0596	·1260	·2348	·3887	·5739	·7582	·9013	·9779	·9988
	5	·0717	·1471	·2639	·4197	·5968	·7653	·8943	·9681	·9953	·9999
	6	·1662	·2902	·4478	·6188	·7752	·8929	·9623	·9917	·9992	1·0000
	7	·3145	·4743	·6405	·7872	·8954	·9598	·9891	·9983	·9999	1·0000
	8	·5000	·6626	·8011	·9006	·9597	·9876	·9974	·9997	1·0000	1·0000

CUMULATIVE BINOMIAL PROBABILITIES—(Continued)

n	x	p=·05	·10	·15	·20	·25	·30	·35	·40	·45	·50
	9	1·0000	1·0000	1·0000	·9995	·9969	·9873	·9617	·9081	·8166	·6855
	10	1·0000	1·0000	1·0000	·9999	·9994	·9968	·9880	·9652	·9174	·8338
	11	1·0000	1·0000	1·0000	1·0000	·9999	·9993	·9970	·9894	·9699	·9283
	12	1·0000	1·0000	1·0000	1·0000	1·0000	·9999	·9994	·9975	·9914	·9755
	13	1·0000	1·0000	1·0000	1·0000	1·0000	1·0000	·9999	·9995	·9981	·9936
	14	1·0000	1·0000	1·0000	1·0000	1·0000	1·0000	1·0000	·9999	·9997	·9988
	15	1·0000	1·0000	1·0000	1·0000	1·0000	1·0000	1·0000	1·0000	1·0000	·9999
	16	1·0000	1·0000	1·0000	1·0000	1·0000	1·0000	1·0000	1·0000	1·0000	1·0000
18	0	·3972	·1501	·0536	·0180	·0056	·0016	·0004	·0001	·0000	·0000
	1	·7735	·4503	·2241	·0991	·0395	·0142	·0046	·0013	·0003	·0001
	2	·9419	·7338	·4797	·2713	·1353	·0600	·0236	·0082	·0025	·0007
	3	·9891	·9018	·7202	·5010	·3057	·1646	·0783	·0328	·0120	·0038
	4	·9985	·9718	·8794	·7164	·5187	·3327	·1886	·0942	·0411	·0154
	5	·9998	·9936	·9581	·8671	·7175	·5344	·3550	·2088	·1077	·0481
	6	1·0000	·9988	·9882	·9487	·8610	·7217	·5491	·3743	·2258	·1189
	7	1·0000	·9998	·9973	·9837	·9431	·8593	·7283	·5634	·3915	·2403
	8	1·0000	1·0000	·9995	·9957	·9807	·9404	·8609	·7368	·5778	·4073
	9	1·0000	1·0000	·9999	·9991	·9946	·9790	·9403	·8653	·7473	·5927
	10	1·0000	1·0000	1·0000	·9998	·9988	·9939	·9788	·9424	·8720	·7597
	11	1·0000	1·0000	1·0000	1·0000	·9998	·9986	·9938	·9797	·9463	·8811
	12	1·0000	1·0000	1·0000	1·0000	1·0000	·9997	·9986	·9942	·9817	·9519
	13	1·0000	1·0000	1·0000	1·0000	1·0000	1·0000	·9997	·9987	·9951	·9846
	14	1·0000	1·0000	1·0000	1·0000	1·0000	1·0000	1·0000	·9998	·9990	·9962
	15	1·0000	1·0000	1·0000	1·0000	1·0000	1·0000	1·0000	1·0000	·9999	·9993
	16	1·0000	1·0000	1·0000	1·0000	1·0000	1·0000	1·0000	1·0000	1·0000	·9999
	17	1·0000	1·0000	1·0000	1·0000	1·0000	1·0000	1·0000	1·0000	1·0000	1·0000

n	r										
19	0	·3774	·1351	·0456	·0144	·0042	·0011	·0003	·0001	·0000	·0000
	1	·7547	·4203	·1985	·0829	·0310	·0104	·0031	·0008	·0002	·0000
	2	·9335	·7054	·4413	·2369	·1113	·0462	·0170	·0055	·0015	·0004
	3	·9868	·8850	·6841	·4551	·2631	·1332	·0591	·0230	·0077	·0022
	4	·9980	·9643	·8556	·6733	·4654	·2822	·1500	·0696	·0280	·0096
	5	·9998	·9914	·9463	·8369	·6678	·4739	·2968	·1629	·0777	·0318
	6	1·0000	·9983	·9837	·9324	·8251	·6655	·4812	·3081	·1727	·0835
	7	1·0000	·9997	·9959	·9767	·9225	·8180	·6656	·4878	·3169	·1796
	8	1·0000	1·0000	·9992	·9933	·9713	·9161	·8145	·6675	·4940	·3238
	9	1·0000	1·0000	·9999	·9984	·9911	·9674	·9125	·8139	·6710	·5000
	10	1·0000	1·0000	1·0000	·9997	·9977	·9895	·9653	·9115	·8159	·6762
	11	1·0000	1·0000	1·0000	1·0000	·9995	·9972	·9886	·9648	·9129	·8204
	12	1·0000	1·0000	1·0000	1·0000	·9999	·9994	·9969	·9884	·9658	·9165
	13	1·0000	1·0000	1·0000	1·0000	1·0000	·9999	·9993	·9969	·9891	·9682
	14	1·0000	1·0000	1·0000	1·0000	1·0000	1·0000	·9999	·9994	·9972	·9904
	15	1·0000	1·0000	1·0000	1·0000	1·0000	1·0000	1·0000	·9999	·9995	·9978
	16	1·0000	1·0000	1·0000	1·0000	1·0000	1·0000	1·0000	1·0000	·9999	·9996
	17	1·0000	1·0000	1·0000	1·0000	1·0000	1·0000	1·0000	1·0000	1·0000	1·0000
20	0	·3585	·1216	·0388	·0115	·0032	·0008	·0002	·0000	·0000	·0000
	1	·7358	·3917	·1756	·0692	·0243	·0076	·0021	·0005	·0001	·0000
	2	·9245	·6769	·4049	·2061	·0913	·0355	·0121	·0036	·0009	·0002
	3	·9841	·8670	·6477	·4114	·2252	·1071	·0444	·0160	·0049	·0013
	4	·9974	·9568	·8298	·6296	·4148	·2375	·1182	·0510	·0189	·0059
	5	·9997	·9887	·9327	·8042	·6172	·4164	·2454	·1256	·0553	·0207
	6	1·0000	·9976	·9781	·9133	·7858	·6080	·4166	·2500	·1299	·0577
	7	1·0000	·9996	·9941	·9679	·8982	·7723	·6010	·4159	·2520	·1316
	8	1·0000	·9999	·9987	·9900	·9591	·8867	·7624	·5956	·4143	·2517
	9	1·0000	1·0000	·9998	·9974	·9861	·9520	·8782	·7553	·5914	·4119
	10	1·0000	1·0000	1·0000	·9994	·9961	·9829	·9468	·8725	·7507	·5881
	11	1·0000	1·0000	1·0000	·9999	·9991	·9949	·9804	·9435	·8692	·7483

CUMULATIVE BINOMIAL PROBABILITIES—(*Continued*)

n	x	$p=.05$	.10	.15	.20	.25	.30	.35	.40	.45	.50
	12	1·0000	1·0000	1·0000	1·0000	·9998	·9987	·9940	·9790	·9420	·8684
	13	1·0000	1·0000	1·0000	1·0000	1·0000	·9997	·9985	·9935	·9786	·9423
	14	1·0000	1·0000	1·0000	1·0000	1·0000	1·0000	·9997	·9984	·9936	·9793
	15	1·0000	1·0000	1·0000	1·0000	1·0000	1·0000	·9999	·9997	·9985	·9941
	16	1·0000	1·0000	1·0000	1·0000	1·0000	1·0000	1·0000	1·0000	·9997	·9987
	17	1·0000	1·0000	1·0000	1·0000	1·0000	1·0000	1·0000	1·0000	1·0000	·9998
	18	1·0000	1·0000	1·0000	1·0000	1·0000	1·0000	1·0000	1·0000	1·0000	1·0000

TABLE 3: CUMULATIVE POISSON PROBABILITIES

$$\text{Entry:} \quad \sum_{r=0}^{x} \frac{\lambda^r e^{-\lambda}}{r!}$$

x	λ=·1	·2	·3	·4	·5	·6	·7	·8	·9	1·0
0	·9048	·8187	·7408	·6703	·6065	·5488	·4966	·4493	·4066	·3679
1	·9953	·9825	·9631	·9384	·9098	·8781	·8442	·8088	·7725	·7358
2	·9998	·9989	·9964	·9921	·9856	·9769	·9659	·9526	·9371	·9197
3	1·0000	·9999	·9997	·9992	·9982	·9966	·9942	·9909	·9865	·9810
4	1·0000	1·0000	1·0000	·9999	·9998	·9996	·9992	·9986	·9977	·9963
5	1·0000	1·0000	1·0000	1·0000	1·0000	1·0000	·9999	·9998	·9997	·9994
6	1·0000	1·0000	1·0000	1·0000	1·0000	1·0000	1·0000	1·0000	1·0000	·9999
7	1·0000	1·0000	1·0000	1·0000	1·0000	1·0000	1·0000	1·0000	1·0000	1·0000

x	λ=1·1	1·2	1·3	1·4	1·5	1·6	1·7	1·8	1·9	2·0
0	·3329	·3012	·2725	·2466	·2231	·2019	·1827	·1653	·1496	·1353
1	·6990	·6626	·6268	·5918	·5578	·5249	·4932	·4628	·4337	·4060
2	·9004	·8795	·8571	·8335	·8088	·7834	·7572	·7306	·7037	·6767
3	·9743	·9662	·9569	·9463	·9344	·9212	·9068	·8913	·8747	·8571
4	·9946	·9923	·9893	·9857	·9814	·9763	·9704	·9636	·9559	·9473
5	·9990	·9985	·9978	·9968	·9955	·9940	·9920	·9868	·9868	·9834
6	·9999	·9997	·9996	·9994	·9991	·9987	·9981	·9974	·9966	·9955
7	1·0000	1·0000	·9999	·9999	·9998	·9997	·9996	·9994	·9992	·9989
8	1·0000	1·0000	1·0000	1·0000	1·0000	1·0000	·9999	·9999	·9998	·9998
9	1·0000	1·0000	1·0000	1·0000	1·0000	1·0000	1·0000	1·0000	1·0000	1·0000

CUMULATIVE POISSON PROBABILITIES—(*Continued*)

x	λ=2.1	2.2	2.3	2.4	2.5	2.6	2.7	2.8	2.9	3.0
0	·1225	·1108	·1003	·0907	·0821	·0743	·0672	·0608	·0550	·0498
1	·3796	·3546	·3309	·3084	·2873	·2674	·2487	·2311	·2146	·1991
2	·6496	·6227	·5960	·5967	·5438	·5184	·4396	·4695	·4460	·4232
3	·8386	·8194	·7993	·7787	·7576	·7360	·7141	·6919	·6696	·6472
4	·9379	·9275	·9162	·9041	·8912	·8774	·8629	·8477	·8318	·8153
5	·9796	·9751	·9700	·9643	·9580	·9510	·9433	·9349	·9258	·9161
6	·9941	·9925	·9906	·9884	·9858	·9828	·9794	·9756	·9713	·9665
7	·9985	·9980	·9974	·9967	·9958	·9947	·9934	·9919	·9901	·9881
8	·9997	·9995	·9994	·9991	·9989	·9985	·9981	·9976	·9969	·9962
9	·9999	·9999	·9999	·9998	·9997	·9996	·9995	·9993	·9991	·9989
10	1·0000	1·0000	1·0000	1·0000	·9999	·9999	·9999	·9998	·9998	·9997
11	1·0000	1·0000	1·0000	1·0000	1·0000	1·0000	1·0000	1·0000	·9999	·9999
12	1·0000	1·0000	1·0000	1·0000	1·0000	1·0000	1·0000	1·0000	1·0000	1·0000

x	λ=3.1	3.2	3.3	3.4	3.5	3.6	3.7	3.8	3.9	4.0
0	·0450	·0408	·0369	·0334	·0302	·0273	·0247	·0224	·0202	·0183
1	·1857	·1712	·1586	·1468	·1359	·1257	·1162	·1074	·0992	·0916
2	·4012	·3799	·3594	·3397	·3208	·3027	·2854	·2689	·2531	·2381
3	·6248	·6025	·5803	·5584	·5366	·5152	·4942	·4735	·4532	·4335
4	·7982	·7806	·7626	·7442	·7254	·7064	·6872	·6678	·6484	·6288
5	·9057	·8946	·8829	·8705	·8576	·8441	·8301	·8156	·8006	·7851
6	·9612	·9554	·9490	·9421	·9247	·9267	·9182	·9091	·8995	·8893
7	·9858	·9832	·9802	·9769	·9733	·9692	·9648	·9599	·9546	·9489
8	·9953	·9943	·9931	·9917	·9901	·9883	·9863	·9840	·9815	·9786
9	·9986	·9982	·9978	·9973	·9967	·9960	·9952	·9942	·9931	·9919

10	·9996	·9995	·9994	·9992	·9990	·9987	·9984	·9981	·9977	·9972
11	·9999	·9999	·9998	·9998	·9997	·9996	·9995	·9994	·9993	·9991
12	1·0000	1·0000	1·0000	·9999	·9999	·9999	·9999	·9998	·9998	·9997
13	1·0000	1·0000	1·0000	1·0000	1·0000	1·0000	1·0000	1·0000	·9999	·9999
14	1·0000	1·0000	1·0000	1·0000	1·0000	1·0000	1·0000	1·0000	1·0000	1·0000

x	λ=4·1	4·2	4·3	4·4	4·5	4·6	4·7	4·8	4·9	5·0
0	·0166	·0150	·0136	·0123	·0111	·0101	·0091	·0082	·0074	·0067
1	·0845	·0780	·0719	·0663	·0611	·0563	·0518	·0477	·0439	·0404
2	·2238	·2102	·1974	·1851	·1736	·1626	·1523	·1425	·1333	·1247
3	·4142	·3954	·3772	·3594	·3423	·3257	·3097	·2942	·2793	·2650
4	·6093	·5898	·5704	·3512	·5321	·5132	·4946	·4763	·4582	·4405
5	·7693	·7531	·7367	·7190	·7029	·6858	·6684	·6510	·6335	·6160
6	·8786	·8675	·8558	·8436	·8311	·8180	·8046	·7908	·7767	·7622
7	·9427	·9361	·9290	·9214	·9134	·9049	·8960	·8867	·8769	·8066
8	·9755	·9721	·9683	·9642	·9597	·9549	·9497	·9442	·9382	·9319
9	·9905	·9889	·9871	·9851	·9829	·9805	·9778	·9749	·9717	·9682
10	·9966	·9959	·9952	·9943	·9933	·9922	·9910	·9896	·9880	·9863
11	·9989	·9986	·9983	·9980	·9976	·9971	·9966	·9960	·9953	·9945
12	·9997	·9996	·9995	·9993	·9992	·9990	·9988	·9986	·9983	·9980
13	·9999	·9999	·9998	·9998	·9997	·9997	·9996	·9995	·9994	·9993
14	1·0000	1·0000	1·0000	·9999	·9999	·9999	·9999	·9999	·9998	·9998
15	1·0000	1·0000	1·0000	1·0000	1·0000	1·0000	1·0000	1·0000	·9999	·9999

CUMULATIVE POISSON PROBABILITIES—(Continued)

x	λ=5·1	5·2	5·3	5·4	5·5	5·6	5·7	5·8	5·9	6·0
0	·0061	·0055	·0050	·0045	·0041	·0037	·0033	·0030	·0027	·0025
1	·0372	·0342	·0314	·0266	·0244	·0244	·0224	·0206	·0189	·0174
2	·1165	·1088	·1016	·0948	·0884	·0824	·0768	·0715	·0666	·0620
3	·2513	·2381	·2254	·2133	·0217	·1906	·1800	·1700	·1604	·1512
4	·4231	·4061	·3895	·3733	·3575	·3422	·3272	·3127	·2987	·2851
5	·5984	·5809	·5635	·5461	·5289	·5119	·4950	·4783	·4619	·4457
6	·7474	·7324	·7171	·7017	·6860	·6703	·6544	·6384	·6224	·6063
7	·8560	·8449	·8335	·8217	·8095	·7970	·7841	·7710	·7576	·7440
8	·9252	·9181	·9106	·9027	·8944	·8857	·8766	·8672	·8574	·8472
9	·9644	·9603	·9559	·9512	·9462	·9409	·9352	·9292	·9228	·9161
10	·9844	·9823	·9800	·9775	·9747	·9718	·9686	·9651	·9614	·9574
11	·9937	·9927	·9916	·9904	·9890	·9875	·9859	·9841	·9821	·9799
12	·9976	·9972	·9967	·9962	·9955	·9949	·9941	·9932	·9922	·9912
13	·9992	·9990	·9988	·9986	·9983	·9980	·9977	·9973	·9969	·9964
14	·9997	·9997	·9996	·9995	·9994	·9993	·9991	·9990	·9988	·9986
15	·9999	·9999	·9999	·9998	·9998	·9998	·9997	·9996	·9996	·9995
16	1·0000	1·0000	1·0000	·9999	·9999	·9999	·9999	·9999	·9999	·9998
17	1·0000	1·0000	1·0000	1·0000	1·0000	1·0000	1·0000	1·0000	1·0000	·9999

x	λ=6·1	6·2	6·3	6·4	6·5	6·6	6·7	6·8	6·9	7·0
0	·0022	·0020	·0018	·0017	·0015	·0014	·0012	·0011	·0010	·0009
1	·0159	·0146	·0134	·0123	·0113	·0103	·0095	·0087	·0080	·0073
2	·0577	·0536	·0498	·0463	·0430	·0400	·0371	·0344	·0320	·0296
3	·1425	·1342	·1264	·1189	·1118	·1052	·0988	·0928	·0871	·0818
4	·2719	·2592	·2469	·2351	·2237	·2127	·2022	·1920	·1823	·1730
5	·4298	·4141	·3988	·3837	·3690	·3547	·3406	·3270	·3137	·3007

(Poisson cumulative distribution. Upper block continues the λ = 6·1 – 7·0 columns from the preceding page; lower block gives λ = 7·1 – 8·0.)

x	λ=6·1	6·2	6·3	6·4	6·5	6·6	6·7	6·8	6·9	7·0
6	·5902	·5742	·5582	·5423	·5265	·5108	·4953	·4799	·4647	·4497
7	·7301	·7160	·7017	·6873	·6728	·6581	·6433	·6285	·6136	·5987
8	·8367	·8259	·8148	·8033	·7916	·7796	·7673	·7548	·7420	·7291
9	·9090	·9016	·8939	·8858	·8774	·8686	·8596	·8502	·8405	·8305
10	·9531	·9486	·9437	·9386	·9332	·9274	·9214	·9151	·9084	·9015
11	·9776	·9750	·9723	·9693	·9661	·9627	·9591	·9552	·9510	·9467
12	·9900	·9887	·9873	·9857	·9840	·9821	·9801	·9779	·9755	·9730
13	·9958	·9952	·9945	·9937	·9929	·9920	·9909	·9898	·9885	·9872
14	·9984	·9981	·9978	·9974	·9970	·9966	·9961	·9956	·9950	·9943
15	·9994	·9993	·9992	·9990	·9988	·9986	·9984	·9982	·9979	·9976
16	·9998	·9997	·9997	·9996	·9996	·9995	·9994	·9993	·9992	·9990
17	·9999	·9999	·9999	·9999	·9998	·9998	·9998	·9997	·9997	·9996
18	1·0000	1·0000	1·0000	1·0000	·9999	·9999	·9999	·9999	·9999	·9999
19	1·0000	1·0000	1·0000	1·0000	1·0000	1·0000	1·0000	1·0000	1·0000	1·0000

x	λ=7·1	7·2	7·3	7·4	7·5	7·6	7·7	7·8	7·9	8·0
0	·0008	·0007	·0007	·0006	·0006	·0005	·0005	·0004	·0004	·0003
1	·0067	·0061	·0056	·0051	·0047	·0043	·0039	·0036	·0033	·0030
2	·0275	·0255	·0236	·0219	·0203	·0188	·0174	·0161	·0149	·0138
3	·0767	·0719	·0674	·0632	·0591	·0554	·0518	·0485	·0453	·0424
4	·1641	·1555	·1473	·1395	·1321	·1249	·1181	·1117	·1055	·0996
5	·2881	·2759	·2640	·2526	·2414	·2307	·2203	·2103	·2006	·1912
6	·4349	·4204	·4060	·3920	·3782	·3646	·3514	·3384	·3257	·3134
7	·5838	·5689	·5541	·5393	·5246	·5100	·4956	·4812	·4670	·4530
8	·7160	·7027	·6892	·6757	·6620	·6482	·6343	·6204	·6065	·5925
9	·8202	·8096	·7988	·7877	·7764	·7649	·7531	·7411	·7290	·7166
10	·8942	·8867	·8788	·8707	·8622	·8535	·8445	·8352	·8257	·8159
11	·9420	·9371	·9319	·9265	·9208	·9148	·9085	·9020	·8952	·8881
12	·9703	·9673	·9642	·9609	·9573	·9536	·9496	·9454	·9409	·9362
13	·9857	·9841	·9824	·9805	·9784	·9762	·9739	·9714	·9687	·9658

CUMULATIVE POISSON PROBABILITIES—(*Continued*)

x	$\lambda=7.1$	7·2	7·3	7·4	7·5	7·6	7·7	7·8	7·9	8·0
14	·9935	·9927	·9918	·9908	·9897	·9886	·9873	·9859	·9844	·9827
15	·9972	·9969	·9964	·9959	·9954	·9948	·9941	·9934	·9926	·9918
16	·9989	·9987	·9985	·9983	·9980	·9978	·9974	·9971	·9967	·9963
17	·9996	·9995	·9994	·9993	·9992	·9991	·9989	·9988	·9986	·9984
18	·9998	·9998	·9998	·9997	·9997	·9996	·9996	·9995	·9994	·9993
19	·9999	·9999	·9999	·9999	·9999	·9999	·9998	·9998	·9998	·9997
20	1·0000	1·0000	1·0000	1·0000	1·0000	1·0000	·9999	·9999	·9999	·9999
21	1·0000	1·0000	1·0000	1·0000	1·0000	1·0000	1·0000	1·0000	1·0000	1·0000

x	$\lambda=8.1$	8·2	8·3	8·4	8·5	8·6	8·7	8·8	8·9	9·0
0	·0003	·0003	·0002	·0002	·0002	·0002	·0002	·0002	·0001	·0001
1	·0028	·0025	·0023	·0021	·0019	·0018	·0016	·0015	·0014	·0012
2	·0127	·0118	·0109	·0100	·0093	·0086	·0079	·0073	·0068	·0062
3	·0396	·0370	·0346	·0323	·0301	·0281	·0262	·0244	·0228	·0212
4	·0940	·0887	·0837	·0789	·0744	·0701	·0660	·0621	·0584	·0550
5	·1822	·1736	·1653	·1573	·1496	·1422	·1352	·1284	·1219	·1157
6	·3013	·2896	·2781	·2670	·2562	·2457	·2355	·2256	·2160	·2068
7	·4391	·4254	·4119	·3987	·3856	·3728	·3602	·3478	·3357	·3239
8	·5786	·5647	·5507	·5369	·5231	·5094	·4958	·4823	·4689	·4557
9	·7041	·6915	·6788	·6659	·6530	·6400	·6269	·6137	·6006	·5874
10	·8058	·7955	·7850	·7743	·7634	·7522	·7409	·7294	·7178	·7060
11	·8807	·8731	·8652	·8571	·8487	·8400	·8311	·8220	·8126	·8030
12	·9313	·9261	·9207	·9150	·9091	·9029	·8965	·8898	·8829	·8758
13	·9628	·9595	·9561	·9524	·9486	·9445	·9403	·9358	·9311	·9261
14	·9810	·9791	·9771	·9749	·9726	·9701	·9675	·9647	·9617	·9585

x										
15	·9908	·9898	·9887	·9875	·9862	·9848	·9832	·9816	·9798	·9780
16	·9958	·9953	·9947	·9941	·9934	·9926	·9918	·9909	·9899	·9889
17	·9982	·9979	·9977	·9973	·9970	·9966	·9962	·9957	·9952	·9947
18	·9992	·9991	·9990	·9989	·9987	·9985	·9983	·9981	·9978	·9976
19	·9997	·9997	·9996	·9995	·9995	·9994	·9993	·9992	·9991	·9989
20	·9999	·9999	·9998	·9998	·9998	·9998	·9997	·9997	·9996	·9996
21	1·0000	1·0000	·9999	·9999	·9999	·9999	·9999	·9999	·9998	·9998
22	1·0000	1·0000	1·0000	1·0000	1·0000	1·0000	1·0000	1·0000	·9999	·9999

x	λ=9·1	9·2	9·3	9·4	9·5	9·6	9·7	9·8	9·9	10·0
0	·0001	·0001	·0001	·0001	·0001	·0001	·0001	·0001	·0001	·0000
1	·0011	·0010	·0009	·0009	·0008	·0007	·0007	·0006	·0005	·0005
2	·0058	·0053	·0049	·0045	·0042	·0038	·0035	·0033	·0030	·0028
3	·0198	·0184	·0172	·0160	·0149	·0138	·0129	·0120	·0111	·0103
4	·0517	·0486	·0456	·0429	·0403	·0378	·0355	·0333	·0312	·0293
5	·1098	·1041	·0986	·0935	·0885	·0838	·0793	·0750	·0710	·0671
6	·1978	·1892	·1808	·1727	·1649	·1574	·1502	·1433	·1366	·1301
7	·3123	·3010	·2900	·2792	·2687	·2584	·2485	·2388	·2294	·2202
8	·4426	·4296	·4168	·4042	·3918	·3798	·3676	·3558	·3442	·3328
9	·5742	·5611	·5479	·5349	·5218	·5089	·4960	·4832	·4705	·4579
10	·6941	·6820	·6699	·6576	·6453	·6329	·6205	·6080	·5955	·5830
11	·7932	·7832	·7730	·7626	·7520	·7412	·7303	·7193	·7081	·6968
12	·8684	·8607	·8529	·8448	·8364	·8279	·8191	·8101	·8009	·7916
13	·9210	·9156	·9100	·9042	·8981	·8919	·8853	·8786	·8716	·8615
14	·9552	·9517	·9480	·9441	·9400	·9357	·9312	·9265	·9216	·9165
15	·9760	·9738	·9715	·9691	·9665	·9638	·9609	·9579	·9546	·9513
16	·9878	·9865	·9852	·9838	·9823	·9806	·9789	·9770	·9751	·9730
17	·9941	·9934	·9927	·9919	·9911	·9902	·9892	·9881	·9870	·9857
18	·9973	·9969	·9966	·9962	·9957	·9952	·9947	·9941	·9935	·9928

COMULATIVE POISSON PROBABILITIES—(*Continued*)

x	λ=9·1	9·2	9·3	9·4	9·5	9·6	9·7	9·8	9·9	10·0
19	·9988	·9986	·9985	·9983	·9980	·9978	·9975	·9972	·9969	·9965
20	·9995	·9994	·9993	·9992	·9991	·9990	·9989	·9987	·9986	·9984
21	·9998	·9998	·9997	·9997	·9996	·9996	·9995	·9995	·9994	·9994
22	·9999	·9999	·9999	·9999	·9999	·9998	·9998	·9998	·9997	·9997
23	1·0000	1·0000	1·0000	1·0000	·9999	·9999	·9999	·9999	·9999	·9999
24	1·0000	1·0000	1·0000	1·0000	1·0000	1·0000	1·0000	1·0000	1·0000	1·0000

x	λ=11	12	13	14	15	16	17	18	19	20
0	·0000	·0000	·0000	·0000	·0000	·0000	·0000	·0000	·0000	·0000
1	·0002	·0001	·0000	·0000	·0000	·0000	·0000	·0000	·0000	·0000
2	·0012	·0005	·0002	·0001	·0000	·0000	·0000	·0000	·0000	·0000
3	·0049	·0023	·0011	·0005	·0002	·0001	·0000	·0000	·0000	·0000
4	·0151	·0076	·0037	·0018	·0009	·0004	·0002	·0001	·0000	·0000
5	·0375	·0203	·0107	·0055	·0028	·0014	·0007	·0003	·0002	·0001
6	·0786	·0458	·0259	·0142	·0076	·0040	·0021	·0010	·0005	·0003
7	·1432	·0895	·0540	·0316	·0180	·0100	·0054	·0029	·0015	·0008
8	·2320	·1550	·0998	·0621	·0374	·0220	·0126	·0071	·0039	·0021
9	·3405	·2424	·1658	·1094	·0699	·0433	·0261	·0154	·0089	·0050
10	·4599	·3472	·2517	·1757	·1185	·0774	·0491	·0304	·0183	·0108
11	·5793	·4616	·3532	·2600	·1848	·1270	·0847	·0549	·0347	·0214
12	·6887	·5760	·4631	·3585	·2676	·1931	·1350	·0917	·0606	·0390
13	·7813	·6815	·5730	·4644	·3632	·2745	·2009	·1426	·0984	·0661
14	·8540	·7720	·6751	·5704	·4657	·3675	·2808	·2081	·1497	·1049
15	·9074	·8444	·7636	·6694	·5681	·4667	·3715	·2867	·2148	·1565
16	·9441	·8987	·8355	·7559	·6641	·5660	·4677	·3751	·2920	·2211

17	·2970	·3784	·4686	·5640	·6593	·7489	·8272	·8905	·9370	·9678
18	·3814	·4695	·5622	·6550	·7423	·8195	·8826	·9302	·9626	·9823
19	·4703	·5606	·6509	·7363	·8122	·8752	·9325	·9573	·9787	·9907
20	·5591	·6472	·7307	·8055	·8682	·9170	·9521	·9750	·9884	·9953
21	·6437	·7255	·7991	·8615	·9108	·9469	·9712	·9859	·9939	·9977
22	·7206	·7931	·8551	·9047	·9418	·9673	·9833	·9924	·9970	·9990
23	·7875	·8490	·8989	·9367	·9633	·9805	·9907	·9960	·9985	·9995
24	·8432	·8933	·9317	·9594	·9777	·9888	·9950	·9980	·9993	·9998
25	·8878	·9269	·9554	·9748	·9869	·9938	·9974	·9990	·9997	·9999
26	·9221	·9514	·9718	·9848	·9925	·9967	·9987	·9995	·9999	1·0000
27	·9475	·9687	·9827	·9912	·9959	·9983	·9994	·9998	·9999	1·0000
28	·9657	·9805	·9897	·9950	·9978	·9991	·9997	·9999	1·0000	1·0000
29	·9782	·9881	·9941	·9973	·9989	·9996	·9999	1·0000	1·0000	1·0000
30	·9865	·9930	·9967	·9986	·9994	·9998	·9999	1·0000	1·0000	1·0000
31	·9919	·9960	·9982	·9993	·9997	·9999	1·0000	1·0000	1·0000	1·0000
32	·9953	·9978	·9990	·9996	·9999	1·0000	1·0000	1·0000	1·0000	1·0000
33	·9973	·9988	·9995	·9998	·9999	1·0000	1·0000	1·0000	1·0000	1·0000
34	·9985	·9994	·9998	·9999	1·0000	1·0000	1·0000	1·0000	1·0000	1·0000
35	·9992	·9997	·9999	1·0000	1·0000	1·0000	1·0000	1·0000	1·0000	1·0000
36	·9996	·9998	·9999	1·0000	1·0000	1·0000	1·0000	1·0000	1·0000	1·0000
37	·9998	·9999	1·0000	1·0000	1·0000	1·0000	1·0000	1·0000	1·0000	1·0000
38	·9999	1·0000	1·0000	1·0000	1·0000	1·0000	1·0000	1·0000	1·0000	1·0000
39	·9999	1·0000	1·0000	1·0000	1·0000	1·0000	1·0000	1·0000	1·0000	1·0000

TABLE 4: NORMAL PROBABILITY TABLE

Entry : $\displaystyle\int_0^x \frac{e^{-t^2/2}}{\sqrt{2\pi}}\,dt$

x	·00	·01	·02	·03	·04	·05	·06	·07	·08	·09
0·0	·0000	·0040	·0080	·0120	·0160	·0199	·0239	·0279	·0319	·0359
0·1	·0398	·0438	·0478	·0517	·0557	·0596	·0636	·0675	·0714	·0753
0·2	·0793	·0832	·0871	·0910	·0948	·0987	·1026	·1064	·1103	·1141
0·3	·1179	·1217	·1255	·1293	·1331	·1368	·1406	·1443	·1480	·1517
0·4	·1554	·1591	·1627	·1664	·1700	·1736	·1772	·1808	·1844	·1879
0·5	·1915	·1950	·1985	·2019	·2054	·2088	·2123	·2157	·2190	·2224
0·6	·2257	·2291	·2324	·2357	·2389	·2422	·2454	·2486	·2518	·2549
0·7	·2580	·2612	·2642	·2673	·2704	·2734	·2764	·2794	·2823	·2852
0·8	·2882	·2910	·2939	·2967	·2996	·3023	·3051	·3079	·3106	·3133
0·9	·3159	·3186	·3212	·3238	·3264	·3290	·3315	·3340	·3365	·3389
1·0	·3414	·3438	·3461	·3485	·3508	·3531	·3554	·3577	·3599	·3621
1·1	·3643	·3665	·3686	·3708	·3729	·3749	·3770	·3790	·3810	·3830
1·2	·3849	·3888	·3888	·3906	·3925	·3943	·3962	·3980	·3997	·4015
1·3	·4032	·4049	·4066	·4082	·4099	·4115	·4131	·4146	·4162	·4177
1·4	·4192	·4207	·4222	·4236	·4251	·4265	·4278	·4292	·4306	·4319
1·5	·4332	·4345	·4357	·4370	·4382	·4394	·4406	·4418	·4429	·4441
1·6	·4452	·4453	·4474	·4484	·4495	·4505	·4515	·4525	·4535	·4545
1·7	·4554	·4564	·4573	·4582	·4591	·4599	·4608	·4610	·4625	·4633
1·8	·4641	·4648	·4656	·4664	·4671	·4678	·4686	·4693	·4699	·4706
1·9	·4713	·4719	·4726	·4732	·4738	·4744	·4750	·4756	·4762	·4767

	0	1	2	3	4	5	6	7	8	9
2·0	·4773	·4778	·4783	·4788	·4793	·4798	·4803	·4808	·4812	·4817
2·1	·4821	·4826	·4830	·4834	·4838	·4842	·4846	·4850	·4854	·4857
2·2	·4861	·4865	·4868	·4871	·4875	·4878	·4881	·4884	·4887	·4890
2·3	·4893	·4896	·4898	·4901	·4904	·4906	·4909	·4911	·4914	·4916
2·4	·4918	·4920	·4922	·4925	·4927	·4929	·4931	·4933	·4934	·4936
2·5	·4938	·4940	·4941	·4943	·4945	·4946	·4948	·4949	·4951	·4952
2·6	·4953	·4955	·4956	·4957	·4959	·4960	·4961	·4962	·4963	·4964
2·7	·4965	·4966	·4967	·4968	·4969	·4970	·4971	·4972	·4973	·4974
2·8	·4974	·4975	·4976	·4977	·4977	·4978	·4979	·4979	·4980	·4981
2·9	·4981	·4982	·4982	·4983	·4984	·4984	·4985	·4985	·4986	·4986
3·0	·4986	·4987	·4987	·4988	·4988	·4988	·4989	·4989	·4990	·4990

TABLE 5: STUDENT-t TABLE

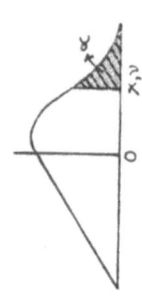

Entry: $t_{a,\nu}$ where $\int_{t_{a,\nu}}^{\infty} f(t_\nu)\,dt_\nu = x$

and $f(t_\nu)$ is the density function of a Student-t with ν degrees of freedom.

ν	$a=\cdot10$	$a=\cdot05$	$a=\cdot025$	$a=\cdot01$	$a=\cdot005$
1	3·078	6·314	12·706	31·821	63·657
2	1·886	2·920	4·303	6·965	9·925
3	1·638	2·353	3·182	4·541	5·841
4	1·533	2·132	2·776	3·747	4·604
5	1·476	2·015	2·571	3·365	4·032
6	1·440	1·943	2·447	3·143	3·707
7	1·415	1·895	2·365	2·998	3·499
8	1·397	1·860	2·306	2·896	3·355
9	1·383	1·833	2·262	2·821	3·250
10	1·372	1·812	2·228	2·764	3·169
11	1·363	1·796	2·201	2·718	3·106
12	1·356	1·782	2·179	2·681	3·055
13	1·350	1·771	2·160	2·650	3·012
14	1·345	1·761	2·145	2·624	2·977
15	1·341	1·753	2·131	2·602	2·947

16	1·337	1·746	2·120	2·583	2·921
17	1·333	1·740	2·110	2·567	2·898
18	1·330	1·734	2·101	2·552	2·878
19	1·328	1·729	2·093	2·539	2·861
20	1·325	1·725	2·086	2·528	2·845
21	1·323	1·721	2·080	2·518	2·831
22	1·321	1·717	2·074	2·508	2·819
23	1·319	1·714	2·069	2·500	2·807
24	1·319	1·711	2·064	2·492	2·797
25	1·316	1·708	2·060	2·485	2·787
26	1·315	1·706	2·056	2·479	2·779
27	1·314	1·703	2·052	2·473	2·771
28	1·313	1·701	2·048	2·467	2·763
29	1·311	1·699	2·045	2·462	2·756
inf.	1·282	1·645	1·966	2·326	2·576

TABLE 6: THE χ^2 TABLE

Entry: $\chi^2_{a, \nu}$ where $\displaystyle\int_{\chi^2_{a, \nu}}^{\infty} f(\chi^2_{\nu}) d\chi^2_{\nu} = a$ and $f(\chi^2_{\nu})$ is the density function of a chi-square variable with ν degrees of freedom.

ν	$a=\cdot995$	$\cdot99$	$\cdot975$	$\cdot95$	$\cdot10$	$\cdot05$	$\cdot025$	$\cdot01$	$\cdot005$	$\cdot001$
1	0·04393	0·03157	0·03982	0·00393	2·71	3·84	5·02	6·63	7·88	10·83
2	0·100	0·0201	0·0506	0·103	4·61	5·99	7·38	9·21	10·60	13·81
3	0·0717	0·115	0·216	0·352	6·25	7·81	9·35	11·34	12·84	16·27
4	0·207	0·297	0·484	0·711	7·78	9·49	11·14	13·28	14·86	18·47
5	0·412	0·554	0·831	1·15	9·24	11·07	12·83	15·09	6·75	20·52
6	0·676	0·872	1·24	1·64	10·64	12·59	14·45	16·81	8·55	22·46
7	0·989	1·24	1·69	2·17	12·02	14·07	16·01	18·48	20·28	24·32
8	1·34	1·65	2·18	2·73	13·36	15·51	17·53	20·09	21·95	26·12
9	1·73	2·09	2·70	3·33	14·68	16·92	19·02	21·67	23·59	27·88
10	2·16	2·56	3·25	3·94	15·99	18·31	20·48	23·21	25·19	29·59
11	2·60	3·05	3·82	4·57	17·28	19·68	21·92	24·73	26·76	31·26
12	3·07	3·57	4·40	5·23	18·55	21·03	23·34	26·22	28·30	32·91
13	3·57	4·11	5·01	5·89	19·81	22·36	24·74	27·69	29·82	34·53
14	4·07	4·66	5·63	6·57	21·06	23·68	26·12	29·14	31·32	36·12
15	4·60	5·23	6·26	7·26	22·31	25·00	27·49	30·58	32·80	37·70
16	5·14	5·81	6·91	7·96	23·54	26·30	28·85	32·00	34·27	39·25
17	5·70	6·41	7·56	8·67	24·77	27·59	30·19	33·41	35·72	30·79
18	6·26	7·01	8·23	9·39	25·99	28·87	31·53	34·81	37·16	42·31
19	6·84	7·63	8·91	10·12	27·20	30·14	32·85	36·19	38·58	43·82

20	7·43	8·26	9·59	10·85	28·41	31·41	34·17	37·57	40·00	45·31
21	8·03	8·90	10·28	11·59	29·62	32·67	35·48	38·93	41·40	46·80
22	8·64	9·54	10·98	12·34	30·81	33·92	36·78	40·29	42·80	48·27
23	9·26	10·20	11·69	13·09	32·01	35·17	38·08	41·64	44·18	49·73
24	9·89	10·86	12·40	13·85	33·20	36·42	39·36	42·98	45·56	51·18
25	10·52	11·52	13·12	14·61	34·38	37·65	40·65	44·31	46·93	52·62
26	11·16	12·20	13·84	15·38	35·56	38·89	41·92	45·64	48·29	54·05
27	11·81	12·88	14·57	16·15	36·74	40·11	43·19	46·96	49·64	55·48
28	12·46	13·56	15·31	16·93	37·92	41·34	44·46	48·28	50·99	56·89
29	13·12	14·26	16·05	17·71	39·09	42·56	45·72	49·59	52·34	58·30
30	13·79	14·95	16·79	18·49	40·26	43·77	46·98	50·89	53·67	59·70
40	20·71	22·16	24·43	26·51	51·81	55·76	59·34	63·69	66·77	73·40
50	27·99	29·71	32·36	34·76	63·17	67·50	71·42	76·15	79·49	86·66
60	35·53	37·48	40·48	43·19	74·40	79·08	83·30	88·38	91·95	99·61
70	43·28	45·44	48·76	51·74	85·53	90·53	95·02	100·4	104·2	112·3
80	51·17	53·54	57·15	60·39	96·58	101·9	106·6	112·3	116·3	124·8
90	59·20	61·75	65·75	69·13	107·6	113·1	118·1	124·1	128·3	137·2
100	67·33	70·06	74·22	77·93	118·5	124·3	129·6	135·8	140·2	149·4

TABLE 7: 5% POINT OF F-DISTRIBUTION

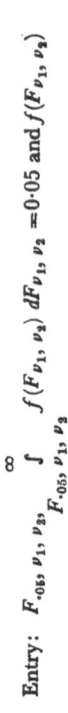

Entry: $F_{.05, \nu_1, \nu_2}, \quad \int_{F_{.05, \nu_1, \nu_2}}^{\infty} f(F_{\nu_1, \nu_2})\, dF_{\nu_1, \nu_2} = 0.05$ and $f(F_{\nu_1, \nu_2})$ is the density function of a F-variable with ν_1 and ν_2 degrees of freedom.

ν_2	$\nu_1=1$	2	3	4	5	6	7	8	10	12	24	∞
1	161·4	199·5	215·7	224·6	230·4	234·0	236·8	238·9	241·9	243·9	249·0	254·3
2	18·5	19·0	19·2	19·2	19·3	19·3	19·4	19·4	19·4	19·4	19·5	19·5
3	10·13	9·55	9·28	9·12	9·01	8·94	8·89	8·85	8·79	8·74	8·64	8·53
4	7·71	6·94	6·59	6·39	6·26	6·16	6·09	6·04	5·96	5·97	5·77	5·63
5	6·61	5·79	5·41	5·19	5·05	4·95	4·88	4·82	4·74	4·68	4·53	4·36
6	5·99	5·14	4·76	4·53	4·39	4·28	4·21	4·15	4·06	4·00	3·84	3·67
7	5·59	4·74	4·35	4·12	3·97	3·87	3·79	3·37	3·64	3·57	3·41	3·23
8	5·32	4·46	4·07	3·84	3·69	3·58	3·50	3·44	3·35	3·28	3·12	2·93
9	5·12	4·26	3·86	3·63	3·38	3·37	3·29	3·23	3·14	3·07	2·90	2·71
10	4·96	4·10	3·71	3·48	3·33	3·22	3·14	3·07	2·98	2·91	2·74	2·54
11	4·84	3·98	3·59	3·36	3·20	3·09	3·01	2·95	2·85	2·79	2·61	2·40
12	4·75	3·89	3·49	3·26	3·11	3·00	2·91	2·85	2·75	2·69	2·51	2·30
13	4·67	3·81	3·41	3·18	3·03	2·92	2·83	2·77	2·67	2·60	2·42	2·21
14	4·60	3·74	3·34	3·11	2·96	2·85	2·76	2·70	2·60	2·53	2·35	2·13
15	4·54	3·68	3·29	3·06	2·90	2·79	2·71	2·64	2·54	2·48	2·29	2·07
16	4·49	3·63	3·24	3·01	2·85	2·74	2·66	2·59	2·49	2·42	2·24	2·01
17	4·45	3·59	3·20	2·96	2·81	2·70	2·61	2·55	2·45	2·38	2·19	1·96
18	4·41	3·55	3·16	2·93	2·77	2·66	2·58	2·51	2·41	2·34	2·15	1·92
19	4·38	3·52	3·13	2·90	2·74	2·63	2·54	2·48	2·38	2·31	2·11	1·88

20	4·35	3·49	3·10	2·87	2·71	2·60	2·51	2·45	2·35	2·28	2·08	1·84
21	4·32	3·47	3·07	2·84	2·68	2·57	2·49	2·42	2·32	2·25	2·05	1·81
22	4·30	3·44	3·05	2·82	2·66	2·55	2·46	2·40	2·30	2·23	2·03	1·78
23	4·28	3·42	3·03	2·80	2·64	2·53	2·44	2·37	2·27	2·20	2·00	1·76
24	4·26	3·40	3·01	2·78	2·62	2·51	2·42	2·36	2·25	2·18	1·98	1·73
25	4·24	3·39	2·99	2·76	2·60	2·49	2·40	2·34	2·24	2·16	1·96	1·71
26	4·23	3·37	2·98	2·74	2·59	2·47	2·39	2·32	2·22	2·15	1·95	1·69
27	4·21	3·35	2·96	2·73	2·57	2·46	2·37	2·31	2·20	2·13	1·93	1·67
28	4·20	3·34	2·95	2·71	2·56	2·45	2·36	2·29	2·19	2·12	1·91	1·65
29	4·18	3·33	2·93	2·70	2·55	2·43	2·35	2·28	2·18	2·10	1·90	1·64
30	4·17	3·32	2·92	2·69	2·53	2·42	2·33	2·27	2·16	2·09	1·89	1·62
32	4·15	3·29	2·90	2·67	2·51	2·40	2·31	2·24	2·14	2·07	1·86	1·59
34	4·13	3·28	2·88	2·65	2·49	2·38	2·29	2·23	2·12	2·05	1·84	1·57
36	4·11	3·26	2·87	2·63	2·48	2·36	2·28	2·21	2·11	2·03	1·82	1·55
38	4·10	3·24	2·85	2·62	2·46	2·35	2·26	2·19	2·09	2·02	1·81	1·53
40	4·08	3·23	2·84	2·61	2·45	2·34	2·25	2·18	2·08	2·00	1·79	1·51
60	4·00	3·15	2·76	2·53	2·37	2·25	2·17	2·10	1·99	1·92	1·70	1·39
100	3·92	3·07	2·63	2·45	2·29	2·18	2·09	2·02	1·91	1·83	1·61	1·25
∞	3·84	3·00	2·60	2·37	2·21	2·10	2·01	1·94	1·83	1·75	1·52	1·00

TABLE 8: THE F-DISTRIBUTION

Entry: $F_{\cdot01, \nu_1, \nu_2}$ where $\int_{F_{\cdot01, \nu_1, \nu_2}}^{\infty} f(F_{\nu_1, \nu_2})\, dF_{\nu_1, \nu_2} = 0\cdot01$, and $f(F_{\nu_1, \nu_2})$ is the density function of an F with ν_1 and ν_2 degrees of freedom.

ν_2	$\nu_1=1$	2	3	4	5	6	7	8	10	12	24	∞
1	4052	4999·5	5403	5625	5764	5859	5928	5981	6056	6106	6235	6366
2	98·5	99·0	99·2	99·2	99·3	99·3	99·4	99·4	99·4	99·4	99·5	99·5
3	34·1	30·8	29·5	28·7	28·2	27·9	27·7	27·5	27·2	27·1	26·6	26·1
4	21·2	18·0	16·7	16·0	15·5	15·2	15·0	14·8	14·5	14·4	13·9	13·5
5	16·26	13·27	12·06	11·39	10·97	10·67	10·46	10·29	10·05	9·89	9·47	9·02
6	13·74	10·92	9·78	9·15	8·75	8·47	8·26	8·10	7·87	7·72	7·31	6·88
7	12·25	9·55	8·45	7·85	7·46	7·19	6·99	6·84	6·62	6·47	6·07	5·65
8	11·26	8·65	7·59	7·01	6·63	6·37	6·18	6·03	5·81	5·67	5·28	4·86
9	10·56	8·02	6·99	6·42	6·06	5·80	5·61	5·47	5·26	5·11	4·73	4·31
10	10·04	7·56	6·55	5·99	5·64	5·39	5·20	5·06	4·85	4·71	4·33	3·91
11	9·65	7·21	6·22	5·67	5·32	5·07	4·89	4·74	4·54	4·40	4·02	3·60
12	9·33	6·93	5·95	5·41	5·06	4·82	4·64	4·50	4·30	4·16	3·78	3·36
13	9·07	6·70	5·74	5·21	4·86	4·62	4·44	4·30	4·10	3·96	3·59	3·17
14	8·86	6·51	5·56	5·04	4·70	4·46	4·28	4·14	3·94	3·80	3·43	3·00
15	8·68	6·36	5·42	4·89	4·56	4·32	4·14	4·00	3·80	3·67	3·29	2·87
16	8·53	6·23	5·29	4·77	4·44	4·20	4·03	3·89	3·69	3·55	3·18	2·75
17	8·40	6·11	5·18	4·67	4·34	4·10	3·93	3·79	3·59	3·46	3·08	2·65
18	8·20	6·01	5·09	4·58	4·25	4·01	3·84	3·71	3·51	3·37	3·00	2·57
19	8·18	5·93	5·01	4·50	4·17	3·94	3·77	3·63	3·43	3·30	2·92	2·49

$F_{\cdot01, \nu_1, \nu_2}$

20	8·10	5·85	4·94	4·43	4·10	3·87	3·70	3·56	3·37	3·23	2·86	2·42
21	8·02	5·78	4·87	4·37	4·04	3·81	3·64	3·51	3·31	3·17	2·80	2·36
22	7·95	5·72	4·82	4·31	3·99	3·76	3·59	3·45	3·26	3·12	2·75	2·31
23	7·88	5·66	4·76	4·26	3·94	3·71	3·54	3·41	3·21	3·07	2·70	2·26
24	7·82	5·61	4·72	4·22	3·90	3·67	3·50	3·36	3·17	3·03	2·66	2·21
25	7·77	5·57	4·68	4·18	3·86	3·63	3·46	3·32	3·13	2·99	2·62	2·17
26	7·72	5·53	4·64	4·14	3·82	3·59	3·42	3·29	3·09	2·96	2·58	2·13
27	7·68	5·49	4·60	4·11	3·78	3·56	3·39	3·26	3·06	2·93	2·55	2·10
28	7·64	5·45	4·57	4·07	3·75	3·53	3·36	3·23	3·03	2·90	2·52	2·06
29	7·60	5·42	4·54	4·04	3·73	3·50	3·33	3·20	3·00	2·87	2·49	2·03
30	7·56	5·39	4·51	4·02	3·70	3·47	3·30	3·17	2·98	2·84	2·47	2·01
32	7·50	5·34	4·46	3·97	3·65	3·43	3·26	3·13	2·93	2·80	2·42	1·96
34	7·45	5·29	4·42	3·93	3·61	3·39	3·22	3·09	2·90	2·76	2·38	1·91
36	7·40	5·25	4·38	3·89	3·58	3·35	3·18	3·05	2·86	2·72	2·35	1·87
38	7·35	5·21	4·34	3·86	3·54	3·32	3·15	3·02	2·83	2·69	2·32	1·84
40	7·31	5·18	4·31	3·83	3·51	3·29	3·12	2·99	2·80	2·66	2·29	1·80
60	7·08	4·98	4·13	3·65	3·34	3·12	2·95	2·82	2·63	2·50	2·12	1·60
120	6·85	4·79	3·95	3·48	3·17	2·96	2·79	2·66	2·47	2·34	1·95	1·38
∞	6·63	4·61	3·78	3·32	3·02	2·80	2·64	2·51	2·32	2·18	1·79	1·00

TABLE 9: RANDOM NUMBERS

6920	9487	6408	5896	2305	8201	0507	8709
6757	6240	2997	9238	2236	1474	3711	5185
4246	1390	5636	7026	2663	9690	2354	2044
8969	2519	1488	4008	5497	9505	5002	4507
8308	7999	6307	4307	0614	4921	5536	0458
1864	3473	5338	8812	4150	2962	7113	0076
1986	6069	8055	4124	2179	6304	8484	4789
1000	1867	2867	4734	7602	2336	9938	2275
6081	3070	9152	2222	1375	3598	4973	8571
5863	0514	6378	6892	3271	0163	3434	3598
4493	4722	9215	3937	3152	7090	0242	7333
4427	1580	6007	7587	3595	1183	4778	5962
3070	0402	3472	3874	7347	1222	8570	9792
2930	9993	2923	2916	5840	8757	4598	3355
0663	0913	1577	2491	4068	6559	0628	7187
7570	6676	4246	0923	5170	6093	1263	7356
6652	7564	4217	1781	5998	7779	3777	1556
2249	4043	6293	0337	6630	6968	3598	0566
4848	3537	8386	1923	0309	2232	2542	4775
5741	9068	4810	3879	8690	2570	1260	3831
6747	3600	0348	3948	4296	8244	2540	0784
0003	9497	9500	8997	8498	7496	5994	3490
4577	5816	0394	6210	6605	2816	9421	2237
3924	3389	7313	0702	9016	8719	6735	5455

6936	9065	6001	5066	1068	6134	7202	3337
4744	8474	3219	1694	4913	6608	1522	8130
8390	8911	7301	6213	3515	9728	3244	2972
9198	3799	2997	6797	9795	6592	6387	2979
5444	1054	6498	7553	4052	1605	5647	7262
5725	1780	7505	9286	6791	6077	2869	8946
4445	9629	4075	3704	7779	1483	9263	0747
9770	5783	5554	1337	6891	8228	5120	3349
1412	8317	9729	8047	7776	5824	3601	9425
6091	0841	6933	7775	4708	2484	7192	9676
0068	6792	6860	3652	0513	4165	4679	8845
8996	5573	4569	0142	4712	4855	9567	4423
1318	8591	9909	8500	8410	6911	5321	2233
7688	3769	1457	5227	6684	1912	8596	0508
7836	0282	8119	8402	6521	4923	1445	6369
6916	9467	6384	5851	2235	8086	0321	8407
2620	3366	5986	9352	5339	4691	0030	4722
1206	0776	1983	2759	4742	7502	2244	9746

RANDOM NUMBERS—(*Continued*)

9216	7925	7141	5066	2208	7274	9482
8896	4082	2979	7061	0040	7102	7143
4399	6444	0843	7287	8131	5418	3550
9509	4017	3527	7545	1072	8617	9690
5995	6453	2449	8903	1352	0256	1608
7190	7266	4456	1723	6179	7903	4083
3273	8062	1335	9398	0734	0132	0867
2214	4489	6704	1193	7897	9091	6989
3544	2115	5660	7776	3436	1213	4650
7033	0632	7666	8298	5965	4263	0229
7576	4909	2485	7394	9879	7273	7153
0741	6703	7444	4147	1591	5739	7331
8363	8155	6518	4674	1193	5868	7062
7954	1310	9264	0574	9838	0412	0250
7815	5003	2819	7822	0641	8464	9106
8619	5976	4596	0572	5169	5741	0911
5333	6890	2224	9115	1339	0454	1794
4165	4732	8898	3630	2529	6159	8688
7317	2092	9409	1502	0912	2414	3327
5092	8923	4016	2939	6956	9895	6852
3324	4108	7433	1542	8975	0518	9493
9485	2976	2461	5438	7900	3338	1238
1658	3896	5555	9451	5006	4458	9465
2191	7646	9837	7464	7322	4806	2129

0540	3878	4419	8297	2716	1013	3730
9652	7782	7434	5217	2651	7869	0521
6216	9189	5406	4595	0002	4598	4600
9367	2346	1713	4060	5774	9835	5609
2919	0182	3101	3283	6385	9669	6055
1816	0762	2578	3341	5920	9262	5183
0002	0750	0752	1502	2255	2757	6012
8469	1819	0289	2108	2398	4507	6905
3026	2452	5478	7930	2409	1340	4750
6869	6546	3416	9963	3380	3343	6724
3524	2369	5893	8263	4156	2419	6576
3991	8414	2405	0820	3226	4046	7272
7555	9788	7343	7131	4474	1606	6081
9105	9614	8720	8335	7055	5390	2446
7814	4184	1999	6183	8183	4366	2550
8729	7137	5866	3004	8870	1874	0745
4753	9475	4228	3704	7932	1636	9569
1991	1737	3729	5466	9195	4662	3858

Index